D0857739

HARVARD EAST ASIAN MONOGRAPHS

145

Technology and Investment
The Prewar Japanese Chemical Industry

Subseries on the History of
Japanese Business and Industry

Japan's rise from the destruction and bitter defeat of World War II to its present eminence in world business and industry is perhaps the most striking development in recent world history. This did not occur in a vacuum. It was linked organically to at least a century of prior growth and transformation. To illuminate this growth a new kind of scholarship on Japan is needed: historical study *in the context of a company or industry* of the interrelations among entrepreneurs, managers, engineers, workers, stockholders, bankers, and bureaucrats, and of the institutions and policies they created. Only in such a context can the contribution of particular factors be weighed and understood. It is to promote and encourage such scholarship that this series is established, supported by the Reischauer Institute of Japanese Studies and published by the Council on East Asian Studies at Harvard.

Albert M. Craig

Technology and Investment
The Prewar Japanese Chemical Industry

BARBARA MOLONY

PUBLISHED BY
COUNCIL ON EAST ASIAN STUDIES
HARVARD UNIVERSITY

Distributed by
Harvard University Press
Cambridge, Massachusetts, and London
1990

Printed in the United States of America

The Council on East Asian Studies at Harvard University publishes a monograph series and, through the Fairbank Center for East Asian Research and the Reischauer Institute of Japanese Studies, administers research projects designed to further scholarly understanding of China, Japan, Korea, Vietnam, Inner Asia, and adjacent areas.

Library of Congress Cataloging-in-Publication Data

Molony, Barbara.
Technology and investment : the prewar Japanese chemical industry / Barbara Molony.
p. cm. – (Harvard East Asian monographs ; 145)
Includes bibliographical references.
ISBN 0-674-52160-9
1. Chemical industry–Japan–History–20th century. I. Title.
II. Series.
HD9657.J32M65 1990
338.4'566'095209041–dc20 89-48815
CIP

To my parents
Dorrit and Donald Molony

Acknowledgments

I take great pleasure in thanking the people who have been so helpful in guiding and facilitating my work on this book. This study began as a dissertation, and I therefore owe my greatest debt to my dissertation advisor, Professor Albert M. Craig, who first suggested that I study the prewar Japanese economy and who patiently saw the thesis and book through several transformations. Professor Edwin O. Reischauer's incisive comments helped give cohesion and direction to my ideas during the thesis preparation stage of my writing. Professor Hara Akira of Tokyo University also generously offered his time and shared his expertise in the wartime economy as I was preparing the dissertation.

Turning the dissertation into a book required substantial rethinking of major issues in business history. I could not have succeeded in that task without the exceptionally generous help of Professor Morikawa Hidemasa of Hōsei University and Professor Nakagawa Keiichirō of

Aoyama Gakuin University. These two scholars, together with Professor Alfred Chandler of the Harvard Business School, raised penetrating questions about the content and theory of my work at a workshop on Japanese business history organized by Professor William Wray at the University of British Columbia in 1982. Other members of that workshop, especially Professors Steven Ericson (Dartmouth College), Mark Fruin (Euro-Asia Centre, France), Andrew Gordon (Duke University) and William Wray have read selections from this work and offered gratefully accepted comments over a period of several years.

During the summer of 1983, Professor Ōshio Takeshi of Meiji Gakuin University gave me access to some of his extensive documents on the Nippon Chisso Hiryō company, which have been supplemented since that time. He also kindly introduced me to Mr. Kamata Shōji, who has been, for the last ten years, assembling both rare documents from the Nitchitsu company and an incomparable collection of written memoirs of people involved in that company. Professor Peter Duus of Stanford University has discussed Japanese business imperialism in Korea with me, and has introduced me to the Chūō Nikkan Kyōkai (Central Japan-Korea Association). This group of "old boys" recounted invaluable tales of life in colonial Korea, and introduced me to Mr. Tamaki Shōji, former President of the Japan Consulting Engineers Association. Mr. Tamaki graciously granted me an extensive interview about the years of his involvement in Chōsen Chisso and informed me of the continuing role of prewar Japanese businesses in post-colonial East Asia. Professor Udagawa Masaru of Hōsei University shared many of his resources with me, and introduced me to other scholars working on the chemical industry, including Professor Shimotani Masahiro of Kyoto University, from whom I received much helpful information by letter and telephone. Professor Kūdō Akira of Tokyo University, a discussant for a paper I presented in 1988 based on material in this book, made insightful comments which helped me focus my analysis.

My father, Professor Donald Molony of Rutgers University, helped me create the figures for this volume, and my mother, Dorrit Molony, gave valuable help in text preparation. Professor Catherine Bell and Sandy Choi, both of Santa Clara University, assisted me in assigning the proper readings to Chinese and Korean place names. My sister, Kathleen Molony, and my colleague, Professor Mary Gordon, have given me excellent advice on structuring my material.

This book could not have been completed without the superb editing of Florence Trefethen, Executive Editor for the Harvard Council on East Asian Studies.

I am grateful to a number of institutions for their funding support during the preparation of this book. As a graduate student beginning my research, I received National Defense Foreign Language grants and a Fulbright fellowship from the Department of Health, Education and Welfare. I completed the dissertation with a Japan Institute grant from Harvard University. Santa Clara University granted me a Presidential Research Grant to collect additional materials in Japan in 1983.

Finally, I thank my husband, Thomas Turley, a historian of medieval Europe, who not only read and commented on the entire manuscript but also shouldered more childcare responsibilities than he may have anticipated while I was preparing this book.

The help of all these people is greatly appreciated and, of course, the responsibility for all errors is entirely my own.

Contents

Tables

Figures

Technology and Investment
The Prewar Japanese Chemical Industry

Introduction

Many in the West equate contemporary Japanese industrial development with high technology. Advances in computers, sophisticated automobiles, biotechnology, ceramics, optoelectronics, and consumer electronics have led the growth of an economy that was leveled by the destruction of World War II. Japan's remarkable redevelopment has not gone unnoticed in the West. Scholars have used a variety of approaches to examine Japan's economy, society, and political structure to find an explanation for its apparent successes. These approaches include studies of the government's guiding role in formulating and applying industrial policy;[1] discussions of the role of innovation and development of technology;[2] analyses of commercial applications of technology;[3] didactic accounts of the techniques of Japanese management;[4] a thoughtful new interpretation of labor-management relations;[5] and an emphasis on entrepreneurial firms, rather than firms associated with the prewar zaibatsu, as innovators in high tech-industries.[6]

Studies using these approaches illuminate, with greater or lesser success, noteworthy aspects of Japan's modern high-technology economy. But only a few are consciously historical, and some are ahistorical in their attribution of contemporary conditions to tradition. To be sure, the roots of Japan's modern high-technology economy are to be found in the prewar era. But these roots are based neither on some allegedly unique cultural tradition nor, as some would have it, on the concentration of technology in a few firms like the large zaibatsu, a supposedly "inevitable" consequence of industrial development in a "backward country" like Japan.[7] Rather, it was managers and entrepreneurs in the prewar technology-intensive industries—many of which were not zaibatsu-related—who pioneered most of the well-known areas of Japan's modern high-tech economy. These efforts were aided by certain macroeconomic policies like the encouragement of scientific education.

Strategies planned by entrepreneurs are central to an understanding of technological innovation. Technology does not advance without some use for it. Thousands of patents expire each year without ever having been used for commercial production; thus, most never serve as the basis for developing a new line of technology. The invention of a new idea is the fortuitous reordering of old ideas; but those old ideas, already developed technologies, must be available and widely disseminated for imaginative minds to reorder them. Invention of new ideas does not guarantee their adoption, of course; consider Leonardo da Vinci's helicopter or the nineteenth-century ballpoint pen. Both these "inventions" became new technologies only in the twentieth century when process technology was adequate for their commercial production and when someone perceived a need for them. Thus, entrepreneurs deciding to commercialize new ideas that they either generate themselves or acquire from another imaginative thinker are integral to the process of technology development. High technology is not particularly meaningful without its industrialization. The study of technology in Japan, as elsewhere, must be the study of businesses devoted to its advancement.

This book will examine the investment decisions of some of the managers and entrepreneurs in the archetypal prewar high-technology industry, the chemical industry. Through this examination, it will attempt to demonstrate the relation of technology to investment and to the development of individual firms; the role of entrepreneurs and the state in economic development; and the economic component of Japanese imperialism.

Investment decisions included choices about which technologies to develop and whether to invest in expansion of existing productive capacity, in diversification of product, or in coordination of the major factors of production. Investment in the development and exploitation of technology occurred only when certain encouraging conditions obtained within the firm and its environment. Although successful investors therefore tended to share company characteristics and values, the types of investments they made determined more than just the shape of their firms. Rather, specific decisions made by entrepreneurs and managers determined the formation of the high-technology sector of the economy. Alfred Chandler's assertion in his penetrating analysis of modern business enterprises in the United States—that "the visible hand of management replaced the invisible hand of market mechanisms"[8]—applies equally well in the case of technology-intensive industries in Japan. Management decisions had far-reaching effects. Certainly, the market continued to be significant, but it was less the prime mover of investment than one factor encouraging or discouraging investment.

At the same time, a different type of "visible hand," that of government industrial policy, was also less significant than many Western analysts have thought, at least until 1937 when Japan was mobilizing for war. Before then, government espousal of a macroeconomic policy was historically more significant than what is commonly called industrial policy or targeting. Hugh Patrick has defined macroeconomic policy in a useful way as policy designed to "increase the quantity and especially the quality of the factors of production—labor, capital, and natural resources—and the general level of technology. This definition incorporates educational policy as an important element."[9] That is, the economy in general is assisted; specific industries are not targetted. In its emphasis on macroeconomic policy to aid development, Japan was little different from most other industrial countries.

These government policies changed after 1937, however. Managerial autonomy in decision making decreased as government agencies used pressure as well as incentives like huge munitions requisitions to encourage diversification in new strategic products. The increasing involvement of government agencies in the affairs of businesses had long-term effects. Most important, the precedent for continuing government involvement was established, although the differences between wartime and postwar industrial policy are extensive.[10] There are some interesting similarities in the responses of entrepreneurs to industrial policies during and after the war. For example, those product lines less affected by government direction were often a company's more profitable ones during the war, and several high-tech companies that were more independent of bureaucratic direction tended to be the most innovative after the war.

The highest levels of technology—both in terms of sophistication of basic science and in terms of methods of its application—were attained in the prewar period in the electrochemical industry. This industry was both commercially successful (it produced, among other things, fertilizers for Japan's largely agrarian society) and strategically important (it also produced explosives). Though surpassed in importance by petrochemicals in the postwar period, and today profoundly depressed as higher energy costs and competition from newly industrialized countries have made much of Japan's electrochemical industry redundant,[11] electrochemicals were Japan's most advanced industry before World War II. It is on this industry, therefore, that the present study centers. This book analyzes the origins of Japan's high-technology industry by focusing on the prewar chemical industry, especially electrochemicals, with shorter descriptions of other related branches. In doing so, it will confront some of the major issues of Japanese industrialization.

CHEMICALS AND INDUSTRIALIZATION

In Japan, as in other industrial countries, chemical industries have been central to industrial development. While the hazards and pollution of chemical products have come to symbolize the excesses of modern industrialization, economic growth and technological development have, to a great extent, been dependent on chemicals. Indeed, from the 1880s to the 1940s, chemical development lay at the cutting edge of technology.

The vital role of chemicals in development is apparent even in the earliest stages of the European Industrial Revolution, which preceded industrialization in Japan. Operation of Europe's spinning mills in the eighteenth century demanded the release of rural labor, and their continued success depended on an increasing population of consumers. But without increasing the productivity of the land to feed this growing population, the requirements for industrial development would have succumbed to Malthus's dire predictions of the limitations of growth. A key response to this need was the development of chemical fertilizers. By increasing the productive capacity of the land that fed an emerging labor force, chemical fertilizers contributed significantly to otherwise unrelated industrial developments. Furthermore, the methods of production of chemical fertilizers required research and development of more efficient machinery. Thus, even in the earliest years of chemical-fertilizer production, the industry simultaneously increased the productivity of the less technologically advanced industry which had spurred fertilizer development in the first place—agriculture—and stimulated the development of a new, higher level of technology outside the field of chemicals—advanced industrial machinery. In Japan, the pattern of industrial development was similar. The early chemical-fertilizer industry proved pivotal in industrialization, stimulating industrial development both "backward" and "forward."

As the manufacturing industries in the West grew during the nineteenth century, they gained sophistication in production techniques, spawning new industries and accelerating the process of development in chemicals. Fancy new fabrics, for instance, called for more abundant and sophisticated dyes, at prices within reach of a growing number of urban consumers. Clever use of by-products from other new industrial processes brought down prices for colorful textiles and stimulated demand for new products. Simultaneously, the transportation industry's demand for steel and widespread need for energy and lighting engendered modern coking furnaces whose "wastes" could be profitably used as the basis for dyes. Subsequent research at the end of the nineteenth century, in the West as well as in Japan, revealed additional dramatic uses for new chemical technologies that few governments could afford to ignore. Powerful explosives, produced by methods similar to those used in making fertilizers or dyes, could release the mineral treasures stored in a nation's soil or could help construct its infrastructure for transportation and communication. Furthermore, they could supplement the arms deemed necessary to preserve the nation's wealth and people and, in some cases, to expand its frontiers beyond the seas. Explosives filled both these functions in Japan's case. The development of explosives by private-sector manufacturers contributed, at first, to the building of the infrastructure of Japan's colony, Korea, and later to the production of armaments.

Certain problems of industrialization were amenable only to technological solutions. In Japan and the West, the fight against both tropical diseases encountered with colonial expansion and bacteria and pests which accompanied urbanization in industrializing countries stimulated growth in the pharmaceutical industry. Low-cost and replenishable substitutes for hard-to-obtain natural products—some of which, like rubber, did not come into demand before the exploitation of colonial resources—led to the creation of the resin, celluloid, and plastics industries.

None of these chemicals was produced totally independently of other chemicals. Increasing sophistication of technology meant that some chemicals would be used as intermediaries in some processes and that others would spin off new products. The search for synthetic quinine to cure colonial malaria, for instance, yielded not the drug but a new branch of the chemical industry, that of aniline dyes.[12] The need for synthetic ammonia to produce nitric acid for explosives during World War I led several governments to support research in ammonia which, after cessation of hostilities, could be turned to peaceful ends in the production of nitrogenous fertilizers.

Governments in several countries supported private investors in particular types of chemical technology with research assistance, tax benefits, licensing privileges, and development grants. Most governments, however, had no consistent policies to support private investors in all areas of technological development. Many consumer products, for example, underwent decades of development without a glimmer of official interest. Japan was no exception. William Wray has written that the Japanese government designated certain industries as "strategic" between the 1880s and the 1920s and granted them "huge subsidies."[13] This was not true of the electrochemical industry. Unquestionably, the government played a large role during that period in influencing how business leaders chose to run their firms. But, although the electrochemical industry was deemed strategic—and the government expended great sums on public scientific education and the creation of research laboratories—no direct subsidies were given to the pioneering entrepreneurs in this industry.

Nevertheless, no expensive technology could be developed without a business climate conducive to risky investment. Whether in Germany, America, or Japan, the scientifically trained entrepreneurs who invested their carefully accumulated capital in chemical technology and manufacture had to be convinced that the initial risk involved would eventually yield worthwhile results. Certain preconditions were deemed essential. A market for investors' products had to develop, permitting profits for reinvestment, accommodation of shareholders' interests, and repayment of high start-up costs. Furthermore, governments had to create favorable conditions. These eased the entry of private investors in high-technology areas through various types of support for specific projects, subsidies for scientific education, and aid in acquisition of technology.

Market creation, government assistance, and the results of both careful research efforts and fortuitous breakthroughs in technology set the stage for development of a sophisticated chemical industry. But without interested investors the embryonic industry would have failed to mature. In all nations where major chemical industries developed, private investors were largely responsible for establishing the industry. And, in each economy, prospective investors had to be satisfied that preconditions for investment in a high-technology enterprise had been met.

INVESTMENT STRATEGY AND THE DEVELOPMENT OF JAPAN'S CHEMICAL INDUSTRY

The conditions for investment were adapted to individual needs of investors in different countries, but there were, in the main, common requirements for all investment. They were:

1. *Access to capital.* Differences in types and sources of capital could affect the structure of the enterprise. Having to pay dividends to shareholders, for instance, limited funds available for reinvestment. Using loans or bonds was sometimes cheaper. On the other hand, if those loans originated in government institutions, managers' ability to make business decisions might be circumscribed by government pressure.

2. *Access to resources and labor.* Different technologies required particular resources and application of labor. A chemical producer whose preferred technology demanded much electricity sought that resource, while a possessor of coal would find investment in chemicals a wise option only when coal-based technologies became available. Furthermore, investors would have to be assured that appropriately skilled white- and blue-collar workers were available and that the conditions of their employment were sufficiently advantageous to retain them. Labor was unique as a commodity, and securing access to labor was far less simple than securing access to other inputs. Workers had wills and desires that restricted the autonomous decision making of managers.

3. *Certain environmental conditions, such as creation of a market and a political climate conducive to new investment.* Generally, government interest or interference was but one of several considerations in

the formation of business strategy. In extraordinary times, however, such as in Japan after 1937, the government could and did interfere so blatantly that businessmen's autonomy was severely limited.

4. The existence among the developers of any new technology of *skilled professional managers* with an understanding of the technology, a willingness to take risks, and an ability to coordinate their activities with the other functions of the enterprise. Many of the original developers of chemicals were scientists-cum-industrialists, men who managed their own enterprises for several decades after founding them. Although some later developers missed the early opportunities available to pioneering entrepreneurs, they shared with their predecessors a strong interest in taking risks to master new technological production. In Japan, as elsewhere, this latter group included fewer scientist-entrepreneurs than the first; more commonly they were professional managers and researchers, usually at the middle-management level, supported by visionary top management in the large corporations.

Business strategies are devised in non-static environments. Technologies are derived from specific existing technologies; business leaders cannot simply decide to develop certain products if available technologies are inadequate to the task. Government intervention may severely limit managerial autonomy, and employees' needs and desires may be more influential in forming business strategy than, say, access to capital or resources. Nevertheless, it is business leaders—scientists, investors, and professional managers—who typically determine what best advances the interests of their companies. This, in turn, has a profound effect on the development of the economy. And in no industries is the impact of business decision making greater than in high-technology industries which produce new industries and help direct the future structure of the economy.

The men who pursued the development of Japan's chemical industry, like their counterparts in Europe and the United States, had to find methods of attaining these conditions. During the first decades of the Meiji period (1868–1912), however, few private investors responded to opportunities to manufacture chemicals or other high-risk products using new technologies. Few in Japan had the scientific knowledge to understand the new technology; consequently, private investors were not interested in complicated chemicals. The earliest chemicals produced in Japan were either manufactured under government auspices or were so simple technologically that private investors were assured that they were investing in products they could understand. These early investors also had access to capital and resources, and operated under improving market conditions with some government encouragement. But most early chemical firms were led by men untrained in science who operated their low-capital plants as unsophisticated traditional firms. Few used complex machinery or sponsored continuing laboratory research. No new patented processes were generated, nor was related technology spun off.

But Japan soon began to catch up with the technologically more advanced West. By the turn of the century, Japanese schools of higher education had begun to produce a scientifically literate group of officials and industrial scientists. Foreign studies also aided a significant number of students. Soon these engineers and scientists were applying their knowledge to industrial production. The cleverest among them found capital, negotiated for resources applicable to the available technology, profited from a growing consumer market, and, as scientists, had a strong desire to develop a technologically advanced product.

Several types of chemicals requiring the highest levels of technological sophistication developed after the turn of the century. These included dyes and fertilizers, both products with large domestic markets in Japan, a country still largely agrarian but with a growing textile industry. Because of the productive superiority of German dye makers and the inability of Japanese dye makers to take advantage of economies of scale (dyes were made in small lots), the dye industry began haltingly before receiving government aid around the time of World War I.

The fertilizer industry had a better start. Fertilizers were produced entirely for the consumer market by private entrepreneurs. In addition, government aided the industry in more indirect ways, stimulating the market, training scientists, controlling distribution of fertilizers, and, most important, providing access to hydroelectric generating rights. In time, fertilizers became one of Japan's largest industries. Its developers–scientists often working independently of large corporations–carved an important niche for themselves in Japan's economy. Until the 1930s, the larger and wealthier companies–zaibatsu like Mitsui and Mitsubishi–lagged behind these scientific entrepreneurs in high-technology investment, despite their enormous capital resources.

The most important type of technology used by Japan's early independent investors was electrochemistry. Basically, these scientists used electricity to "fix" atmospheric nitrogen; that is, they removed nitrogen from the air and combined it with other elements in a form usable in fertilizers and explosives. Within a few years of Western discovery of methods to fix nitrogen, several young Japanese engineers, most notably Noguchi Jun (1873–1944), perfected similar processes in Japan and decided to begin commercial production of chemicals. Noguchi's story highlights how investment in an industry dependent on technological advances can be analyzed in terms of the conditions for development outlined above.

Noguchi was one of the first professionals in science to emerge from the excellent program at the nationally run Tokyo Imperial University. Early in his career, he perceived the entrepreneurial niche opening up in nitrogenous chemicals. Through close ties to political leaders who granted him water-power development rights, Noguchi gained access to the necessary resource, hydroelectricity. His only real problem, access to capital, was solved when Mitsubishi Bank granted him a series of loans. As his technology developed and became more sophisticated, it not only produced changes in his company's organizational structure but also necessitated finding cheaper energy sources. Just as the requirements of his new technology led him to seek cheaper colonial sources of electricity, political problems associated with fertilizer distribution to farmers encouraged him also to expand his sales to the colonies. Changes in government relations subsequently produced changes in investment decisions by his firm, one of the "new zaibatsu." After 1937, wartime conditions significantly altered his company's ability to promote its own interests. From its beginnings as a relatively freewheeling high technology company, it found—to paraphrase William Wray in his study of N.Y.K—that its autonomy was circumscribed by its close ties to the fabric of Japan's political and military institutions.[14]

It would be inaccurate to imply that decisions made in Noguchi's enterprise, or in any business for that matter, had ever been based entirely on the four rational criteria noted earlier. Personal whims—in Noguchi's case, his often-cited and therefore probably atypical patriotism—influenced his evaluation of optimal strategies. "Rational" business decisions always had their "irrational" human components. Economic rationality was also compromised by adherence to contemporary social values. Managers were never objective in assessing even the seemingly value-free environment, because their perceptions of it were influenced by their own values. For instance, it was taken for granted in the prewar period that effective employee interaction would be impeded by bringing women or Koreans into management, although bias against these two groups was economically irrational. When we eliminate these admittedly extensive areas of irrational bias, however, we find that business decisions were made for otherwise rational reasons.

The efforts of chemical companies like Noguchi's caught the attention of the established zaibatsu, who subsequently began to pursue technological advances themselves. By the 1930s, Mitsui, Mitsubishi and Sumitomo had plunged into electrochemical manufacture and, within a few years, had some of the largest plants in Japan. They, too, were transformed from cautious financial giants to tough competitors in the chemical industry. In examining this development, it is imperative that the zaibatsu not be viewed as monoliths, but rather as organizations made up of subsidiaries, some involved only in manufacturing. The professional managers and researchers managing these subsidiaries planned strategy to advance their own specific businesses. They were given significant latitude to do so.

Like the earliest investors, the zaibatsu managers had to be satisfied that certain conditions for investment were met. They were. New production technologies, available by the late 1920s, gave them access to an important resource, coal. Capital was available for establishing specialized, professionally staffed research facilities. Government monetary policies of the early 1930s effectively removed the threat of foreign imports in Japan. Finally, a new group of professional managers with ideas about scientific development similar to those of men in existing chemical companies had risen to the top levels of management in both the zaibatsu and their subsidiaries. Top-level management in the zaibatsu was able to direct resources, both material and financial, to the innovative professional managers of their technology-intensive subsidiaries. When the conditions for investment were met, advanced chemical technology was developed by some subsidiaries of the zaibatsu. The issues that have usually intrigued historians of the zaibatsu—like ownership of capital and the role of family members as major shareholders—were less important than the role of management professionalism, technology, and resources in determining how the zaibatsu acted as manufacturers.

It is noteworthy that the zaibatsu lagged behind the earliest innovators in investment in electrochemicals. But this lag was not unique to Japan. In other economies as well, inventors led in commercialization of new processes and products. When their innovations were successful, they created a new market, which gradually enticed larger firms to invest in the new products. These larger firms often had greater resources for experimentation, testing, and test-marketing, and therefore began themselves to develop, and often improve on, the new technologies.[15] In some cases, "imitation" (first use of a technology after its innovation elsewhere) was undertaken because companies wished to respond quickly to market opportunities. In other cases, these large firms innovated new technologies derived from earlier discoveries because their research scientists were as enthralled by the technological possibilities of experimentation as the originators of those discoveries had been. It is difficult, and perhaps inappropriate, to separate the effects of demand-pull and technology-push in attributing cause for investment in innovative technology. They were both significant, and must be considered together.

THEMES OF THIS STUDY

This study analyzes the development of the Japanese chemical industry in the pre-World War II period. The first chapter covers the industry's historical background. The rest of the book focuses on the electrochemical branch of the industry. Because of their extensive production, widespread use, and demanding technology, electrochemicals, the branch of the chemical industry derived from fertilizer production, will be used to represent development in the industry. This is an area that has not been studied by Western scholars to date. Western writers have tended to focus on the role of government in Meiji development or the rise of the zaibatsu and their position in the economy, and so have generally overlooked some central questions about economic development.

The recent surge in scholarly interest in postwar Japanese enterprises, however, has suggested new areas for historical study in continuities (and discontinuities) within prewar and postwar enterprises. One such area is decision making. Contemporary entrepreneurs in non-traditional and risky but potentially profitable new industries make decisions based on their best understanding of present and future conditions. Their industries are shaped in their capital structure, labor relations, resource allocations, and a variety of other defining characteristics in ways consistent with prevailing practices and the individuals' assessments of risks.[16] Prewar investors faced a different set of conditions, whether in the market, in the political economy, or in the location of capital and resources, but their different decisions about investment were based on questions similar to those asked by all entrepreneurs. The preconditions for investment—access to capital and resources, appropriate management, and encouraging external factors in the political world and the marketplace—influenced the timing of investment, the types of technologies used, the justification for production, and the organizational structure of chemical firms. This process was not particularly mystifying or peculiarly Japanese. One goal of this book is to help demystify the story of Japan's high technology development. And the story's most informative chapter involves the chemical industry.

The development of the chemical-fertilizer industry, the backbone of Japan's prewar chemical industry, progressed through several levels of technological improvement. As the technology became more advanced, investment decisions were increasingly determined by the requirements of the technological methods. This induced the earliest investors to seek both new markets for chemical products and new sources for necessary raw materials.

As the first important investor in electrochemical technology, Noguchi Jun undertook several major changes in his company's structure and product line. When he upgraded his technology for nitrogenous fertilizers (the principal electrochemical product) from a relatively uncomplicated method to a sophisticated synthetic method in the years after World War I, he found his need for electricity increasing. He handled this growing requirement through expansion to Korea, structural changes in his company, and product diversification to lessen dependence on nitrogenous fertilizers. Unfortunately, the last response—diversification—was less than profitable in many cases and required, in other cases, manufacturing processes that, willfully uncontrolled, produced deadly effluents over a three-decade period.

Noguchi's success with electrochemicals, analyzed in Chapters 2 through 5, encouraged other investors to enter the field. These later investors are discussed in Chapter 6. Those who used electrochemical methods shared Noguchi's problems and often his solutions. Some of Japan's largest chemical companies, including Shōwa Fertilizer and Nissan Chemical, grew by using electrolytic manufacturing processes during a second wave of chemical-industry development in the late 1920s.

In contrast to these entrepreneurial companies, the zaibatsu were relatively slow to invest in electrochemicals, despite their apparent access to resources and capital. During the World War I era, however, the zaibatsu began to make important changes in their corporate structures which permitted greater flexibility for their subsidiaries' managers to develop new technologies and processes. By the late 1920s, as they became informed of changing methods of production, top zaibatsu managers found the risks of investment lower than they originally believed. Several invested in the latest production methods and initiated a third wave of industrial development and technological innovation. The new methods produced electrochemicals by a "coal-gas" technique which used coal rather than electrochemistry. Access to a resource they already possessed–coal–made the zaibatsu less dependent on the cooperation of government officials than were the earlier investors who needed official approval to harness hydroelectricity for electrochemistry. In the last analysis, the zaibatsu scientists were creative, but they were unable to lead the electrochemical industry in its earliest innovations because of the initially more conservative policies of the large companies' top managers. This pattern may also be seen in the technology-intensive sector in other contemporary advanced industrial economies, especially those with highly developed capital markets permitting start-up companies to flourish.

This book focuses on one branch of one industry in pre-World War II Japan, and is most concerned with one company in that branch. But it is more than a case study of an interesting company. This analysis of the electrochemical industry helps to explain the development and diffusion of technology in Japan, the nature of prewar entrepreneurship, the role of private capital in Japan's development, the relationship of technology and investment, the nurturing of scientist-entrepreneurs, and the economic components of imperialism.

ONE

Early Development of the Chemical Industry in Japan

The Japanese countryside was untouched by technology-intensive industry in 1868 as the Meiji period began; no chemical plants, nor any other plants, concentrated labor and capital in towns and villages. Farmers had been accustomed to buying fertilizers—oil cakes, dried fish, and night soil—but none of these required manufacturers to be technologically literate or farmers to be sophisticated in fertilizer application. The opening of Japan to imports from the West introduced consumers to new products requiring more advanced processing methods. Growing demand for these products led both the government (concerned about serious trade imbalances) and producers (interested in profits) to invest in production requiring some technological input. Chemical fertilizers were one potentially profitable area for investment, though not the first to be pursued.

The chemical industry in its early stages was still fairly primitive. Its technology encouraged little diversification and demanded few skilled technicians. It benefited from direct public investment in production. And it had yet to develop a market for its products. In short, it bore little resemblance to the dynamic chemical industry of the twentieth century. But the achievements of the last third of the nineteenth century provided the foundation for a rapidly growing high technology sector. Indeed, the robust structure of Japan's modern chemical industry was determined by these early developments. This chapter will trace the origins of this sophisticated, technology-driven industry.

Three subjects will be considered here. First, the earliest chemical products will be discussed in order to place the later growth of the electrochemical industry in its proper perspective. The earliest chemicals were developed in both the public and private sector; in the latter case, the government often assisted by purchasing the finished products. Most early products entered the market as substitutes for items already being imported. Second, the initial stages of the development of the chemical-fertilizer industry will be analyzed. The technology was simple, and the firms were not modern, multi-unit, vertically integrated companies. But they were instrumental in creating and expanding the market for more advanced-process fertilizers. And third, the revolution in scientific education and research will be described. Technology cannot be absorbed if it is not understood. But, once understood, it can be used as a springboard for further innovation. Without innovation, an industry stagnates, and, in time, so does the economy in which it is situated. The Japanese government fostered technological competence in two ways—through schools and through research institutes. Both were important parts of the macroeconomic policy encouraging technological parity with the West.

SERVING THE NATION AND THE CONSUMER: THE FIRST CHEMICALS

Japan's first chemical plants were of two types: advanced factories run under government authority because there were few private entrepreneurs able to invest in the sophisticated products demanded by national security; and primitive factories differing little from traditional workshops. The former type—more advanced factories—began production of two chemicals, sulfuric acid and soda, in the 1870s. Sulfuric acid was needed by the government to print paper money and to refine metal, and soda was needed by the developing textile industry for use as a bleach. The latter type—more primitive factories—manufactured matches, paper, and ceramics, all chemical products having a ready consumer market and requiring negligible capital investment. Their production contributed little to the advancement of technology in Japan during the nineteenth century. Technological advances in the private sector would come only with fertilizers, which were not among these earliest manufactures. But the earliest products were nevertheless important because they accustomed consumers to chemical products and because their manufacture helped disseminate production technology. Fertilizer manufacture, requiring some new technology and undertaken by private entrepreneurs, constituted a third, and later, type of investment.

The desire to develop science was not, however, the principal reason for the advance of the fertilizer industry. The need to raise agricultural output supplied the justification for investment in fertilizers. Agriculture was the most important sector of the economy throughout the nineteenth century and employed the majority of the people. Agriculture also supplied most government revenues through the land tax. To obtain sufficient revenues to fund industrial development, the government encouraged farmers to maintain and increase land productivity. Responsibility for producing the surplus was, in the time-honored and universal pattern of government policy, that of individual farmers who paid taxes. Because it was therefore in the farmers' interest to increase their yields, popular interests and public policy joined in encouraging the development of chemical fertilizers, although only gradually and with much government prodding.

That fertilizer became more important than sulfuric acid or soda in the Japanese chemical industry during the Meiji era is significant for several reasons. Most fertilizers were primarily produced by private investors, and most, though not all, required a higher level of technological sophistication than sulfuric acid or soda. Japan, as a late-developing country, could import the most efficient modern technology for the advanced fertilizers without immediately having to upgrade outmoded technology. The technology for sulfuric acid, on the other hand, was improving rapidly at the end of the nineteenth century, and soon required investment in improvements. This is not to underplay the importance of production of sulfuric acid in Japan. In addition to its strategic uses, large quantities of sulfuric acid were needed in fertilizer manufacture, and its growth paralleled that of fertilizers.

Acids and alkalis shared with explosives the honor of being the first chemicals produced in Japan; these products were given government support. But, whereas officially run production continued in the area of explosives, which had strategic and military significance, the government did not continue investment for long in acids and alkalis, which had various industrial uses. Private manufacture of sulfuric acid and soda rapidly outstripped official operations in importance.

The first production of both types of chemicals was under government auspices, however, because of the chemicals' political significance in policies of national unification in the first decades of the Meiji period. A national currency had to be established; sulfuric acid was used for refining gold and silver, while bleach and caustic soda figured in the manufacture of paper currency.[1] Sodium products and sulfuric acid were, therefore, soon produced by the Printing and Minting Offices of the Ministry of Finance. When no other facilities existed, Roland Finch, a British citizen in Japanese government employ from 1872 to 1875, established sulfuric acid production. But operation of factories by government officials was more the result of necessity than of an official preference for nationalized industry. By

1879, Toyohara Hyakutarō had begun private production, and, a few years later, the government left the business. Government support for the sulfuric acid industry was a matter of timing; there had been no Japanese producers at the time the products were initially needed for currency manufacture. Though in the hands of private investors by the end of the century, the sulfuric acid industry continued to develop slowly. The major retardant was the continuing presence of British exports to Japan which amounted, at times, to 10 times the domestic output.[2]

Chemical products were not the only ones first developed in government-operated factories.[3] The Ministry of Technology (Kōbushō) operated facilities in shipbuilding, mining, telegraphy, and machinery as well. Most of these, like those for sulfuric acid, soda, and products classified as chemicals like cement and glass, passed into private hands within a few years. The Ministry of Agriculture and Commerce (Nōshōmushō) was involved in cattle raising, sugar manufacture, and textiles. One area of official production retained by its responsible ministry was munitions. Government arsenals initially operating during the last years of the Tokugawa shogunate were allowed to continue production by the Military Ministry (Heibushō). That ministry's successors, the Army and Navy Ministries, quickly moved to build new munitions factories, starting with the Army's Itabashi Arsenal in 1875.[4] As in some other countries, explosives for military use remained the only chemical products consistently produced in large quantities by public enterprise.[5]

In sum, sulfuric acid and soda are interesting examples of early government enterprises in the chemical industry; fertilizers are important as a sophisticated leading sector of the industry; and the munitions industry offers an insight into national perceptions of strategic necessity.

It is important to note, however, that other branches of the chemical industry far outweighed these three in value of output and in number of employees during the first decades of the Meiji period. These were products used by consumers, such as matches, iodine, soap, paint, and celluloid. Imports of these consumer products in the first years of the Meiji period climbed quickly, worsening Japan's

TABLE 1 Representative Chemical Processes, Capital Investment in 1896

Product	*No. of Firms*	*No. of Factories*	*Capital Investment (yen)*	*No. of Workers*	*Investment per Worker (yen)*
Matches	31	307	565,253	46,326	12
Sugar	9	4	1,474,106	98	15,042
Wax	4	6	100,00	98	1,020
Cement	14	15	1,330,000	1,858	716
Paper	28	97	2,872,793	4,909	585
Coke	6	21	248,700	512	486
All Chemicals	353	1,601	11,509,100	86,101	134

Source: Calculated from data presented in Watanabe Tokuji, *Gendai Nihon sangyō hattatsushi,* p. 121.

Note: Totals for the representative products are less than the figures given for "all chemicals." By 1896, at least 15 additional chemical products were manufactured in Japan, and the sums listed under "all chemicals" include data for these chemicals.

TABLE 2 Use of Inanimate and Human Energy in the Chemical Industry, 1903

Industry Branch	*No. of Factories*	*Mechanized Factories*	*Male Workers*	*Female Workers*	*Total Workers*
Ceramics	457	49	11,958	1,956	13,914
Paper	88	44	3,592	2,332	5,924
Matches	239	49	6,946	13,492	20,438
Fertilizers	8	8	550	8	558
Pharmaceutical	65	35	1,342	427	1,769

Source: Table adapted from Nakamura Chūichi, *Nihon sangyō no kigyōshiteki kenkyū*, p. 20.

balance of international payments. Soap alone accounted for 9,000 yen in imports in 1868, rising quickly to over 31,000 yen in 1871, and 85,000 yen in 1876.[6] Imports stimulated consumer demand, and import substitution appeared a reasonable way for enterprising investors to use their money. Domestic production of consumer products did little to alter consumer tastes. It simply made desirable products more readily available from local sources. Imports had created a market; early manufacture of these products—matches in 1872, soap in 1873, paint in 1881, and celluloid in 1889—testified to investor confidence in their marketability. This confidence was well placed. Domestically produced soap, for example, overtook imported soap by 1877.[7] For some unexplained reason, however, the Finance Ministry's Printing Office began producing soap, thereby creating new competition for Japan's private soap manufacturers, the following year. Fortunately, the private makers survived, and the government dropped soap production by 1886.[8]

The match industry employed the vast majority of workers in the Japanese chemical industry, and matches constituted the industry's single largest export item around the turn of the century. Since the match industry differed in several respects from most other branches of the chemical industry, it can hardly be considered representative. Yet its early position earns it a brief description.[9] Begun by foreigners in Yokohama in 1872, small-scale match factories and home workshops multiplied throughout Japan in the next five years. The initial French production methods were supplemented by Swedish methods as soon as Sweden developed the safety match in the 1870s. The composition of the labor force, mostly female and often young, was significantly different from that of later chemical industries, but similar to other export-oriented light industries such as textiles. Capital investment was extremely low, and labor intensity high (see Tables 1 and 2).[10]

Chiefly produced for export, Japanese matches found their way into Southeast Asia and the United States by the turn of the century. Because manufacturers wished to maintain their share of these foreign markets in the face of stiff international competition, however, they were forced to squeeze extra work from their laborers. International competition had forced down Japanese prices (see Table 3).

The particularly severe pressure on the unskilled female and child labor in match production was not, for the most part, replicated in other branches of the chemical industry, although conditions for all workers, male and female, were, as in other countries' chemical industries, intrinsically hazardous to health.[11]

TABLE 3 Decline in Unit value of Exports in the Match Industry, 1881–1912, selected years

Year	*Value of Exports (yen)*	*Quantity of Exports (gross)*	*Value per gross*
1881	249,758	566,324	.44
1887	941,276	3,384,296	.28
1899	5,890,666	19,628,134	.30
1912	12,043,784	44,871,921	.27

Source: Watanabe Tokuji, *Gendai Nihon sangyō hattatsushi,* p. 123.

Of the remaining building blocks of the prewar Japanese chemical industry—synthetic resins, fats and oils, plastics, petrochemicals, and dyestuffs—only dyestuffs and fats and oils were products of the developmental stage of the industry to 1920. Dyes were developed because of the importance of textiles in Japan's prewar exports, and received government support in research and development during World War I similar to the research aid in nitrogenous fertilizers.

THE FIRST CHEMICAL FERTILIZERS: SUPERPHOSPHATES

Superphosphate of lime, one of three important classes of fertilizers, was the first major chemical product bought by Japanese consumers. But the fertilizer's advance into the market initially proved difficult, and investors were somewhat wary of beginning research and development in superphosphates. As in other nations, the passage of time, government incentives to use artificial fertilizer, and advertising overcame the apprehension farmers felt about using the new chemical products. With market and government conditions eventually stabilized, entrepreneurs could have confidence in their investments.

The manufacturing method was simple; the only resources necessary for superphosphate production were phosphate rock and sulfuric acid. The former was supplied entirely by imports. Producers of the fertilizer, therefore, established close ties with the Japanese importing firms (mainly Mitsui Bussan); this permitted the trading companies to influence the activities of some of the companies in that industry. The other resource, sulfuric acid, was initially purchased from domestic suppliers outside the fertilizer firms, but superphosphate manufacturers saw the benefit of supplying their own acid. The resultant expansion into sulfuric acid production boosted company profits, but it did not resemble the sophisticated diversification later undertaken by electrochemical companies. Superphosphate technology was so simple as to preclude the possibility of spin-off technology. The superphosphate industry, therefore, never manifested a tendency to create vertically integrated, complex firms making interrelated products. Instead, expansion within the superphosphate industry meant creating a larger company, with more factories making the same two products: sulfuric acid, both used as a component in fertilizer manufacture and marketed as a finished product; and, of course, superphosphate fertilizer.

The sources of capital in the superphosphate industry may have affected this proclivity toward growth without diversification. The largest of the superphosphate firms, Tokyo (later Dai Nihon and later yet Nissan) Artificial Fertilizer, had more than half the market. Its original investors were mostly not scientists but financiers like Shibusawa Eiichi. Their goals and subsequent decisions about investing showed a certain interest in advancing Japan's chemical technology; investors were even more interested, however, in preserving the profitability of their shareholdings. Expansion of the firm's capacity in proven areas of production rather than risky investment in experimental products offered investors greater security. Merger, monopoly, and market restraint followed as production rose to meet demand for superphosphates. The superphosphate industry is notable as a contrast to the electrochemical industry.

Japanese farmers had for centuries relied on nutritive additives to the soil, using such natural fertilizers as dried fish, manures, and compost for the necessary nitrogen, potassium, and phosphate. Farmers' demand for fertilizers increased during the early years of the Meiji period when government agronomists, attempting to raise the productivity of existing fields, urged village leaders throughout the coun-

TABLE 4 Comparative Nutrient Content in Fertilizers Used in Japan before World War II (% of nutrient present)

Fertilizer	*Nitrogen*	*Phosphate*	*Potassium*	*Total*
	Traditional			
Night soil	0.57	0.13	0.27	0.97
Barnyard manure	0.58	0.30	0.50	1.38
Soy meal	6.20	1.20	1.40	8.80
Tea meal	5.00	2.00	1.00	8.00
Cottonseed meal	6.00	3.00	1.60	10.60
Rice bran	2.00	3.80	1.40	7.20
Dried sardines	7.50	3.50	0.70	11.70
Dried herring	6.60	2.30	0.70	9.60
Sardine-oil cake	10.00	4.00	0.60	14.60
Herring-oil cake	9.00	5.00	0.70	14.70
	Chemical			
Ammonium sulfate	20.8	0.0	0.0	20.8
Calcium cyanamide	19.0	0.0	0.0	19.0
Sodium nitrate	15.7	0.0	0.0	15.7
Superphosphate	0.0	19.5	0.0	19.5
Potassium sulfate	0.0	0.0	48.0	48.0

Source: Kameyama Naoto, *Kagaku kōgyō gairon*, p. 212.

try to develop new strains of rice particularly responsive to fertilizer.[12] Because traditional organic fertilizers, as Table 4 indicates, failed to offer particularly high levels of the three nutrients, the demand arose for more effective artificial fertilizers.[13]

As in the West, superphosphate fertilizers were the first inorganic fertilizers to appear on the market in Japan. Acceptable traditional nitrogenous fertilizers could already be purchased in Japan, so investors were not initially motivated to introduce technology for nitrogenous fertilizers. Furthermore, the imported form of nitrogenous fertilizer most common in Europe, guano, was inappropriate to

Japanese wet-field cultivation. Potassium fertilizers were still something of a novelty in the 1880s, even in Germany where they were first used in large quantities.[14] It was, therefore, logical that the simple superphosphate technology would be the first introduced. Takamine Jōkichi (1854–1922), an engineer in the Ministry of Agriculture and Commerce who would later make major international breakthroughs in science (he discovered adrenalin, among other accomplishments), was responsible for introducing the superphosphate technology he had viewed at the New Orleans Exhibition of 1884.[15]

Soon thereafter, in 1886, the Sulfuric Acid Company of Ōsaka (Ryūsan Seizō Kaisha, later Ōsaka Alkali) set up pilot production facilities and succeeded in making Japan's first superphosphate fertilizer from phosphate rock. Though, for a short period of time, it produced only a small amount, sufficient to carry out experimentation in a test market in Shikoku,[16] its results impressed Takamine Jōkichi and Yoshida Kiyonari, Vice-Minister of Agriculture and Commerce. They wished to establish a full-scale pilot plant under government auspices. Unfortunately, their enthusiasm sparked just as the government was divesting itself of enterprises set up by the former Ministry of Technology.[17]

Undaunted by the government's reluctance to set up a pilot plant, Takamine gathered an "elite" group of wealthy investors willing to put up the money to start a private plant.[18] He first contacted Shibusawa Eiichi, who believed fertilizer production to be potentially profitable. Shibusawa convinced Mitsui's Masuda Takashi to cooperate, thereby gaining support in capital and management from Mitsui.[19] In April 1887, the group established Tokyo Artificial Fertilizer (Tōkyō Jinzō Hiryō).

In contrast to the enthusiasm of the firm's small group of initial investors, farmers were wary of the product.[20] Government officials, anxious to promote agricultural productivity through improved fertilization, were as unhappy with the farmers' reluctance as were the company's shareholders. Though anxious to help the new firm, they hesitated to grant direct subsidies to manufacturers. Instead, they offered indirect forms of aid. They sold surplus sulfuric acid to the

TABLE 5 Sales at Tokyo Artificial Fertilizer, 1888–1898

1888	184
1889	465
1890	390
1891	1,560
1892	1,853
1893	1,571
1894	3,176
1895	4,013
1896	7,013
1897	11,096
1898	16,395

Source: Shimotani Masahiro, "Dai Nihon Jinzō Hiryō," p.36.

fledgling company from the Finance Ministry's Printing Office at Ōji at bargain prices and, more important, undertook an educational program to encourage chemical-fertilizer use. Officials at the Ministry of Agriculture and Commerce appealed to farmers by advertising in newspapers, by disseminating thousands of explanatory pamphlets replete with fancy wood blockprints, and by sending out graduates of agricultural colleges to instruct farmers in new methods of fertilization.[21] Moreover, Tokyo University's agricultural school at Komaba continued its experimentation in the 1890s under the direction of an international team.[22] The team attempted to mix superphosphate with a number of substances, including herring, rice bran, and night soil, and had particular success with the last. This was fortunate, because farmers were less reluctant to try a new product if it was mixed with an old familiar product.

In addition to publicizing information about research to gain the confidence of farmers in chemical fertilizers, the government dispensed research assistance to individual firms. Some were permitted to use government laboratories, which were far superior to their company labs.[23]

Not only did the government *promote* the use of artificial fertilizer; the Ministry of Agriculture and Commerce also attempted to *control*

TABLE 6 Comparative Unit Prices of Organic and Inorganic Fertilizers, 1896–1920 (yen per ton)

	Nitrogen			*Phosphate*			*Potassium*		
Year	*a*	*b*	*c*	*a*	*b*	*c*	*a*	*b*	*c*
1896*	706	719	–	201	205	–	208	–	–
1897*	708	719	–	202	205	–	216	–	–
1898	839	718	–	236	205	–	245	–	–
1899	808	718	–	231	205	–	240	–	–
1900	767	729	–	229	213	–	207	–	–
1901*	679	740	–	203	210	–	175	–	–
1902*	642	786	–	176	204	–	167	–	–
1903*	724	773	–	198	200	–	189	–	–
1904	898	753	–	244	200	–	245	–	–
1905	869	769	–	264	220	–	219	208	–
1906	822	649	–	274	205	–	272	222	–
1907	873	613	–	283	183	–	238	176	–
1908	723	646	–	217	175	–	219	205	–
1909*	620	647	628	186	173	168	186	204	198
1910	743	638	626	220	172	169	227	204	200
1911	745	713	649	204	170	155	203	205	187
1912	856	737	611	206	158	131	205	189	157
1913	828	732	616	198	155	130	199	185	156
1914	782	665	642	220	173	167	236	207	200
1915*	669	751	659	168	168	147	170	200	175
1916*	753	1048	793	165	202	153	165	242	183
1917*	995	1392	910	189	240	157	198	285	186
1918*	1255	1764	1148	245	302	197	247	359	234
1919	1589	1559	1260	471	400	323	468	476	385
1920	1634	1228	1638	781	555	740	869	662	883

Source: Chōki keizai tōkei: nōringyō, Vol. IX, Tables 23–25, pp. 202–207, were used in preparing this table.

Notes: a. Nutrient source is an organic fertilizer.

b. Nutrient source is an inorganic fertilizer.

c. Nutrient source is a compound fertilizer.

Asterisk (*) denotes years in which organic fertilizers tended to be cheaper than inorganic fertilizers. Through 1908, unit values for nutrient content in compound fertilizers are not separately calculated, but rather are included in the price calculations for the other two types. Values are for the appropriate component in all types of fertilizers, although the nutrients cannot be used in an "unfixed" state.

TABLE 7 Sources of Phosphate Fertilizer Used in Japan, 1881–1915 (1,000 metric tons)

Year	*Animal* [a]	*Vegetable* [b]	*Mineral* [c]
1881–1885	37.6	9.4	–
1886–1890	30.9	8.9	–
1891–1895	36.2	10.3	–
1896–1900	39.1	12.1	14.7
1901–1905	53.3	26.0	57.8
1906–1910	59.7	50.6	162.9
1911–1915	70.3	76.1	226.9

Source: Calculated from *Chōki keizai tōkei: noringyō*, Vol. IX, Table 21, pp. 198–199. Note that superphosphate supplied, after the turn of the century, a larger percentage of demand than animal or vegetable sources. Though fish cakes appear more popular than soy cakes, this is only as a source of phosphate; soy cakes supplied more nitrogen by 1900, and the net weight used was higher than that of fish cakes. Furthermore, a large part of column a is not made up of fish cakes but of bone meal.

Notes: a. Dried fish, fish meal, bone meal.
b. Soybean cake, rapeseed oil cake, cottonseed oil.
c. Superphosphates.

its quality. The State Agricultural Experimental Station, established in 1893, included an analysis department. Furthermore, the Fertilizer Control Law of 1899 proscribed adulteration of fertilizers and required licensing and inspection of manufacturers and vendors by prefectural governors. The Law was strengthened in 1901 to include inspection of component ingredients in fertilizers as well.[24] These laws were helpful in gaining consumers' trust and were, therefore, an important part of the government's policy to encourage development of the industry.

At Tokyo Artificial Fertilizer, Takamine tried to overcome the problem of the farmers' insecurity by mixing the superphosphate with materials they understood, thereby creating compound fertilizers, which did sell.[25] Clever fertilizer merchants soon realized that there was no reason Tokyo Artificial Fertilizer should earn most of the profits from mixing fertilizer types. It was a simple matter to buy the constituent parts, mix them, and pocket the middleman's percentage.

Surprisingly, these same clever merchants were stymied when it came to carrying out the central part of the manufacturing operation—the production of superphosphate. The production method was actually quite simple and should not have required special equipment or knowledge. Sulfuric acid was poured over phosphate rock and reaction time allowed. This produced the superphosphate fertilizer.[26] But general technical knowledge among farmers and average commercial people at the time was so meager that even this simple technique was beyond them.

Despite the efforts of both government and industry, chemical fertilizers did not really achieve great popularity until after the China War (1894–1895). Hostilities temporarily impeded imports of the popular soybean-cake fertilizer from China.[27] The temporary setback in imports, simultaneous with an unusually poor catch by Hokkaido fishermen, gave chemical manufacturers the opportunity to establish a toehold in a fertilizer market dominated by fish and soybean cakes. Farmers had little choice but to try the chemical fertilizers: An unnatural fertilizer was better than an insufficient harvest. Increased agricultural productivity beginning in the 1880s (due, in part, to application of fish and soybean cakes) had accustomed farmers to high yields unattainable with application of night soil, compost, and other traditional fertilizers alone. Once a significant number of farmers had overcome their initial reluctance to experiment with non-traditional substances, demand for new fertilizers of every type increased quickly. This increased production and, in time, reduced prices dramatically (see Tables 5, 6, 7).

MARKETS, MERGERS AND MONOPOLIES: COMPANY ORGANIZATION IN SUPERPHOSPHATES

Like superphosphate manufacturers in the United States, Italy, and France, Tokyo Artificial Fertilizer and other Japanese superphosphate firms were, despite their simple processes, the precursors of a diversified chemical industry.[28] One of the reasons for the central position of Tokyo Artificial was its large share of the Japanese chemical industry. Absorbing other fertilizer firms throughout Japan after the Russo-Japanese War (1904–1905), Tokyo Artificial Fertilizer came to control 63 percent of the Japanese market by 1913, making it the first of the immense firms characteristic of the chemical industry in the twentieth century.[29] As nitrogenous fertilizers began to be produced in larger quantities during World War I, however, Tokyo Artificial, still producing only superphosphates, began to lose its large market share in a fertilizer market no longer dominated by superphosphates.

Tokyo Artificial spent the first eight years of its existence struggling to survive. In 1895, its management observed that all European fertilizer makers made their own sulfuric acid[30]—and conversely, that many of the sulfuric acid makers had added fertilizer to their lines.[31] Thus, it made plans to start production of the important acid.

Tokyo Artificial Fertilizer's decision to supply its own sulfuric acid was well timed. Within two years of beginning manufacture, the Tokyo company was one of the largest producers of sulfuric acid in a market increasingly dominated by fertilizer makers. Indeed, in 1898, fertilizer production required 37 percent of Japan's output of sulfuric acid, surpassing its consumption in soda manufacturing (13.4 percent), bleach making (15 percent), oil refining (4.6 percent), and all other uses (30 percent).[32] By that year, Tokyo Artificial Fertilizer's ability to produce 1.2 million pounds of sulfuric acid per month made it one of Japan's largest producers, surpassed only by Ōsaka Sulfuric Acid and Soda (Ōsaka Ryūsō), Ōsaka Alkali, and Nihon Chemie.[33] However, despite the increase in productive capacity of sulfuric acid, Japan's output continued to trail world production at the turn of the century (see Table 8).

The decision to begin production of sulfuric acid for the company's own use was a wise financial decision for Tokyo Artificial, but it failed to lead to a complex, diversified chemical company. It represented no technological advance; the decision was made more for financial than scientific reasons. Thus began the process of Tokyo Artificial's expansion without substantial upgrading of technology, a process that would continue until the company's important decision to invest in electrochemical production in the mid-1920s.

TABLE 8 Estimated Production of Sulfuric Acid, 1900 (1,000 metric tons)

Great Britain	1,100
United States	940
Germany	850
France	625
Austrian Empire	200
Belgium	160
Russia	125
Japan	50

Source: Haber, *The Chemical Industry 1900–1930,* p. 11.

The years after the Russo-Japanese War finally brought an end to Tokyo Artificial Fertilizer's ability to function in a market relatively free of domestic competition. Better technology had generated cheaper production of the superphosphate fertilizer, farmers had changed their farming habits to demand increasing amounts of the chemical, and the war itself, like the Sino-Japanese War ten years earlier, had temporarily impeded imports of organic fertilizers interchangeable with superphosphates. Naturally, demand was stimulated. Manufacturers overresponded to this demand and flooded the market with superphosphates. The number of producers had gradually increased since the turn of the century from 11 in 1898, to 21 in 1902, and to 44 in 1906. One hundred and fifty-nine domestic manufacturers were selling superphosphate fertilizers in the Japanese market by 1907.[34]

The sudden increase in the number of manufacturers had a deleterious effect on the profitability of Tokyo Artificial Fertilizers. Profits plummeted about 38 percent between 1905 and 1910, despite increases in the size of the company. Manufacturers, led by Tokyo Artificial Fertilizer, felt that their overproduction made their firms grossly unprofitable. Many sought to limit production through formation of cartels and merging of resources. In the event, cartels had limited success in restraining trade; mergers, however, served to increase Tokyo Artificial Fertilizer's control of the market.

Precedents already existed in Japan's young industrial economy for restraining trade. Furthermore, the chemical industry had been among the first to generate inter-firm agreements to limit competition and stabilize prices. Manufacturers of paper products had been in communication with each other as early as 1880, and entrepreneurs in the sulfuric acid industry tried their hands at establishing a marketing cooperative in 1900.[35] In 1904, Kansai area producers attempted a regional agreement, but their tentative gropings came to nothing. Similarly, attempts a few years later to regulate production and sales nationwide failed to bridge the distance between producers in the Kansai and Kantō regions.[36]

Cartelization attempts generated in the Kansai area failed to organize producers on a national scale, and regional organizations in the Kantō area had only limited success. The Artificial Fertilizer League (Jinzō Hiryō Rengōkai), formed in 1907, brought together ten Kantō producers and four importers of the raw material phosphate rock.[37] The group was eventually able to attract members from both the eastern and western regions of Japan, but it was not particularly effective. Some important manufacturers of superphosphate were not included in the League, thereby diminishing its ability to regulate prices. Furthermore, the presence of importers and suppliers of phosphate rock was inappropriate, since they shared few of the manufacturers' concerns. But, just as the cartel was reorganizing to provide for potentially more effective joint action, the need for a cartel like the League became moot. Tokyo Artificial Fertilizer's transformation that year into Dai Nihon Artificial Fertilizer (Dai Nihon Jinzō Hiryō KK) and its sustained growth so increased the firm's dominance in the Kantō region that the need for a nationwide cartel was obviated.[38]

Its large share of the Kantō market allowed Tokyo Artificial Fertilizer to begin acquiring other firms in 1908, despite poor market conditions. Although the firm's high profit rates decreased, it continued to make money and to pay dividends. In 1909, Tokyo Artificial bought out Settsu Oil's fertilizer operation in Ōsaka, earning Tokyo a foothold in the Kansai area. This opening was followed in 1910 by an even greater acquisition, that of Ōsaka Sulfuric Acid and Soda.

Thereafter, mergers continued apace; only Taki and Niigata Sulfuric Acid remained independent of either Dai Nihon or one of its related subsidiaries.[39] Thus, mergers and not cartels succeeded in regulating Japan's superphosphate industry before World War I.

A second round of mergers among superphosphate manufacturers occurred after prices plummeted following World War I, most notably the merger of Dai Nihon with Nihon Chemical and Kantō Sulfuric Acid in November 1922 (see Figure 1). At one turn, Dai Nihon regained the 60 percent of market share it had held before the war. Higher profits permitted expansion into areas of more highly sophisticated technology. The company almost doubled its capital before becoming Nissan Chemical in 1936.[40] Accompanying these mergers were more successful inter-firm agreements concerning pricing and production; increasing concentration of capital in the fertilizer industry allowed for easier, although not automatic, cooperation among producers.

This cooperation occurred within the private sector. The government's role in regulating prices and distribution was minor until after 1930. Existing government regulations, such as requirements for proper labeling of products (1908) and the intensification of research and standardization of fertilizer use through the Regulations for the Encouragement of Fertilizer Production (1921), generally aided the industry. But self-regulation by industry members, especially through the threat of stiff penalties by the East-West Monthly Society (Tōshi Tsukinamikai), a manufacturers' association, was more instrumental than government regulation in aiding recovery after 1920.[41]

Repeated mergers allowed Dai Nihon to grow to the immense proportions that would become characteristic of the chemical industry as a whole. But Dai Nihon, until the late 1920s, resembled other conglomerates in size only. Dai Nihon was simply Japan's single largest producer of superphosphates, and, until the 1920s, made large quantities of only one other chemical—the sulfuric acid needed for fertilizer production. It did not diversify in related product areas as did the electrochemical firms, nor did it have their vertically integrated organization.[42] The artificial nitrogenous fertilizer produced by the electrochemical firms, on the other hand, was the object of rapid

FIGURE 1 Growth of Tokyo Artificial Fertilizer through Mergers and Acquisitions, 1887–1930

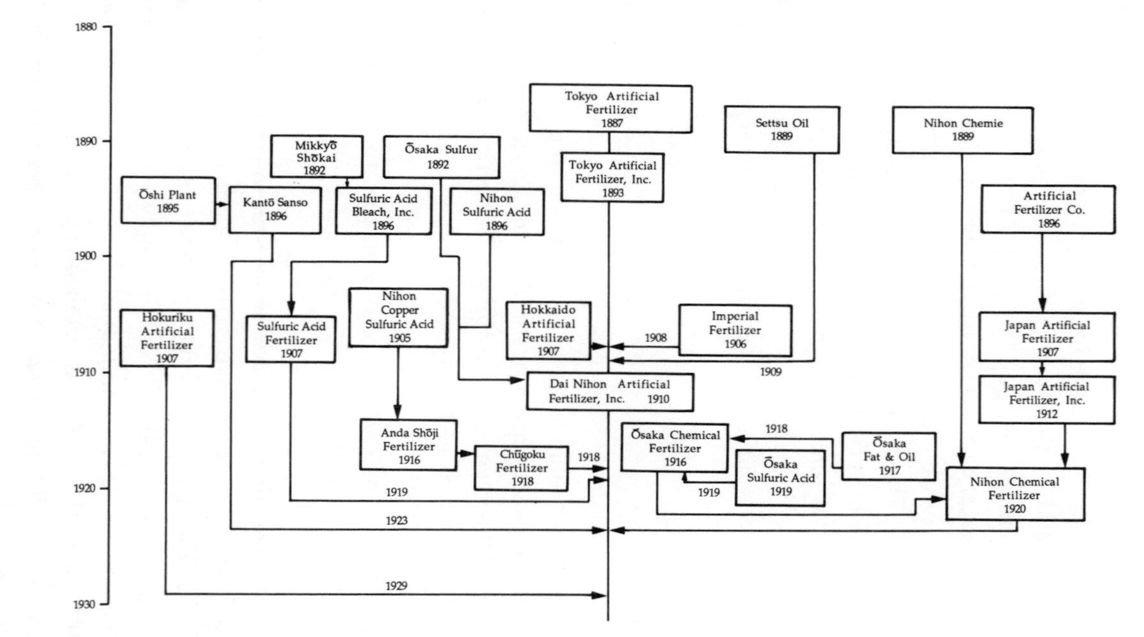

Sources: Nissan, pp. 569–570; Watanabe Tokuji, *Gendai Nihon sangyō hattatsushi*, p. 88; Nakamura Chūichi, *Nihon sangyō*, p. 102.

Note: Of the remaining firms producing superphosphate, only Sumitomo Chemical, Rasa Industries, Niigata Sulfuric Acid, and Taki Fertilizer remained completely independent.

transfer of sophisticated technology and the catalyst for further technological development and transmission. Technological advances in these firms pushed them to diversify widely.

JAPAN'S REVOLUTION IN SCIENTIFIC EDUCATION AND RESEARCH

Sulfuric acid, soda, matches, and superphosphates were important early products of Japan's chemical industry, but it was the development of the electrochemical industry that spurred the nation's advances in chemical technology. The growth of electrochemistry in Japan in the first two decades of the twentieth century was extraordinarily rapid. This quick development had several requirements: investors able to understand technology; the effective growth of electrical generation; investors with access to capital; and the maturation of creative management capable of effective response to competition through organizational and managerial solutions.

The most basic of these developments was the improvement of technical competence permitting technology transfer and innovation. Technology cannot be absorbed, even by a nation with financial resources to purchase it, if educated engineers and technicians are incapable of applying it to industry. A purchasing nation that fails to innovate as it purchases will see its level of technology lag behind the forefront of scholarship, always a step behind the most productive leaders making the product with the highest quality and lowest cost. Productivity increases derive from technological breakthroughs, either in application of tested basic research or in new basic research itself. New engineering techniques can make labor more productive and products therefore cheaper. Innovative discoveries can lead not only to new manufacturing methods but also to original products for which a market can profitably be created.

Progress in the ammonium sulfate industry offers a clear example of the stages of Japan's technological maturation. When the fertilizer first entered the country, scientists examined it and considered methods of import substitution. Simple technology of production by the by-product method was acquired from overseas.[43] A market developed but continued to be supplied mainly from overseas. Japanese manufacturers sought ways of improving their product, but there were limits to the efficiency of the technology of by-product ammonium sulfate.

Technological breakthroughs in Europe and North America during the 1890s and 1900s pushed the level of technology in ammonium sulfate production far beyond the level possible with the old by-product technology. To remain abreast of international progress in chemicals, interested Japanese would have to absorb improved technology. By the turn of the century, thirty years of public policy encouraging scientific education had prepared a number of young scientists capable of comprehending and therefore replicating foreign technology. Especially important was the replication of the method of fixation of atmospheric nitrogen, widely used in producing two nitrogenous fertilizers, calcium cyanamide and ammonium sulfate.

Until the 1920s, this technology was purchased overseas and modified by knowledgeable engineers for domestic use. In a few cases, Japanese producers showed enough creativity to allow their products international competitiveness. They knew they could not remain competitive without contributing to international technology, for, without domestic innovation, all their technology would have to be imported, generally after its successful application in the West. Although innovation carried greater risks than acquisition of foreign technology, it also paid greater rewards in the form of higher retained profits.

The ability to innovate was, furthermore, an index of the degree to which technology transfer was effective. Occasional innovation was an indication of the ability and willingness of Japanese scientists to keep up with, understand, and absorb advances in the West. Japanese research in and development of the continuous method of cyanamide production to replace Europe's batch method in 1912 was the first example of an engineering breakthrough pioneered in Japan and exported to the world.[44] Improvements in methods of manufacturing calcium cyanamide had a direct bearing on production of ammonium sulfate, as the former was not only a finished fertilizer itself, but was also the main component used in manufacture of ammonium sulfate in Japan until after World War I.

Despite occasional innovations, Japan's technological level had not caught up to that of the West by World War I. For example, Japanese entrepreneurs were surprisingly unable to absorb already developed Western know-how for ammonia synthesis during the war. Although technological development was somewhat uneven, however, progress both in academic research and in its industrial applications eventually allowed the growth of the nitrogenous fertilizer industry. Moreover, the growth of that industry stimulated further research. Ready industrial uses for skills acquired in universities, colleges, and specialty schools made scientific studies attractive. Indeed, a widening pool of technically trained people meant that all types of undertakings requiring scientific skills would benefit, not only firms making nitrogenous fertilizers. Japan's development as a fertile medium for transfer, absorption, and regeneration of technology is one of the pivotal chapters in the history of the integrated development of its society, economy, and science.

Technological developments in electrochemicals dictated certain structural elements inherent in modern chemical firms in any society; for example, technical training even for upper management—managers in other types of firms were typically trained in finance or law—and expenditures for research and development. Other factors related to the growth of the industry—those concerned with energy resources, capital, and management—affected how that technology would be developed. Generation of the firms' own supplies of electricity was a common feature of nitrogenous-fertilizer plants. Because electrochemical reactions are produced by electricity, cheap and reliable sources of electricity constituted a second development, almost as important as technological advancement, encouraging growth of the nitrogenous-fertilizer industry.

A third interesting development in the growth of chemical firms was their type of financing. Relationships with zaibatsu banks as well as with official sources of funding, such as the Industrial Development Bank (Kōgyō Ginkō), were often cultivated. Technological considerations may have impelled chemical firms to lean toward management boards dominated by engineers and scientists, but financial considerations reminded industry leaders that investors and financiers had to be included as well.

These three developments necessary for the growth of the fertilizer branch of electrochemicals—sustained educational and scientific advances sufficient for technological transfer, growth of the electrical industry, and acquisition of capital for high technology industries—were well underway during the first decade of this century.

DEVELOPMENT OF THE INFRASTRUCTURE FOR TECHNOLOGY TRANSFER AND INNOVATION

Creating an infrastructure in which high-level, state-of-the-art technology could be understood and developed was a difficult task for a developing country like Japan in the nineteenth century. Being able to produce domestically the tools of modern economic and military strength was an important goal of the Meiji leadership, however, and so they made diffusion of the scientific knowledge necessary to replicate those tools one of their priorities. There were two mutually reinforcing ways to accomplish this priority, and the Japanese pursued both: First, they established a system of scientific education at various levels, from vocational-technical schools to universities; and, second, they created several public and private research institutions and laboratories. Successful researchers were graduates of Japan's technical colleges and universities. Many had additional training in Europe or the United States. Other alumni, though not necessarily affiliated with prestigious laboratories, formed the intellectual backbone of Japan's high-technology chemical firms. Still others worked in privately funded laboratories, although this phenomenon was more common in Japan's mining and related coal-tar industries. Because the school system was most responsible for scientific training, its establishment was a necessary condition for scientific advance.[45]

The Japanese worked long and hard to establish such a system, a key element in their macroeconomic policy. Technical education in

Western subjects had been established in Japan even before the Meiji Restoration (1868), but only sporadically at a few locations. A small number of students received some training at two shipyards set up in the 1850s and 1860s at Nagasaki and Yokosuka.[46] In the 1870s, a clearly defined, formalized system of scientific education was established. At the pinnacle of the system of scientific education during the 1870s were the Kaisei School (later part of Tokyo University) and the College of Technology (Kōbu Daigakkō). Other schools existed, such the Tokyo Telegraph School, run by the Ministry of Communications, and several army and navy schools, all of which taught technical subjects; education in these tended to be specialized, reflecting the purposes of the particular ministry running the school.[47] The Kaisei School and the College of Technology, although run by the Ministry of Education and the Ministry of Technology respectively, prepared their graduates for scientific careers in government, business, and academia and, therefore, had a more important role in diffusion of technology than the other schools during the Meiji period.

The Kaisei School, a descendant of an old shogunal institution, taught a number of subjects including science and technology. Supplementing the education offered by a large contingent of foreign teachers, trips to the United States and Europe rounded out the educational program of the best students at Kaisei.[48] These trips abroad were prerequisites for excellence and leadership in science and technology as the young students matured. Among the earliest graduates were men who would become Japan's leading academic chemists and researchers.

The College of Technology, managed by the Ministry of Technology (1870–1885), was more oriented than the Kaisei School toward practical applications of scholarship. Itō Hirobumi, the first Minister of Technology, was much taken with the idea of establishing an institute devoted to engineering and technological studies to replace the more informally organized Institute of Engineering (Kōgakuryō, established 1871).[49] He approached representatives of the trading firm Jardine Matheson, who did business in Yokohama, for help in finding suitable professors for a polytechnic institute in Japan. As a result, Henry Dyer, a 25-year-old disciple of the Scottish professor William Rankine (known for his advocacy of technology as an academic discipline), arrived in Japan in 1873 to create a new college for the Ministry of Technology.[50] With him arrived a large staff of foreign professors of science, most from England, Scotland, or Ireland. Among them, chemistry professor Edward Divers left a major imprint on the teaching of chemistry in Japan. He stayed at the College of Technology until it was absorbed by Tokyo Imperial University in 1885.[51]

"Dyer's College," as foreign residents in Japan called the College of Technology, offered instruction in seven major areas: industry, shipbuilding, mining, applied chemistry, engineering, electricity, and architecture.[52] Students received a rigorous, practical education. After six years of study—two years of preparatory education and four of specialization—the first class graduated in 1879, the year the school finally obtained a chemistry laboratory. Of the 33 completing the six-year course, 11 found themselves, within a few months of graduation, enroute to Europe and the United States for further study.[53] Takamine Jōkichi, founder of the superphosphate firm Tokyo Artificial Fertilizer, was one of the 11, earning a trip to Scotland and England, and later New Orleans, where he was introduced to superphosphates. One of 6 graduates in chemistry in the Class of 1879, Takamine was followed by 24 other graduates in chemistry (of a total of 212 graduates of the College of Technology between 1879 and 1885). Although not all would become as famous as Takamine, almost all who had completed the arduous course took leadership positions in government agencies and ministries.[54] A decade later, as government positions were increasingly filled, graduates began to venture into private industry, eventually expanding the scope of Japan's chemical industry.

While the College of Technology was producing chemists and other technically trained scholars, Tokyo University was also producing scientists. Formed in April 1877 by the merger of the Kaisei School and the Tokyo Medical College, Tokyo (later Imperial) University offered scientific education in various fields of science and engineering. The 1885 dissolution of the Ministry of Technology and resultant transfer of responsibility for the College of Technology to the Ministry of Education permitted Tokyo University's absorption of the College in that year.[55] Absorption of the College, with its more practical and less theoretical bent, produced some structural changes in the University. For instance, it meant the University would establish a Faculty of Engineering separate from the Faculty of Science. Stressing both theoretical and applied science, Tokyo University was moving in the direction of German-style technical education. Far in advance of any scientific education elsewhere in the world in the 1880s, the German system of universities and technical colleges (technische Hochschulen) produced scientific, and particularly chemical, scholarship worthy of international emulation.[56] Not surprisingly, the Japanese adopted many features of German technical studies. There was a major shift during the 1880s from sending Japanese students to study primarily in the United States and Great Britain to sending them to Germany, especially for chemistry.[57] Less obvious as a measure of Japanese adoption of German scientific education, but in the long term more important, was the creation of a multi-tiered system of scientific education based on the German model.

The multi-tiered system, in theory, would allow not only the dissemination of skills used in production, that is, skills necessary for commercial application of basic technology, but also for the creation of original theoretical science. Lower vocational schools in locations throughout Japan would create a skilled work force capable of dealing with the manufacture of high-technology products. Higher-level technical colleges would presumably train business leaders in sophisticated engineering and applied chemistry. And the Imperial Universities would train students in pure science. Logically, all levels of education should have been established simultaneously, but the lower level, the vocational schools, were set up later than the universities (after the turn of the century). People were being trained to conduct research in pure science before a skilled work force could be trained to manufacture the products of that research. It was not until after 1900 that skilled workers were to be employed in the chemical trades, and, even then, the work force at some plants had to be gathered from all over Japan.

The middle and upper levels of the chemical education hierarchy also failed to fulfill the original purposes for which they were created; graduates were, until the 1890s, absorbed by government agencies and given good jobs, which made those students who chose entrepreneurial careers instead seem to be taking unnecessary risks. An official career was attractive, and there simply were insufficient university graduates to fill all posts. The result was the siphoning off of most graduates of universities and colleges into government service rather than business.

Tokyo University took its chemical training quite seriously, as evidenced by its establishment of a separate department of applied chemistry in 1883. This was particularly important for the industry, as the department taught the kinds of studies entrepreneurs would need.[58] Kyoto Imperial University was founded in 1897 under government auspices and with government money, and had departments in pure and applied chemistry by 1898.[59] Tōhoku and Kyūshū Imperial Universities, which were particularly strong in the applied sciences, were established in 1911. Kikuchi Dairoku, the Minister of Education at the time, advocated funding higher technical schools rather than imperial universities. Tōhoku University was, therefore, not built with central-government money but was supported by grants from the Miyagi prefectural authorities and the Furukawa family, the latter wishing by its generosity to dispel public antagonism for Furukawa's tragic pollution of the area surrounding its copper mine at Ashio.[60] Tōhoku University included both a science faculty and an agricultural faculty. The science faculty differed from those of Tokyo and Kyoto in its stronger emphasis on pure science and its admission of women students.

Scientific education at the university level continued to expand. Although graduates in engineering consistently outnumbered science students 4 to 1—and some of the most influential early chemical manufacturers were trained as engineers—students of pure and applied chemistry also increased. By 1917, 1,182 students of chemistry had graduated from Tokyo Imperial University, 220 from Kyoto, 52 from Kyūshū, and 31 from Tōhoku, in addition to the 24 pioneering students in chemistry at the old College of Technology.[61] The effect of the increasing numbers of graduates was compounded by their high degree of occupational mobility from government jobs to private enterprise to teaching and research. Of all graduates of the Faculty of Technology at Tokyo University before 1890, approximately equal numbers entered government and private firms; firms claimed 56 percent of all who had graduated by 1921, however, while government employed only 34 percent. Almost two-thirds of all graduates of the College of Engineering and the Tokyo University Faculty of Engineering before 1905 had moved from the sector in which they found initial employment.[62] Such high inter-sectoral mobility increased the rate of diffusion of technological information, since interested scientists could share their knowledge with colleagues in various fields. Dissemination of knowledge received further impetus from the early formation of professional societies by graduates of programs in technology. The first chemical association, Tokyo Chemical Association (Tōkyō Kagakukai), was founded in 1877 by the earliest graduates of Tokyo University.[63]

Below the university level were several other strata of educational institutions sponsored by the Ministry of Education. The top of the middle tier of the multi-tiered system was occupied by the Tokyo Workers School (Tōkyō Shokkō Gakkō), founded in 1881.[64] Within a few years, the school had been upgraded to the level of an industrial school (Tōkyō Kōgyō Gakkō, 1890), and then to the level of a technical college (Tōkyō Kōtō Gakkō, 1903). Moving away from its initial purpose of training industrial workers, Tokyo Technical College eventually graduated a large number of managers for Japan's chemical firms. Ōsaka, Nagoya, and other cities emulated Tokyo in setting up their own technical colleges (called *kōtō gakkō* after the German technische Hochschulen).[65] Blue-collar technical workers were trained throughout Japan in a large number of vocational schools by the Ministry of Education under its Apprentice School Rules (1894), Technical School Ordinance (1899), and Professional College Ordinance (1902).[66] Supplementing these formal educational institutions under the Ministry of Education were the training facilities attached to research centers of the Ministry of Agriculture and Commerce after 1901. By the end of the Meiji period, 17 centers had been created, either for research or for training of operatives for technical and chemical plants, although government subsidization for these facilities was not approved until 1906. Even after the government agreed to fund the centers, subsidies failed to cover all expenses. Local funding sources had to make up the difference. That they continued to be established attests to the importance of the centers in the minds of the businessmen and officials who supported them financially, though they were far from the academic and commercial mainstream of Tokyo.[67]

Indeed, by the beginning of the twentieth century, the value of a scientific education had become clear to many Japanese. Individuals of various educational backgrounds could use the skills in their occupations; employers sought trained workers, engineers, and research scientists; and many within the government acknowledged the importance of a technically trained population able to develop products that were then being imported in large quantities. Although it produced a temporary glut of chemists during the 1920s, expanding the scope of chemical education through a multi-tiered system based on the German model was a major goal of officials in the late Meiji period.

Fostering education was one government policy used to achieve the goal of replicating foreign technology in Japan. The other was establishment of government-sponsored research laboratories. The contribution of publicly funded research to the chemical industry, especially to the nitrogenous fertilizer and electrochemical sectors, should not be underestimated. Japan had few role models when considering ways to set up research laboratories divorced from a university setting. German universities had the world's most advanced chemical laboratories in the late nineteenth century, and particularly productive professors were assigned lighter teaching responsibilities to be able to carry out research. By contrast, Britain's laboratories were outmoded, and America had just two modern facilities, at two new universities, Johns Hopkins and the University of Chicago. Around the turn of the century, a number of German chemists urged establishment of a chemical research institute; by 1912, the newly founded Kaiser Wilhelm Gesellschaft promoted independent laboratory work by chemists outside the university. As a result, scholars conducted excellent research in Germany's universities and private enterprises. The Carnegie Institute in the United States similarly promoted research after the beginning of the twentieth century.[68] Research laboratories were established in other countries and improvements of those in the United States and Germany were carried out under the pressure of World War I. Japan, on the other hand, did not have to wait for World War I to feel pressure to advance its sophistication in chemical research. Far behind in technology, Japanese scientists were well aware of the need for a research institute.

The first institution conducting research of benefit to the Japanese chemical industry was the Agricultural Experimental Station, the progenitor of branch stations throughout Japan. While studying the effectiveness of various fertilizers, researchers there carried out Japan's first scientific experiments in agricultural chemistry. This government-supported research had commercial applications, contributing to the development of Japan's phosphatic and nitrogenous fertilizer industries. Similarly, the Tokyo Industrial Experimental Laboratory (Tōkyō Kōgyō Shikenjo, established 1900, hereafter TIEL) and the later regional test laboratories used government-funded staffs and equipment to aid private industry. Public policy aided the scientists' attempts to catch up rapidly with Western technology and commercial applications.

Some large private companies also financed research, but on a limited scale.[69] The early mining and metallurgy companies, including Mitsui, Mitsubishi, Furukawa, and Sumitomo, all set up small research laboratories during the 1890s and 1900s. The scope of their experiments at that time, however, was limited to investigating segments of their current operations. Little attention was given to what would seem a natural object for investigation for mining companies—the study of the highly profitable field of chemical coal-tar derivatives. Other firms with research facilities, such as Tokyo Electric Company's light bulb laboratory, did not focus on chemical research. With the exception of the agricultural experimental stations and university laboratories, there were no organized facilities to which aspiring chemical producers could turn for technical assistance during the nineteenth century. As the experience of the pioneers in electrochemical nitrogenous fertilizers indicates, some of the best original research was conducted by individuals, either in the inventor's home or in a bit of work space solicited from his employer. The tinkerers' fortuitous success in electrochemistry earned them profits with which to try out new methods of production; at the same time, the founding of the Tokyo Industrial Experimental Laboratory opened up additional resources for scientific investigations of industrial processes.

By the end of the ninteenth century, it had become obvious to officials in the Ministry of Agriculture and Commerce that even the groundbreaking efforts by private entrepreneurs investigating some rather simple electrochemicals failed to go far enough in advancing the level of technology in Japan. Some even felt that Japan would never be able to substitute domestic products for technologically sophisticated imports. To counter this situation, the Ministry of Agriculture and Commerce held a three-part series of conferences from 1896 to 1898 that vitally affected the status of basic research.[70] Although the major reason for the conferences was to plan how Japan could deal with its foreign trade problems, the question of scientific research was a crucial consideration. Japan would have to learn to produce for itself the high-technology goods it was importing in vast quantities; for that reason, facilities accessible to all manufacturers would have to be set up to do basic research as well as investigate imported products.[71] These conferences attracted the attention of businessmen, academics, and officials in the Ministry of Agriculture and Commerce.[72]

The necessity of centralized research facilities had also become increasingly obvious to the government officials who would have to fund any such expensive venture. By 1900, the Cabinet of Yamagata Aritomo threw its support behind the idea of establishing research facilities, and both houses of the Diet approved it. The Tokyo Industrial Experimental Laboratory was officially founded on 2 June 1900.[73] In 1903, the central laboratory in Tokyo was ready for operation. Chemical analysis was only one of a number of areas of scientific study that received the attention of the scientists there.

Because the laboratory had been established to aid private entrepreneurs in their research, it had the responsibility of carrying out experiments on request. During 1906, the chemical laboratory's third year of operation, 77 percent of the requests for research assistance were for help in chemical analysis of minerals or metals. Only 8 percent of the requests that year concerned chemical products themselves.[74] At that time, the laboratory was not yet equipped to carry out a wide range of chemical tests; dye research was begun only that same year, and electrochemical research had yet to commence.[75] But the number of experiments conducted increased rapidly from 275 in 1903, the laboratory's first year, to over 1,000 in 1907.

The Tokyo Laboratory was fortunate to have several outstanding leaders in chemical research. The first director was Takayama Jintarō, one of Japan's first graduates in chemistry (Tokyo Imperial University, Class of 1878), a long-time student in Germany, geologist in the Ministry of Agriculture and Commerce, and Professor at Tokyo Imperial University. He was also the laboratory's major proponent before its establishment. Kodera Fusajirō, head of the electrochemical division (founded in 1909), was a talented scientist with foreign training and long experience with high-pressure machinery.[76] His background made him exceptionally adept at maximizing advances in mechanical engineering for the benefit of another field—chemistry. Understanding of mechanics and machinery was fundamental to advances in electrochemistry, as was demonstrated later when limitations in Japan's machinery industry impeded development of commercial production of nitrogenous fertilizers after World War I. Under Kodera's directorship, the electrochemical division of TIEL began to do basic research in areas left untouched by the laboratory before 1909. Carbide, potassium chloride, and calcium cyanamide, all fertilizers or components of fertilizers, were developed for several commercial firms (Denka, Tōyō Soda, and others).

By the end of the Meiji period, TIEL was already deeply involved in advanced research in Japan. During World War I, it expanded considerably, and an affiliated nitrogen section added at that time conducted even more highly sophisticated research. Furthermore, TIEL was joined by other government-sponsored institutions during the war. Although the laboratory's level of research was somewhat primitive compared to the level achieved after the war, it laid the groundwork for systematic research and, together with the diffusion of science through the educational system, created the infrastructure for Japan's chemical industry.

Thus, the climate for the investment in electrochemicals had been created by the turn of the century. Less sophisticated chemicals, produced under government auspices or by private investors, had inaugurated the chemical industry. Chemical fertilizers had become acceptable to farmers. The chemical-fertilizer industry centering on superphosphates bore little resemblance to the later electrochemical industry, either technologically or institutionally, but its profitability encouraged later investment in other chemical fertilizers. And men with sound scientific training were able to benefit from the new environment to become the pioneers in the high technology of electrochemicals.

TWO

Pioneers in Electrochemicals: The First Wave

The name of Noguchi Jun figures prominently in all manner of Japanese studies dealing with the pre-World War II period: studies of the chemical industry, sketches of captains of industry, analyses of investors in colonialism, discussions of pioneers in electrical generation. By the 1930s, Noguchi's reputation was already larger than life, since he was the most successful of the early pioneers in electrochemicals. Although he was not alone in the early years of development, Noguchi and his company, Nippon Chisso Hiryō Kabushiki Kaisha (Japan Nitrogenous Fertilizers, hereafter referred to as Nitchitsu) dominated the first wave of high-technology development in the Japanese chemical industry.

Actually, there were two preeminent Japanese scientists in chemical production in the early years of the twentieth century: Fujiyama Tsuneichi as well as Noguchi Jun. The two worked first together and then as rivals to advance the Japanese electrochemical industry. Both were recent graduates of the Faculty of Engineering at Tokyo Imperial University and were the most creative early pioneers in industrial application of chemical principles in Japan.[1] Well read in the latest discoveries in Western technology, these young electrical engineers were fascinated by the use of electricity in the field of applied chemistry. Electrochemistry was a burgeoning field of study in the West in the 1890s, so germinal articles in Western journals were readily available to interested readers. Commercial applications of the theories expounded in these articles were uncommon, however; the young Japanese scientists were intrigued and interested in carrying out further studies of electrochemical processes.

By the beginning of the century, graduates of Japan's still relatively new technical educational system had advanced to the point where they could adopt and even modify some types of foreign technology. Increasingly, they had adequate power sources to use as they tinkered with new processes. Electrical generation, essential to the production of electrochemicals, was developing smoothly, one step ahead of chemical manufacture. And investors' increasing interest in advanced technology industries and government formation of credit institutions to further those important industries made it easier to accumulate capital, the other requirement for starting the electrochemical industry.

The last years of the Meiji period, when wars limited imports of organic fertilizers and farmers' increasing demands for soil additives spurred a growing market, were years of opportunity for young, highly educated graduates of Tokyo Imperial University. Their ability to innovate and absorb technology, increasingly apparent in the electrochemical field, testifies to the success of the policy encouraging scientific training and to the increasing science-mindedness of the Japanese people. Many were excited by the possibility of innovating new processes. In addition, their advances in electrochemical technology were stimulated by the existence of a market for ammonium sulfate fertilizer made by the "by-product" method, although this method itself was technologically unrelated to the synthetic, electrochemical method. Young enthusiastic researchers, then, became young entrepreneurs, enjoying a ready market for their electrochemically produced fertilizers, the springboard for all advanced chemical technology. How and why this by-product ammonium sulfate was introduced, and how young researchers built on early electrochemical discoveries to arrive at the more sophisticated synthetic method of ammonium sulfate manufacture will be examined in this chapter.

THE "NITROGEN PROBLEM" IN THE WEST AND JAPAN

Fertilizers with a high nitrogen content were of great value to Japanese farmers. Intensive agriculture requires nitrogenous fertilizers, a fact not lost on farmers who had been spending hard-earned cash on nitrogen-rich soy cakes. But nitrogen fertilization remained inadequate, and, according to German scientist Justus Liebig's Law of the Minimum (a restatement for fertilizers of the law of diminishing returns), the "plant nutrient in least supply determined the yield of the crop."[2] Finding adequate sources of nitrogenous fertilizers independent of foreign sources like the Manchurian suppliers of the ever-popular soy cakes would, therefore, be of great benefit to Japanese agriculture. At the same time, entrepreneurs involved in producing a chemical in as much demand as fertilizers could turn a tidy profit.

Japanese farmers understood in practice what was scientifically articulated in the West around the turn of the century: that nitrogen was the "key nutrient, determining plant growth and setting the level of consumption for other fertilizers."[3] High consumption of soy cakes and fish fertilizers indicated Japanese farmers' desire to obtain the soil-improving qualities of nitrogen—even if few knew what the element was—as these organic substances are particularly high in nitrogen. Japanese demand for commercial fertilizer reached 100,000,000 yen before World War I. This included both organic and synthetic products, but growth in demand was primarily for artificial fertilizers.[4]

During the first decade of the twentieth century, Japanese agronomists at Komaba and Nishigahara in Tokyo corroborated rural wisdom by carrying out sophisticated research in the comparative benefits of various fertilizers, including nitrogenous fertilizers. Scientists conducted parallel research on sturdy plant types capable of withstanding heavy applications of fertilizer without collapsing.[5] They found sodium nitrate, a nitrogenous fertilizer popular in Europe, to be poisonous when used under the normal conditions of Japan's paddy agriculture. Other studies indicated that calcium cyanamide and ammonium sulfate were effective nitrogenous fertilizers and that the latter was more beneficial than the former if the farmer was unable to afford more than a small amount.[6]

The research with sodium nitrate (Chilean nitrate) was particularly interesting, since it revealed one of the problems Japanese farmers had to consider when assessing relative values of fertilizers. Although safe in Europe's dry-field agriculture, nitrate fertilizer broke down in Japan's wet fields, producing poisonous nitrites and allowing beneficial nitrogen to escape.[7] Unable to use that source of nitrogen and dependent on soy cakes, the supply of which could be cut in time of war with Asia, Japan's agricultural policymakers thought it important to find other sources of nitrogenous fertilizer.

Locating these sources was a problem not only in Japan but also in the West. Indeed, commonly held neo-Malthusian concerns that the world's supply of natural sources of nitrogen would soon be depleted produced talk of a "Nitrogen Problem."[8] Western farmers using Chilean nitrates for fertilizer competed with explosives companies and sulfuric acid makers who also needed the mineral for their products. During World War I, the international shortage of nitrogen became so acute that governments around the world made nitrogen research a matter of top priority. Actually, European concern for increasing supplies of nitrogen had manifested itself as early as the 1780s, when Henry Cavendish (1731–1810) worked out a simple way of combining nitrogen and oxygen in the laboratory.[9] Over a hundred years later, Sir William Crookes, an eminent British scientist, reiterated the worry about the Nitrogen Problem:

> England and all civilised nations stand in deadly peril of not having enough to eat. As mouths multiply sources dwindle. Land is a limited quantity; it is the chemist who must come to the rescue of the threatened

> communities. It is through the laboratory that starvation may ultimately be turned to plenty.[10]

Eventually, the British developed another source of nitrogenous fertilizer, although in potentially limited quantities. During the 1870s, producers of gas and iron and steel realized that they had the valuable resource at their fingertips.[11] Officials of municipal-gas companies were the first to discover ways to use otherwise wasted ammonia produced in their coking furnaces. The process of converting the waste to ammonium sulfate, a nitrogenous fertilizer, was simple. The major additional cost was for sulfuric acid, necessary for conversion of ammonia to ammonium sulfate. As producers more than recouped their expenses for the sulfuric acid by sales of fertilizer, the incentive for manufacturing by-product ammonium sulfate—so called because it was a by-product of gas production or steel refinery—was strong. Furthermore, ammonia was not the only waste product of the coking procedure for which furnace operators found a purpose; tar, another by-product, produced pitch for naval vessels. Managers at gas works and steel mills began to invest in fertilizer production, although many had to upgrade their technology to use special recovery ovens for coke production. They found it was worth the cost. Soon other Europeans copied the British. German producers followed suit, and, by 1907, 85 percent of German ammonium sulfate was produced by the by-product method.[12] For several decades, the by-product method appeared to supply sufficient quantities of nitrogenous fertilizers to supplement Chilean nitrates (see Table 9). Indeed, British gas companies and steel mills produced so much ammonium sulfate (430,000 tons in 1913) that three-quarters of it was exported, primarily to Spain and Japan.[13] By-product ammonium sulfate was eventually manufactured in Japan, too. The development of the by-product industry was a significant stage in the advancement of electrochemical technology and in the creation of a market for its products.

TABLE 9 World Nitrogenous Fertilizer Use, 1880–1913, selected years (1,000 metric tons)

a. Chilean Exports of Sodium Nitrate

1880	*1885*	*1890*	*1895*	*1900*	*1905*	*1910*	*1913*
224	436	1,063	1,239	1,454	1,650	2,336	2,738

b. Estimated Production of Ammonium Sulfate

	1900	*1905*	*1910*	*1913*
Germany	108	268	373	549
Britain	217	273	374	440
United States	n.a.	59	105	177
France	37	45	57	75
Belgium	20	18	36	49
Austria-Hungary	18	20	29	35
Other (of which Japan)	24	43 (.47)	30 (3.5)	76 (7.3)
Total ammonium sulfate	424	726	1004	1401

Source: L. F. Haber, *The Chemical Industry 1900–1930,* pp. 100–101; Nakamura Chūichi, *Nihon sangyō no kigyōshiteki kenkyū,* p. 13. Note that, although sodium nitrate imports continued to surpass significantly ammonium sulfate production, the amount of contained nitrogen supplied by ammonium sulfate was fast approaching the level supplied by nitrates, since the nitrogen content of ammonium sulfate was higher.

NITROGEN FERTILIZERS BEFORE ELECTROCHEMISTRY

Japan's first sample of by-product ammonium sulfate came not from England but from the Sydney, Australia, municipal gasworks in 1896; subsequent imports were predominantly British in origin before World War I. Within a few years, the process was copied by Japanese gas producers, but imports continued to dominate the Japanese ammonium sulfate market.[14]

Import substitution was one consideration when Japanese investors in gasworks were planning their companies. Although Japanese producers clearly were unable to make much of a dent in the massive quantities of imported ammonium sulfate before World War I, they knew that a market for chemical nitrogenous fertilizer existed because farmers were buying the imports (see Table 10).

Ammonium sulfate offered greater profit to manufacturers than superphosphates. In contrast to superphosphates, most of the final value of the product was added by manufacturers. The process was quite simple, requiring only that producers take coal, process it to obtain coke, retain the waste-material ammonia, and add sulfuric acid, which they usually made themselves.[15] Though the process was simple, the producer "did more" than the producer of superphosphate.

The second reason for the profitability of ammonium sulfate was the lucrative effect of the economies of joint cost. That is, a large part of the cost of manufacturing by-product ammonium sulfate could be written off as cost of coke or coal-gas production. What accounting procedures municipal gas companies or coke producers used to attribute costs to which products are not apparent in company histories; suffice it to note that the incremental cost of ammonium sulfate conversion from ammoniacal wastes was sufficiently low to allow profits that encouraged its production.

Despite the obvious profitability of by-product manufacture and despite the relative simplicity of its production, growth of the by-product ammonia industry was initially slow because prospective producers were afraid that recovery ovens would not be able to produce the high-quality coke they needed in their principal operations. It was only in 1898 that Shimomura Kotarō of Ōsaka Chemie imported one of Europe's modern recovery furnaces, the Semmet-Solvay. But, to his company's dismay, the coke it produced was not of high quality, vindicating the fears of conservative producers. The machinery also had a poor breakdown record. As demand for ammonium sulfate was still capable of being satisfied by foreign imports, the cost of using Ōsaka Chemie's furnace appeared to be greater than any benefits of by-product recovery. Nevertheless, the recovery furnace produced Japan's first by-product ammonium sulfate and coal tar in November 1898, laying the foundation for later effective production.[16]

TABLE 10 Ammonium Sulfate Consumption, 1900–1913 (tons)

Year	*Domestic Product*	*Imported Product*
1900	n.a.	1,744*
1902	n.a.	2,523
1903	n.a.	2,902
1904	n.a.	11,228
1905	467	26,012
1906	549	39,778
1907	1,341	62,649
1908	1,935	66,376
1909	2,764	46,202
1910	3,534	69,364
1911	5,699	74,242
1912	7,194	84,599*
1913	7,342	110,140**

Sources: Watanabe Tokuji, *Gendai Nihon sangyō hattatsushi*, Appendix pp.27–35.

Notes: * starred figures from *Meiji kōgyōshi: kagaku kōgyō*, p. 734; ** double-starred figure from Nakamura Chūichi, *Nihon sangyō no kigyōshiteki kenkyū*, p. 13. Nakamura's import figures for years in which there are multiple sources of data tend to be about 1% below other sources, and should probably be adjusted upward by 1% for consistency.

After the rather unspectacular results at Ōsaka Chemie, original Japanese research in organic chemicals including coal-tar derivatives awaited better market conditions.[17] The Russo-Japanese War ended the short period of inactivity in research by stimulating production of pitch, made from coal tar, for Navy vessels. Producers, most of them zaibatsu firms, who had earlier hesitated to switch their furnaces to recovery ovens, came to see the possibilities of immediate profits. Furukawa (in 1905), Hachiman Iron (1907), Miike Coal (1912), and Mitsubishi Mining (1913) bought recovery ovens.[18] The last seven years of the Meiji period saw a proliferation of recovery operations by mining companies and other firms with a ready supply of coal. Production of coal-tar derivatives and ammonia by these recovery ovens was historically significant because it was the zaibatsus' only major contribution to Japanese fertilizer supplies before their large-scale development of synthetic ammonia in the 1930s.[19] Furthermore, coal-tar derivatives spawned a vast array of modern chemicals for industrial, consumer, and military uses.

More important in the development of by-product ammonium sulfate than the establishment of recovery ovens at refineries, however, was the creation of municipal gas companies. As in England, gasworks were responsible for a large part of the fertilizer's production in its first decades in Japan. Gas sent to consumers had to be purified by removing ammonium compounds, and the simplest way was to filter it through sulfuric acid. Tokyo Gas Company was the first of the big gas companies to produce ammonium sulfate on a large scale, although its commencement of fertilizer production in June 1901 lagged behind the company's founding in 1874.

Tokyo's installation of gas service followed that of Yokohama, where foreign residents had encouraged the industry.[20] Viewing the progress of their neighbors to the south, city fathers in Tokyo decided to set up a municipal gasworks as well.[21] In time, the company grew and prospered: By 1901, it increased its capitalization by 1500 percent and paid out high dividends to stockholders. This rapid growth stimulated official researchers at Tokyo University to turn their attention to finding ways of using gas in chemical production and experimentation.[22] The academics' spirit of investigation was contagious; managers at Tokyo Gas made theirs the first Japanese municipal gas company to emulate foreign works and replicate the ammonium sulfate originally brought from Sydney. Thereafter, the company, which had manifested steady growth in employment and wealth, grew at an accelerating pace.[23]

After production of the fertilizer began at the company's Fukugawa plant in 1901, other firms followed suit. Ōsaka Gas, founded in 1897, began to produce by-products in 1906. Nagoya Gas, founded in 1906, retrieved by-products after 1913. Kyoto Gas had a tiny (75 tons) but noteworthy output of ammonium sulfate by 1912. At least 74 municipal gas companies were created during the Meiji period—all

FIGURE 2 Growth in By-Product Ammonium Sulfate Manufacture, 1903–1912

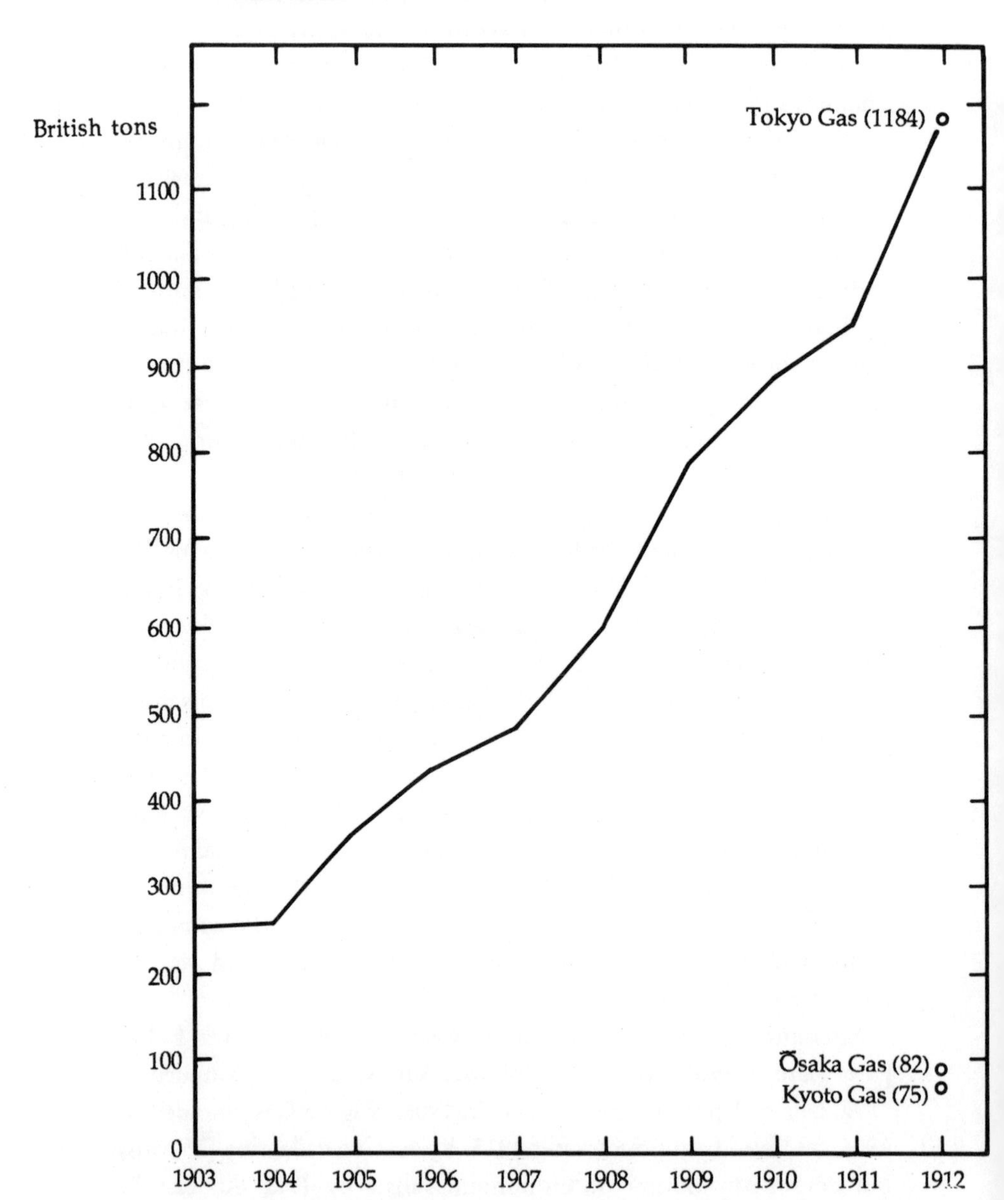

Sources: Figures for Osaka and Kyoto Gas from *Meiji kōgyōshi: kagaku kōgyō*, pp. 176–177; other data from Kantō Taru Seihin KK, p. 30.

TABLE 11 Ammonium Sulfate Produced as Gas By-Product, 1918 (tons)

Tokyo Gas	2,100
Osaka Chemie	600
Nagoya Gas	240
Ueda Gas	240
Osaka Gas	150
Ida Coal Mining*	600
Fuji Paper*	400
Railroad Board*	360

Source: Nihon Ryūan Kōgyō Kyōkai, p. 53.

Note: *Starred producers used the Mond method, which yielded more ammonium sulfate but at greater expense.

but 12 during the last four years—and many found it profitable to collect by-products (see Figure 2 and Table 11).[24] Tokyo Gas continued to be the leader in the field.

By 1918, however, the growth rate of advances in by-product recovery was decelerating. Imports of ammonium sulfate continued to surpass domestic production, although imports had dropped precipitously during World War I from approximately 16 million yen in 1913 to 300,000 yen in 1918.[25] The slowdown in growth rate meant that the output of by-product ammonium sulfate would increase only by a factor of 3 between 1918 and 1930.[26] Instead, newer methods of ammonium sulfate production by electrochemical synthesis were the principal beneficiaries of import restrictions during the war.

Other nations continued to develop the technology of by-product recovery; in 1929, the Germans produced 16 times as much by-product ammonium sulfate as the Japanese and the Americans twice as much as the Germans. But electrochemical methods made better use than the by-product method of Japanese resources in electrical generative capacity and of the Japanese researchers' technological ability. By the 1930s, Tokyo Gas stopped producing ammonium sulfate altogether.[27] Although by-product ammonium sulfate had been superseded by the new synthetics, it had helped stimulate the demand by Japanese farmers for the product and created a market for artificial nitrogenous fertilizer.

The demand for increasingly sophisticated fertilizers and the advances in production technology to manufacture them were incremental, indicating the expansion of Japanese ability to use technology. The Japanese progressed through the same stages of technology development as their counterparts in the West, although at a much accelerated pace because of their late start and the possibility of adopting already painstakingly researched Western ideas. The history of the advancement of technology is like a collapsible telescope. Whether extended or collapsed, the telescope permits light to pass from the external lens to the eye, where it is interpreted. The light follows the same path, whether short or long. Advances in scientific applications follow similar though not identical paths, whether made slowly or adopted quickly, because they are dependent on incremental progress both in intellectual skills and in the infrastructure of technology in industry. From soybean cakes to superphosphates to by-product ammonium sulfate, the manufacturing technology became ever more complex, and the value added by processing ever greater. This is a logical progression for the development of technology in any economy, but it is most clearly shown in the use and manufacture of chemical products. At their most primitive, nitrogen in fertilizers was available in, say, manufactured soy cakes (night soil was not, strictly speaking, manufactured), and the raw material counted for most of the value of the finished product. At their most complex, chemical fertilizers made use of elements as abundant and cheap as the nitrogen in the air; the development of technology and the nurturing of minds able to use it was costly and thus added much value to the finished product. The stages between the simplest and the most complex methods were incremental; thus, the by-product method was significant, even though it was technologically unrelated to the succeeding method.

EXPERIMENTING WITH ELECTROCHEMISTRY, 1900 TO 1908

Noguchi Jun, who along with Fujiyama Tsuneichi was the archetypical Japanese chemical engineer, was born on 26 July 1873 in Kanazawa, Ishikawa prefecture, the son of Noguchi Yukinobu.[28] Soon after Jun's birth, his father, a former samurai of Kaga, uprooted his family and moved them to Tokyo, where he went to work in what would later be the Ministry of Education. The boy attended the best schools in Tokyo and graduated in 1896 from Japan's only school of engineeering at that time, the Faculty of Engineering at Tokyo Imperial University.[29] He also had plenty of practical experience to supplement his formal study. The autumn before his graduation, Noguchi did part-time work (*arubaito*) as an electrical engineer at the Numagami electric generating plant which supplied the Kōriyama Silk Reeling Company. The plant generated far more electricity than was needed for silk reeling, and management therefore sought additional uses. These included long-distance transmission (Numagami was Japan's first such attempt), which Noguchi studied for his senior thesis. The knowledge certainly came in handy; Noguchi later developed into one of Japan's leading producers of electricity. His summer job had further implications; the engineering staff's search for additional uses for electricity led Noguchi to consider, probably for the first time, production of calcium carbide (CaC_2), an electrochemical.[30]

For the next two to three years, young Noguchi continued on as a consultant to the new Kōriyama Electric Light Company in Fukushima prefecture. This was a fruitful time for Noguchi to build important personal ties with later collaborators. At some time during those years in the Tōhoku, he was joined by classmate Ichikawa Seiji, later his most important partner. Toward the end of the period in the northeast, he also began to work with a third graduate of the Faculty of Engineering (Class of 1898), Fujiyama Tsuneichi, on a process for manufacturing the calcium carbide he had thought about while still a student.[31] Recently discovered by Canadian Thomas Willson, calcium carbide had commercial applications of interest to Noguchi and Fujiyama. The two carried out their research in Sankyozawa in Sendai, using a small 50-kilovoltampere electric furnace.

In 1898, Noguchi's father died, leaving his widow with young children to support. Jun, at almost 25, was the eldest son. Suddenly, his life of few resonsibilities and ample opportunity to carry out scientific experiments came to an end. Noguchi had to return to Tokyo to find more lucrative work. While Fujiyama continued his laboratory research in the Tōhoku,[32] Noguchi went to work for an American trading firm in the capital. But Noguchi was ambitious, loath to stay for long in a boring and unpromising job. Neither he nor his contemporary technical friends fit the recent stereotype of dedicated company men who remain throughout most of their working lives with one firm or agency. Noguchi was particularly mobile. Almost immediately after joining the American firm, he quit to open a store selling ink products. He soon abandoned that venture, too, and took a job more appropriate to his training with the Japanese branch of the German Siemens Company, at the time Europe's largest manufacturer of electrical equipment.[33] This time he stayed longer. He enjoyed his job marketing equipment, and remained until 1901. He even found time to continue his carbide research. All the while, he was impressed with what he saw at Siemens, and was determined that, in time, he would buy some electrical machinery, set up his own firm, and become his own boss. That time would arrive within the decade.[34]

While Noguchi was in Tokyo, Fujiyama had taken some quicklime purchased in an ordinary drugstore, smashed it with a hammer and, adding wood charcoal, sent a 30-volt charge through it.[35] Twenty-seven hours later, he had produced Japan's first electrochemical, calcium carbide. This was in 1901.[36] The search for a way of producing calcium carbide had originally interested Fujiyama in electrochemicals; in the long run, calcium carbide was less important in Japan as a finished product than as an ingredient used in the manufacture of more sophisticated chemicals, the first of which was the nitrogenous fertilizer calcium cyanamide. Fujiyama's employer, Itō Seijirō of Miyagi Spinning and Electric Light Company (Miyagi Bōseki Dentō), noticed his employee's diligence and had sufficient foresight to offer him space in the company's warehouse to continue his experiments.

Fujiyama was working on a product that enjoyed a moderate demand following its introduction to Japan after the Sino-Japanese War. Solution of calcium carbide in water produced acetylene, used in lamps on bicycles, on fishing boats, in mining operations, and at festivals and shrines.

Until Noguchi and Fujiyama began their experiments, Japan's demand for acetylene had been supplied by imports of calcium carbide. In January 1902, the Tanaka factory in Koriyama, following plans drawn up by Noguchi, began production of the substance.[37] This particular example of substituting a domestic product for the foreign one was not an obvious solution to the recurrent nineteenth-century problem of excess imports. Cheaper foreign imports made penetration of the market by a domestic product difficult. Furthermore, start-up costs for import substitution were not small. Entrepreneurs considering commercial production had to weigh the high cost of electricity in Japan at the turn of the century. Several decades later, when electricity costs were far lower, energy inputs still accounted for a hefty one-fifth of total cost of the finished product.[38] The high cost of electricity was often a moot point, however, because of the immaturity of the generating industry. Hopeful electrochemical producers, often lacking cheap commercial sources of electricity, had to build their own electrical generating facilities. Self-sustaining chemical development required ready access to all necessary resources, and that often meant adoption and appropriate modification of all aspects of the technology supporting the final product—in this case, the technology of electrical generation. Chemical producers were thus instrumental in the early hydroelectric industry, either as producers of electricity themselves or as the principal consumers of other companies' output. Their demand stimulated both increased capacity and productivity in the industry.

Advances in methods of generation of electricity occurred around the turn of the century. During the 1890s, coal supplied most of the energy to produce electricity. Japan, blessed with abundant though seasonal rainfall as well as varied terrain with its differentials in altitude, was a natural site for hydroelectric facilities.[39] Harnessing Japan's rivers and streams would lower the cost of electricity.

Although capital costs for hydroelectric plants were higher, operating costs, and thus unit costs, were lower than for coal-fired plants, since the generating material (water) was free. Furthermore, what might appear to be a drawback of hydroelectric power in Japan—the seasonal variations in water volume of its rivers—was of aid to the chemical industry. Hydroelectric facilities were often set up to generate the minimal amount of electricity needed at the driest time of the year; this meant that, when water was more plentiful, surplus electricity was generated. This excess production had low marginal cost; fixed capital was in place to supply the minimal demand, and additional water over the dam was nature's gratis gift.[40]

By the turn of the century, hydroelectric plants were beginning to appear, especially in areas where waterfalls were plentiful. Eventually, the energy could be used for new industries like electrochemicals, but it would be several more years before cheap and plentiful electricity would be seen as the crucial element for producing ammonium sulfate, calcium cyanamide, and other electrochemically produced fertilizers.

Young scientists like Noguchi and Fujiyama were more interested, for the time being, in calcium carbide than in fertilizers. And indeed, calcium carbide was important. Its production marked a watershed in technological development in Japan by inaugurating the age of electrochemicals, as important in the years before World War II as petrochemicals would be after the war. In the next few decades, an entire industry originating in calcium carbide production—ammonia, explosives, acetylene, acetic acid vinyls, butanol, esters, acetates, and other synthetic products—came into being. The industry first began to grow as the demand for nitrogenous fertilizers, produced as derivatives of calcium carbide, grew.

The earliest stages in the carbide industry coincided with the growth of hydroelectric power. Not surprisingly, the earliest developers of the chemical product were also interested in generation of electricity. Noguchi was one of these. For him, working on calcium carbide and electricity also offered a degree of personal satisfaction. He had tired of working in Tokyo and tried several jobs in the country, including consulting for the builders of the Enoshima Railroad and managing a lead mine in Miyagi prefecture. None of these jobs suited him, making him recollect more fondly the years of doing research in Sendai with Fujiyama. A trip to Europe and America during the Russo-Japanese War allowed him to defer the question of his future a while longer while he absorbed ideas about Western science and technology.[41] Ten years had passed since his graduation with a small, elite group of scientists from Tokyo University. He felt it was time to settle down.[42]

The climate was right for Noguchi's involvement in electrochemicals. He had mastered the technology required for commercial application of chemical processes. As an electrical engineer, he was soon to show that he could augment the already developing electrical industry in Japan by establishing electrical facilities. What remained to be gathered was investment capital. In that area, he was also prepared; Noguchi had not wasted his years in Tokyo. He had honed his skills at drinking and socializing. His bibulous activities brought him into contact with men who would support his investment efforts. Not only did he strike up friendships with colleagues in engineering and science, but he also apparently impressed at least one financier, Chisawa Heisaburō of the Shimotani Bank. A few years later, in 1906, Chisawa decided to lend his old friend Noguchi Jun 100,000 yen for which he received 1,000 shares of electric-company stock.[43]

Noguchi continued to employ a personal style to persuade investors to part with their money throughout his career, from his days as a struggling young technician to his relationship with the Governor General of Korea as a mature entrepreneur thirty years later. Because he had to rely on loans, investments from large financiers, and sales of shares to finance his early enterprises, Noguchi employed energetic methods to raise capital. At the same time, he and his scientific colleagues were responsible for advancing the level of technology in his firm. The functions of an executive in Noguchi's position were quite eclectic. Noguchi apparently sought similar versatility in his executive staff. Over the years, the most successful managers in his company tended, with a few exceptions, to be technically trained men with a penchant for negotiations and financial matters.

While Noguchi was following a circuitous path toward acquiring the technology, resources, and capital to begin his own electrochemical production, he maintained a consistent interest in that field in the years after his departure from Sendai. Several additional carbide factories had been established by others while he was meeting influential friends and contemplating his future. Noguchi remained involved, to varying degrees, in the affairs of some of these factories. At Sankyozawa, for instance, where Noguchi had originally experimented with electrochemicals in 1898, capital was needed to get Fujiyama's production of calcium carbide underway. Noguchi, Fujiyama's boss Itō, and Ichikawa Seiji each contributed 1,000 yen as soon as Fujiyama was sure of success in the venture.[44] At another site, Yokohama financier and a later supporter of Noguchi Jun, Tanaka Shinshichi, was busily setting up facilities for Kōriyama Carbide Company. With his connections at the German Siemens company, Noguchi was in a position to help Tanaka. The young technician had carefully absorbed wisdom at the feet of Siemens's Herman Kessler (president of Siemens's office in Japan), and Viktor Hermann (head of the technical division of Siemens in Japan), enabling him to help Tanaka design the firm. Furthermore, he had maintained good relations with Siemens, which enabled him to acquire the appropriate machinery from the German firm.[45]

Within a few years, other firms were producing calcium carbide. Tanaka opened another plant at Niigata in 1903, using Hokuriku Hydroelectric Company's surplus electricity, and companies started producing carbide in Nagano (1906), Iwate (1907), and Toyama (1910). To all appearances, Japan had a rapidly maturing carbide industry, even before fertilizer production absorbed a large part of its output. At the same time, the lag between technological innovation in the West and Japan's ability to absorb new technology was narrowing. For example, in 1900, Fujiyama replicated in the laboratory Thomas Willson's process for producing calcium carbide after reading about the discovery in a journal; this occurred only eight years after Willson's experiments.[46] Certainly Fujiyama was a creative genius, but more important for the continued development of science and technology in Japan, his recreation of a modern chemical process within a year of graduation from the university, without having observed the process and without years of on-the-job training, attests to the growing sophistication of scientific training in Japan. Few Japanese at the time shared his level of education, but those who did were beginning to be able to make practical use of the knowledge disseminated by Japan's training facilities.

Fujiyama's business acumen, however, fell far short of his scientific abilities, underscoring the rarity of the combined managerial and technical skills of his partner, Noguchi. In the years before the two men began manufacture of calcium cyanamide, they evinced strong contrasts in ability to manage successfully. When called upon to function in a managerial capacity, Fujiyama was less than outstanding. He had been appointed an administrative manager at the first Nihon Carbide, Inc., founded in 1906 when the workshops at Sankyozawa, Kōriyama, and Nagaoka were merged, but his company had split into its three original parts by November 1907.[47] Fortunately, Fujiyama's stint as manager was short-lived. Soon he accepted Noguchi's invitation to leave the northeast where he had been working since his graduation from college to join Noguchi and Ichikawa in setting up a new carbide firm in Kagoshima.[48]

With his move to Kagoshima after returning from his foreign trip, Noguchi was finally making direct application of his training as an electrical engineer. Two other drinking friends from his Tokyo days, Nagasato Yūhachi and Hino Tatsuji, were developing mines in Kagoshima and apprised him of the need for additional electric generation there.[49] Here was a good opportunity. Noguchi hoped that by turning to these friends for technical help and to Chisawa of Shimotani Bank for financial help, he could harness the water power of the Sogi waterfalls. He succeeded. Established at Ōguchi in Kagoshima on 12 January 1906 and capitalized at 200,000 yen (50,000 paid up), Sogi Electric, Inc. used Siemens equipment to generate electricity for private consumers and nearby gold mines.

Noguchi was more successful than he originally calculated, and found that only half the firm's generative capacity was consumed by its intended users.[50] This offered him, now joined by his friend Fujiyama, an ideal opportunity to use the surplus capacity to produce calcium carbide.

The chemical process for carbide manufacture required more electricity than Noguchi's surplus, however. Ironically, Noguchi would have to increase his generating capacity in order to utilize that surplus. With his original investment, he had been able to buy the equipment to generate 800 kilowatts of energy from the waterfalls; doubling his investment to 400,000 yen would procure the machinery to produce 6,000 kilowatts. This expansion initiated an interesting relationship between chemical production and generative capacity at Noguchi's company. Surplus electricity led Noguchi to increase or diversify production. This, in turn, would lead to a shortage of electricity to match desired production, which would encourage greater output of electricity. A surplus usually ensued. This pattern, in which new products were the result of the quest for ways to use surplus capacity, is discernible until the years of economic mobilization during World War II.

Noguchi followed his decision to expand generative capacity by another bold managerial stroke in 1907. With his friends Fujiyama Tsuneichi and Ichikawa Seiji, Noguchi established Nihon Carbide Company (authorized capitalization: 200,000 yen) at nearby Minamata, an ancient salt-producing village later to suffer horribly from mercury poisoning associated with his company's acetylene manufacture.[51] (This notorious example of excessive enthusiasm for technology-driven growth was caused by the Chisso Company, none other than Nitchitsu, renamed Chisso after World War II. The mercury was dumped over a period of several decades beginning in the 1920s.)

Noguchi's decisions of 1906 and 1907 were bold indeed, and eventually rewarded him and his associates handsomely. In the short run, though, they caused a severe cash-flow problem that almost brought his company to an untimely end. Just as Nihon Carbide was about to succumb because of Noguchi's tying up every last yen in fixed capital, Kagoshima businessman Mutō Kinosuke came through with a welcome loan of 2,000 yen. Though seemingly small, the loan kept the firm solvent at that juncture and earned its granter gratitude disproportionate to the loan's size.[52] Soon thereafter, the facilities at Minamata became Japan's largest production operations for calcium carbide, using in 1907 a 500-kilowatt electric furnace yielding 15 tons per month.[53] After the initial tense moments during the summer of 1907, Nihon Carbide began to earn profits from its production of carbide used for acetylene. These profits were divided with Sogi Electric.

Noguchi proved to be an effective manager, attracting talented men to work in both his electric company and in his carbide company.[54] Though some of his early employees were friends wooed over cups of sake, these colleagues chose to stay with their able leader despite the many job opportunities, especially with prestigious government agencies, that competed for their skills. Engineers at that time were generally highly mobile (although the youthful Noguchi may have been an extreme case!). Company loyalty was not yet a widespread social value. Yet Noguchi's employees chose to stay because of his charisma; salaries were comparable to those elsewhere. In 1909, for example, Ichikawa was asked why he did not seek his fortune in Korea, Japan's colonial equivalent of a "frontier." He replied, "Because it would upset Noguchi's work."[55] Noguchi also showed his effectiveness as a manager by being able to oversee two different though related types of industries—electricity and chemicals—and managing two firms, all while still in his early thirties. Noguchi began to emerge as a leader in a growing industry.

THE FOUNDING OF JAPAN NITROGENOUS FERTILIZERS

As Noguchi Jun and Fujiyama Tsuneichi placed the period of uncertainty in 1907 behind them, they redirected their attention to the process of technology acquisition and development which had initially interested them in science. Intrigued by an interesting European discovery in the field of fertilizers, they sought to adopt it and, in the process, merged their two Kyūshū firms. The young scientists wasted no time; within two years of first commercial production of calcium cyanamide (also called lime nitrogenous fertilizer, $CaCN_2$) at Piano d'Orta, Italy, in 1906, they were sailing for Europe to observe and acquire the process. Vigilant about spotting new trends in science, the two men continued to advance chemical technology in Japan. Calcium cyanamide was still something of an unknown quantity. Because the fertilizer was so new in Europe, it was still unknown to European farmers; only a few companies in Europe and North America had begun to produce the fertilizer between 1906 and 1909.[56] The move by Noguchi and Fujiyama was rather prescient and indicative of their understanding of recent advances in chemistry and technology.

The cutting edge in modern technology in electrochemicals had been honed in Europe and North America through a series of scientific breakthroughs during the 1890s and 1900s. The first was Willson's discovery of calcium carbide in 1892. From 1895 to 1898, Germans Adolf Frank and Nikodem Caro carried out experiments in fixing atmospheric nitrogen at a German-Austrian agricultural experimental station; that is, they separated nitrogen from the air and forced it to combine with other elements in a usable form. Fixation of nitrogen was extremely important. The ability to synthesize nitrogenous fertilizers was crucial, since the world's supply of Chilean nitrate (sodium nitrate) was finite and by-product ammonium sulfate was produced in insufficient quantities. Several governments were also concerned that future wars might impede the flow of nitrates from Chile just at the time those imports were most needed for manufacture of explosives, their other use. Thus, the implications of the Germans' discovery were far-reaching, and strategic applications of nitrogen technology could not help but facilitate development of more peaceful uses of fixed atmospheric nitrogen as a fertilizer. The Frank-Caro cyanamide method of fixing atmospheric nitrogen had the additional virtue of simplicity.[57] The experiments of Frank and Caro with purified nitrogen were themselves made possible by the discovery in 1895 by Karl von Linde of the method for liquifying nitrogen from air. In 1901, Adolf Frank's son Albert publicized his observation that his father's discovery made a good fertilizer. In 1906, the Frank-Caro method was first used for commercial production at Società Generale per la Cianamide in

Piano d'Orta.[58] This brings the story to its Japanese chapter. Reading about the process in a newspaper article, Noguchi Jun decided to expand the possible uses for the calcium carbide his factory produced at Minamata.

Noguchi and Fujiyama went straight to Berlin in 1908 to obtain the patent rights to the Frank-Caro method. The two Japanese, both in their thirties, faced stiff competition from powerful adversaries. Both Mitsui and Furukawa, wealthy zaibatsu, sent skilled negotiators to win the patent rights; Mitsui's Masuda Takashi, with his proven track record in fertilizer investment (superphosphates), went over from London, and Furukawa's Hara Kei, later Prime Minister and perhaps the best-known negotiator in Meiji politics, went from Paris.[59] It seems rather remarkable that the two small investors could best the large zaibatsu. But the young men made a good impression in Germany and were sent on to Italy for the next round. The Società Generale was similarly impressed with them. Noguchi clearly managed his firms well and produced a large amount of calcium carbide, the fundamental raw material for calcium cyanamide. Not only were the Italians aware of Noguchi's personal accomplishments; they also knew of his connections with the German Siemens firm, the Società's principle stockholder and designer of its electrical plant. Moreover, Frank and Caro had conducted their experiments with the backing of Siemens and the Deutsche Bank, which then sold the patents to the Italian firm.[60] Noguchi's Siemens connection offered more than a passive recommendation; Kessler, his former mentor from his days in Tokyo, put in a good word for the young Japanese engineers.[61] Kessler may also have persuaded the Italians that Noguchi planned to develop the license with the financial aid of Mitsui. Noguchi and Fujiyama were thus able to win the licensing rights over their well-established rivals.

Having snared the valuable rights to produce calcium cyanamide in April 1908, the two returned to Japan.[62] In August 1908, they combined their two companies, Sogi Electric and Nihon Carbide, and founded Japan Nitrogenous Fertilizer, Inc. (Nitchitsu). But buying the patent rights and importing the necessary and expensive electrical equipment for the new firm cost 400,000 yen, or about the same amount as the total worth of both Sogi Electric and Nihon Carbide.[63] The engineers clearly needed help.

Mitsui was the obvious choice for such help. It had been understood at the time of Noguchi's negotiations with the Società that Mitsui would cooperate with Noguchi. Noguchi informed his Board of Directors about negotiations with Mitsui on 20 August 1908. Negotiations continued through September and October, when they were broken off because Noguchi clearly could not accede to Mitsui's conditions. These conditions stipulated that Mitsui would own half the shares and would influence management decisions by dominating the Board.[64] A form of financing with fewer strings attached would have to be found.

Noguchi next appealed for assistance to a relative on his mother's side of the family,[65] Hori Tatsu, a member of the Board of Directors of Japan Mail Steamship Co. (Nippon Yūsen Kaisha, N.Y.K.). Hori asked Kondō Renpei, President of N.Y.K., how he might help Noguchi. Kondō referred the question to Toyokawa Ryōhei, head of the banking division (later Mitsubishi Bank) of Mitsubishi Gōshi Kaisha. Toyokawa not only promised to help Noguchi; he also introduced him to Iwasaki Hisaya, President of Mitsubishi, and Nakahashi Tokugorō, President of Ōsaka Merchant Steamship (Ōsaka Shōsen Kaisha, O.S.K.). All these influential men were instrumental in furthering Noguchi's enterprise. Nakahashi became a major investor and President of Nippon Chisso Hiryō.[66] Toyokawa (Nakahashi's successor as President) and Iwasaki were investors. Significantly, both invested as private individuals, not as representatives of Mitsubishi Gōshi or Mitsubishi Bank. Iwasaki was known to use his personal wealth to back creative, strong entrepreneurs.[67] Although venture capitalists are thought to have been rather rare during the early twentieth century in Japan, Iwasaki and Toyokawa were clearly two such investors. The common misconception that they only represented their company probably derives from the belief that the zaibatsu dominated all investments, including those of small, innovative start-up firms. Further studies on investment in other technology-intensive production would indicate whether venture capitalists, zaibatsu-related or not, played an important role in other companies as well.

Noguchi reported to his Board members in August 1908 that he had obtained the Frank-Caro process, that he planned to increase Sogi Electric's capital to 1,000,000 yen (640,500 was paid up initially) and that he planned to merge Sogi Electric and Nihon Carbide. Nitchitsu may thus be said to have been launched in 1908, although Nakahashi, who is often referred to as the first president, was not installed until 1909.[68]

Understandably, the new company's backers expressed a strong interest in what Noguchi and his colleagues were doing with the assistance they received. One way to make sure the company was properly managed was to lend some experienced managers from Mitsubishi. Noguchi and his long-time friends Fujiyama and Ichikawa retained the top managerial positions under President Nakahashi in the fledgling company, as executive director and managing director, respectively. The rest of top management included: Toyokawa (director), Kagami Katsuichirō (auditor), and Shiraishi Naojirō (director), all with Mitsubishi connections; Hori Keijirō (auditor) from O.S.K.; Watanabe Yoshirō (director), President of Aichi Bank and a college friend of Noguchi; and Yamaoka Kuniyoshi, Hino Tatsuji (directors), Iwakiri Tarōkichi and Onoe Hiroshi (auditors), all four of whom had been part of the original management team of Nitchitsu in the months from August 1908 to January 1909.[69] Noguchi, Fujiyama, and Ichikawa were most closely involved in day-to-day operations, although Mitsubishi men occupied many of the company's first management positions. Nitchitsu was Noguchi's company and he retained control over its activities.

In succeeding years, Mitsubishi's influence on the Board of Directors diminished as men with Mitsubishi backgrounds died or left. As the company became better established, men whose primary interest lay with Nitchitsu rose to management positions. Investors had gained greater confidence in Nitchitsu and felt little need to secure their investments by offering either management assistance or by meddling. To be sure, Noguchi availed himself of loans from Mitsubishi Bank, but extending loans did not give Mitsubishi the power to help make management decisions in the way ownership of shares would have done. After World War I, Noguchi was generally free of interference from his "patron" company. Fortunately for Noguchi, Mitsubishi men did not press for additional control in management. Until 1931, Noguchi customarily read Nitchitsu's annual report before a group of Mitsubishi executives, but this was a courtesy rather than a way of incorporating Mitsubishi into Nitchitsu decision making.[70] Mitsubishi executives concerned about Nitchitsu saw the chemical firm more as a potentially profitable personal investment than as one of their manufacturing operations.[71]

Noguchi underscored his respectful distance from Mitsubishi by giving ample recognition to other investors and by borrowing money from the government's Industrial Bank (Kōgyō Ginkō). The latter loan, in the first half of 1910, was particularly helpful, because its large size (500,000 yen) made large-scale capital expansion of Nitchitsu possible without reliance on Mitsubishi alone and because Nitchitsu was close to being unable to meet the monthly payroll in 1910.[72] The large loan from the government is also interesting because it indicates a significant difference from chemical firms in some other industrialized countries. Except for periods when Nitchitsu was unusually profitable, management was, in the early years, willing to borrow amounts that far exceeded the firm's reserves. In contrast, German chemical firms on the eve of World War I tended not to borrow in excess of their reserves.[73] (This pattern changed in later years, however.) Nitchitsu's capital structure was more debt-heavy than most comparable Western firms at that time.

Noguchi had been careful to manage his financial operations, the import of advanced technology, and the integration of carbide manufacture and electrical generation. Yet his success as a manager did not guarantee his company's initial success in marketing nitrogenous fertilizers. The product had serious imperfections. Explosions caused by impurities in the calcium cyanamide rocked the plant. Calcium cyanamide should be approximately 20 percent nitrogen, but the Japanese firm's impure product contained only about 10 percent. This was perceived as an extraordinary problem, as Mitsubishi executives uncharacteristically expressed their concern about the product's quality. Indeed, this appears to be the only case of Mitsubishi interest in micromanagement of Nitchitsu affairs until 1932.[74] Furthermore, farmers found the product unappealing for two reasons. First, it was somewhat foul-smelling and unattractive in color; second, it retained some calcium carbide, harmful to plants, which made application riskier than farmers would have liked.[75] Nitchitsu was faced, then, with two separate problems: improving the technology of manufacture to remove impurities; and changing farmers' attitudes to make the fertilizer more acceptable.[76]

The second problem was easier to solve than the first. A relatively painless way to circumvent farmers' reluctance to use the product was to change it to something they were accustomed to using. That was ammonium sulfate, another inorganic nitrogenous fertilizer, which farmers had been buying from municipal gas companies for several years. The method of production for ammonium sulfate from calcium cyanamide, pioneered in 1900 by Adolf Frank and Nikodem Caro, was simple: Steam was blown over calcium cyanamide, which produced ammonia and limestone, and then the ammonia was taken and added to sulfuric acid to produce ammonium sulfate.[77] This was called the cyanamide process. Until the development of synthetic methods of producing ammonia—that is, by direct combination of atmospheric nitrogen and hydrogen—this process was the most commonly used method for ammonium sulfate production in Japan. It relieved Japan of dependence solely on the by-product process in domestic manufacture.[78]

Noguchi and Fujiyama, always eager to begin new processes that held the promise of profit and scientific excitement, were ready to enter ammonium sulfate production. They set up an experimental pilot plant at Hiejima in Ōsaka, near Nitchitsu's corporate headquarters, and used the calcium cyanamide produced at Minamata. But the problem of quality maintenance plagued the factory there. High-quality cyanamide was necessary as an ingredient in the manufacture of ammonium sulfate, and the operations at Minamata were not yet running smoothly enough to deliver it. It would take time for the technicians at Nitchitsu, and especially Fujiyama, who was beginning to feel somewhat disaffected from his friend Noguchi, to overcome the flaws in the process. Fujiyama claimed to be achieving an 18-percent nitrogen content in his cyanamide, but the product never contained more than 11 percent during these first years. After two years of operation, the Hiejima plant stopped production in May 1912.[79] This closing did not mean that Nitchitsu had given up on ammonium sulfate production for the future. Experiments had been carried out at Minamata, and plans were being laid for a plant at Himekawa in Niigata. Within a few years the company would make a major commitment to ammonium sulfate production by building a large plant in Kyūshū at Kagami. Furthermore, Hiejima continued to be used as a pilot plant.

The Himekawa plant was to be Nitchitsu's most ambitious effort to improve the quality of their fertilizer. The location seemed ideal, close to a source of limestone, a necessary resource for the calcium cyanamide they needed to produce ammonium sulfate. Moreover, Noguchi had access to capital, borrowing 500,000 yen from the Hypothec Bank (Kangyō Ginkō) and using funds later obtained from the Railroad Board's purchase of Nitchitsu's Kyūshū electric facilities.[80] Market conditions were improving gradually, and Nitchitsu's most talented managers personally oversaw construction. The requirements for effective investment were met. Nitchitsu increased capitalization to 2 million yen with its expansion to Himekawa. What no one anticipated was a freak flood in July 1912 that wiped out the entire plant. This was a major setback.[81]

In spite of the tribulations his company suffered, Noguchi's decision in 1909 to use calcium cyanamide as the basis for ammonium sulfate production was probably one of the wisest moves he ever made. In less than a decade, the firm would grow to produce more than twice as much ammonium sulfate as its nearest competitor (Denka) and 9 times as much as Tokyo Gas, the giant among producers of ammonium sulfate by the by-product method.[82] Although the firm eventually diversified, during its first years it concentrated on cyanamide and ammonium sulfate. The latter superseded superphosphate as Japan's top fertilizer after World War I. By that time, the cyanamide process had become less competitive, as cheaper methods using more modern technology were discovered. Rather than clinging to the cyanamide method, Noguchi placed his faith in ammonium sulfate itself and shifted his production method to newer processes as they came along. He eventually stopped making calcium cyanamide altogether. Noguchi's entrepreneurial sense would manifest itself repeatedly over the next three decades. Thus, Noguchi solved his second problem—the inability to market calcium cyanamide to reluctant farmers—by upgrading his technology to use cyanamide as a raw material in ammonium sulfate production. Market conditions in this instance affected his decision to invest in technology.

The first of the two major problems encountered by Noguchi in his early production of nitrogenous fertilizer, that of assuring high nitrogen content in cyanamide produced by the Frank-Caro method, was considerably thornier, and required the expertise of Fujiyama to solve. A number of factors contributed to the dilemma. For one, the carbon source used at Nitchitsu was charcoal rather than high-grade coke. Charcoal was not merely qualitatively inferior; its supply was also apparently uncertain. Managers worried that their operations might be disrupted if charcoal dealers went on strike.[83] This must have occurred with some regularity if managers noted it as a concern. Furthermore, the Siemens machinery used to produce the cyanamide broke down daily. Fujiyama felt it necessary to find a reliable way of producing cyanamide of consistent and high quality.[84]

Noguchi shared his partner's concern, but looked outside the firm for a solution. He invited an Italian technician to work on carbide quality, thereby exacerbating tensions with Fujiyama. Fujiyama eventually succeeded in improving on the European technology, but an unbridgeable chasm was created in his relationship with Noguchi.[85] At the same time, Fujiyama and Noguchi had a fundamental difference of opinion on how to market carbide, the raw material for cyanamide fertilizer. Fujiyama disagreed with the company's policy of relying on sales of carbide, which had been increased in 1909, to maintain company solvency. This disagreement alone may have sufficed to persuade Fujiyama to leave the company.[86] In the event, Fujiyama's departure was imminent by late Meiji. The company's auditor persuaded top management to decide which approach was superior, Noguchi's or Fujiyama's. Technical and managerial employees like Ichikawa were asked to investigate the situation; in the end, internal scrutiny favored Noguchi over Fujiyama.[87]

But Fujiyama continued his research, despite the escalation of tensions with Noguchi. After two years of study, Fujiyama eventually improved production technology which, in turn, permitted improved product quality. On 26 September 1911, Fujiyama's "continuous method" (flow method) replaced the original "alternate method" (batch method) of Frank and Caro.[88] The implications of this discovery were enormous. For the first time, a Japanese engineer had made an important and original breakthrough in a widely used commercial process. Fujiyama had a profound understanding of the original technology, including its faults, and was therefore able to advance it. In recognition of the importance of Fujiyama's contribution, Albert Frank, son of the original discoverer of cyanamide, Adolf Frank, journeyed to Japan to inaugurate the production process at Nitchitsu. Since 1908, Nitchitsu had introduced the cyanamide method to Japan, had constructed a major hydroelectric works, and had made internationally significant advances in technology. Those were remarkable achievements made at breakneck speed.

Frank's visit was the high point in Nitchitsu's first five years. But many problems lingered on. Two months after the ammonium sulfate plant at Hiejima closed down, the Himekawa plant under construction in Niigata prefecture was washed away by the flood.[89] Financial problems nearly crushed the company. Mitsubishi refused to extend Nitchitsu an extra loan in 1912, and things looked particularly bleak. Toyokawa suggested Noguchi petition the Industrial Development Bank instead. The company also tried to sell 1,000 tons of stockpiled cyanamide, but they failed to get the 90 yen per ton they needed. Finally, Mitsui Bank came through with a 150,000-yen loan for two months. When Mitsui refused to extend the repayment period in December 1912, however, Nitchitsu was again close to bankruptcy.[90] And, perhaps of even greater personal significance to Noguchi, he had to run the firm without the ever-ready technical expertise of Fujiyama Tsuneichi. The two college friends had different styles of operation, Noguchi being more the businessman and Fujiyama more the scientist. It is quite possible that their disparate personalities clashed even before Noguchi, by bringing in the Italian technician, demonstrated his lack of faith in Fujiyama's ability to perfect the technology for calcium cyanamide. Furthermore, Noguchi appears to have been more adventuresome, always willing to risk new types of investment. Whatever the cause of the rift, Fujiyama left Minamata in December 1911 and officially quit Nitchitsu in February 1912. He used the technology he had helped implant at Minamata to start a new electrochemical operation at Ōji Paper's factory at Tomakomai.[91]

Fujiyama Tsuneichi's departure from Nitchitsu in the winter of 1912 marked the end of the embryonic stage of Japan's electrochemical industry. Before Fujiyama's efforts to establish an independent firm, Nitchitsu had been the only firm in Japan to struggle with industrial applications of electrochemical technology for nitrogenous-fertilizer production. Fujiyama's creation of production facilities challenged Nitchitsu's dominance. It also expanded Japan's capacity for manufacturing the fertilizer and helped stimulate the market for artificial fertilizers in parts of the country where it had been weak, like Hokkaido. For a short period, the firms of Noguchi and Fujiyama dominated domestic production in the nitrogenous fertilizer industry, but, within five years, other entrepreneurs followed their lead. This competition did not end the originators' dominance in domestic production during the next two decades. But their leadership in production was not replicated in sales in the Japanese market. As was the case with other chemicals, foreign manufacturers enjoyed a large share of the Japanese market, offering stiff competition to Japanese producers.

Japan was part of a large international fertilizer market supplied aggressively by European and American merchants (see Table 12). The existence of a major foreign component in the Japanese market prevented Japanese manufacturers from creating an oligopolistic position for themselves. The price at which Japanese firms could sell their fertilizer was frequently higher than the price accepted by the more efficient European producers who could afford periodic sales below cost in order to maintain their market share. Japanese purchasers of fertilizer, therefore, had cheap alternatives to domestic products. Furthermore, for several years after World War I, a group of Japanese firms preferred earning profits by handling the European imports rather than by using available but expensive German patents

TABLE 12 Sources of Ammonium Sulfate Consumed in Japan, 1912–1930
(in tons; figures in parentheses are percentages of total consumption)

Year	England	Germany	U.S.	All Imports	Nitchitsu	Denka	All Domestic
1912	83,515	51	51	84,600	60 (.06)	–	7,313
1913	109,105	–	–	111,525	–	–	7,463
1914	104,589	71	–	105,638	7,579 (6.2)	–	16,050
1915	17,952	–	–	19,950	17,119(33.0)	3,000 (5.7)	31,838
1916	3,017	–	–	7,163	20,693(46.5)	6,000(13.5)	37,350
1917	11,223	–	102	15,113	20,949(37.5)	10,010(17.9)	40,688
1918	255	–	–	1,088	34,599(64.2)	11,001(20.4)	52,800
1919	25,510	–	64,166	101,213	52,681(29.2)	18,006 (9.9)	78,975
1920	28,068	–	38,792	72,413	51,331(33.6)	19,000(12.4)	80,100
1921	11,867	–	47,673	79,238	54,612(31.4)	17,000 (9.8)	94,763
1922	2,166	–	83,698	93,038	52,977(28.5)	18,997(10.2)	92,963
1923	45,620	1,071	86,454	145,725	55,920(22.4)	21,138 (8.4)	104,213
1924	59,262	54,586	34,052	168,397	61,084(22.0)	30,595(11.0)	108,713
1925	27,089	106,271	52,805	203,550	61,394(18.3)	38,564(11.5)	131,138
1926	39,630	174,785	68,721	296,025	68,756(15.5)	42,799 (9.6)	148,544
1927	60,584	136,512	40,740	250,014	85,081(19.9)	50,221(11.7)	177,950
1928	98,015	149,119	28,416	284,475	109,224(21.0)	64,581(12.4)	234,055
1929	139,087	182,515	53,706	380,658	96,973(15.7)	65,101 (9.6)	236,686
1930	83,769	173,346	41,372	302,905	218,718(32.4)	58,575 (8.7)	371,235

Sources: Figures for imports from data of Agricultural Affairs Bureau, Ministry of Agriculture and Forestry, cited in Hashimoto Jurō, "1920 nendai no ryūan shijō," p. 382. Nitchitsu, Denka, and All Domestic figures are from Shimotani Masahiro, "Nitchitsu kontsuerun to gōsei ryūan kōgyō," p. 75.

Notes: All Imports include imports from non-major suppliers as well (including Kwantung). Denka also made much cyanamide which it did not convert to ammonium sulfate.

for producing synthetic ammonium sulfate. This apparent preference of several Japanese companies for the easier task of importing fertilizer retarded their creative exploitation of patented methods. Though Noguchi and Fujiyama were producers rather than import-

TABLE 13 Percentage of Ammonium Sulfate Produced, by Method, 1917

Method	Percentage
Calcium cyanamide	61.3
Coke by-product	25.1
By-product of gas manufacture	11.2
Uremia	2.4
Total	100%

Source: Nihon Kagakukai, p. 217.

ers, they were affected by the presence of large-scale imports. Neither Fujiyama nor Noguchi was able to exercise total control in the fertilizer market, even during those decades when they led in production.

The prominence of two firms in fertilizer manufacture was nevertheless significant. Their pioneering roles allowed them to retain their leading positions, especially Nitchitsu with its continuing emphasis on innovation. Their large profit margins during World War I eventually persuaded manufacturers of the potential profits in nitrogenous fertilizer production, especially by the newer cyanamide method. In 1917, Hokuriku Electric Industries (Hokuriku Denki Kōgyō) was founded, followed in 1918 by Hokkai Electrochemicals (Hokkai Denka), Ibigawa Electric (Ibigawa Denki), and Daidō Fertilizers (Daidō Hiryō), all using the cyanamide process to make nitrogenous fertilizers.[92] The method became, therefore, the predominant process, and remained most important until it was superseded by the synthetic method in the 1920s (see Table 13). Most important, the leading positions of the two firms indicate that, since different methods of capital acquisition and different types of resources produced different organizational structures and rates of profitability, large-scale investment was possible under various technological conditions.

Nitchitsu's growth, in little more than half a decade, to become one of Japan's most technologically sophisticated companies, manufacturing a product in increasing demand, generating large amounts of electricity, and advancing the international levels of processing technology, appears remarkable. It is significant that this development occurred in a company headed by a risk-taking entrepreneur with scientific training and management skills. These qualities enabled him to make wise investments based on an understanding of technology. As technologies changed, he modified his investments and his company structure to maximize profitability. The company grew rapidly, and, as it grew, it began to alter its environment in perceptible ways—like stimulating greater demand for fertilizer, generating a pool of more talented employees, or encouraging growth in power generation. In time, the environment would not only affect how Nitchitsu managers made decisions, but it would also be affected by those decisions. It was not the zaibatsu alone that ordered the economy, but also, in the area of advanced technology, the dynamic entrepreneurial ventures.

THREE

Nitchitsu's Profitability and Expansion (1912–1923)

Noguchi Jun and Fujiyama Tsuneichi followed different paths to develop their chemical firms during the decade from 1912 to 1922. Each had access to the necessary funding, albeit from different types of sources. Each had access to necessary resources; both used the same raw materials throughout the decade. But, after an initial period of similarity, the two companies began to use vastly different forms of technology. This mattered, because the form of technology was as influential as access to capital in determining the types of decisions the companies' managements made. That is, the cyanamide process or "three-product cycle," (see Figure 3) used by both companies during the 1910s to manufacture carbide, cyanamide, and ammonium sulfate, engendered horizontal expansion where company size was increased by building similar plants. It failed to encourage the spin-off of technology in vertically integrated subsidiaries. By contrast,

Nichitsu's shift, in later decades, to a more sophisticated type of technology encouraged the spin-off of technology.

Noguchi abandoned the cyanamide method because the large profits he earned during World War I permitted him to invest in more advanced technology. Technological innovation, in turn, spawned new processes and products. This had two major effects. First, the production of related but unidentical products led Nitchitsu to become a multifaceted, complex chemical firm with numerous subsidiaries. Second, it "upped the ante"; the technological level of the Japanese economy as a whole was raised, making it necessary for other firms to follow suit to remain competitive. Nitchitsu's innovations therefore affected both company strategy and structure as well as the national economy. The converse was also true. That is, the growth of the economy during World War I had a beneficial effect on Nitchitsu's profitability and therefore its ability to innovate, and Noguchi's business strategy encouraged innovation.

Fujiyama's strategy, on the other hand, was less adventurous. His company diversified like Noguchi's only after its founder was replaced during the 1920s. The attitudes of the two men, both engineers with similar training, helped account for the contrasts in their firms' approaches to management, technology, investment, and expansion. In the end, Fujiyama's company, during his tenure in office, was less competitive and less influential than Noguchi's because it failed to follow the technological imperative; it was less innovative. This was reflected in its company structure and in its role in the industry. Nitchitsu was the leader, coming up with new products ahead of its competitor and manufacturing old products by cheaper, more efficient methods.

NOGUCHI'S DEVELOPMENT OF AMMONIUM SULFATE

It was not clear from the start that Noguchi would later be successful. Fujiyama's departure from Nitchitsu in 1912 coincided with a period of relative financial insecurity for Noguchi. Though concerned about the loss of his top technical man, Noguchi could afford little time to mourn his leaving. The preceding two years had produced a number of managerial problems. Following his 1909 decision to make ammonium sulfate by the cyanamide method, and after receiving a loan of 500,000 yen from the Hypothec Bank in August 1910, Noguchi went to the stockholders and requested an increase in capital investment to 2,000,000 yen to construct the Himekawa plant. They granted this in September 1910.[1] A rare flood hit the Himekawa plant the following year, but Noguchi remained optimistic about the site because little damage was done. Feeling encouraged by Nitchitsu's progress, Noguchi returned to his fellow shareholders for authorization to increase capitalization to 400,000 yen in March 1912 in order to build a new generating plant on the Shirakawa River. Then disaster struck, just as Nitchitsu had extended itself financially. The Himekawa flood of 1910, rare though it was, should have been a warning to Noguchi. In July 1912, an even more destructive flood put the Himekawa plant out of operation completely. This was an extraordinary setback, and it would have seriously damaged many other companies. But Nitchitsu survived. Noguchi continued to make effective use of available capital and resources and to manage his firm's relations with the government adroitly. He even recycled resources from the Himekawa to other units of his firm.

Noguchi's plans in the fall of 1912 for construction of a new electric generating plant on the Shirakawa River as well as a nearby electrochemical plant included using the creative talent he had assembled in Niigata prefecture for the ill-fated plant.[2] With equipment salvaged from Niigata, Noguchi lost no time preparing new operations in Kyūshū which these staff members could run. In addition, conditions outside the plant had changed, creating a better environment for investment. Government decisions in the preceeding months were particularly helpful. Noguchi had not anticipated these environmental changes, but he immediately figured out how to take advantage of them. He was an aggressive entrepreneur who saw change as opportunity to rethink investments. A more reactive investor who responded passively to change might have viewed these developments outside the firm with dread. Not so the proactive Noguchi.

Indeed, at first glance, the proposed changes appeared detrimental to Noguchi. Hoping to electrify all the railroads in Kyūshū, the Railroad Board of the Ministry of Communications determined that additional sources of electricity would be needed, and entered into negotiations with Nitchitsu for government acquisition of Sogi Electric and Minamata.[3] Though loath to part with their facilities, management at Nitchitsu came to see the government's request as a good business opportunity. In reality, Nitchitsu may have had little choice; having initially constructed their facilities without obtaining permission for the water rights at Minamata from the Ministry of Communications, their hold may have been tenuous. So they made the best of the situation. Eventually, they realized it offered an opportunity to profit. Skillful negotiations over a two-month period involving Ichikawa Seiji and chief engineer Nomura Ryūtarō netted 1,570,000 yen for Nitchitsu in exchange for the Railroad Board's acquisition of Sogi Electric and part of the Minamata factory.[4] Furthermore, although the government assumed ownership of much of Nitchitsu's fixed capital, it did not prevent the firm from continuing to develop hydroelectricity on the Shirakawa River.

Ironically, it had not been certain from the start that Nitchitsu would have the right to develop the Shirakawa. Indeed, the company had relinquished all its rights to develop hydroelectricity in Kyūshū through its negotiations with the Railroad Board. Fortunately for Nitchitsu, the Governor of Kumamoto prefecture was attempting to build an opposition political party in his prefecture and found it politically expedient to grant Nitchitsu the rights to develop the Shirakawa. The Railroad Board apparently was not involved with this decision. The rights to the Shirakawa were more likely transferred because of politics at the prefectural level than because officials felt compelled to compensate Nitchitsu for the actions of the Railroad

Board, as it is sometimes asserted.[5] In any case, Nitchitsu proceeded with its expansion of generating capacity as if its operating environment had not been radically altered.

At the same time, production of electrochemicals continued apace. The Railroad Board had acquired title to Nitchitsu's facilities before it was ready to convert the electric generating plant to use in upgrading Kyūshū railroads or in powering trains. While the Railroad Board procrastinated, Nitchitsu won permission to continue conducting business, though at a reduced level in Minamata's case—the plant was forced to reduce its work force to 270—for the acceptable fee of 5.5 percent interest.[6] This indirect relationship between management and Nitchitsu's productive facilities—private management of publicly owned capital—continued for almost three years, from June 1912 until April 1915, when Ichikawa negotiated a return of the generating facilities to Nitchitsu.[7]

The three-year tie with the Railroad Board offered Nitchitsu several benefits. The first was the timely receipt of government funds permitting continued solvency and expansion to new facilities, including eventual construction of a large plant at Kagami to use the Shirakawa's hydroelectric power. The second was the close personal ties that developed between Nitchitsu management and the Railroad Board. Nakahashi Tokugorō, then President of Nitchitsu and Noguchi's mentor, had once been head of the Railroad Board and may have convinced Noguchi of the benefits of agreeing to government takeover. Moreover, the President of the Railroad Board during its tenure as owner of Nitchitsu's facilities, Sengoku Mitsugu (1857–1931), so obviously enjoyed his connection with the industrialists from Kyūshū that he chose to maintain his ties with Nitchitsu by becoming a member of Nitchitsu's Board of Directors when the facility reverted to the company's ownership.[8] A third factor encouraging Nitchitsu's leaders to relinquish their facilities to the government was the opportunity to put intangible sentiment into actual practice. Ichikawa and Noguchi, the pivotal men in managing the transfer to the Railroad Board's hands, claimed to be pleased to aid national goals through company policy. The two explained their patriotic dedication during the period of outside control of part of their business by stating: "We have always felt that we have worked, at the risk of our own lives, for an enterprise benefiting the nation. Therefore, we have persevered."[9] In this case, it is difficult to differentiate the sentimental from the rational. In any case, Nitchitsu's dealing with the Railroad Board was as profitable as it was patriotic.

With careful planning, Noguchi was eventually able to implement his 1909 decision to invest in large-scale commercial production of ammonium sulfate. His most important initial responsibility was generating funds to support new facilities. Conditions outside company control clearly aided him. The Railroad Board had bailed him out in 1912; timely loans from other sources also helped. The loss of the Niigata factory was somewhat mitigated by the loan from the Hypothec Bank.[10] He planned to recycle some of the fixed capital, ready for installation at Himekawa, in the plant to be constructed at Kagami in Kyūshū. Building the complex at Kagami from December 1912 to May 1914, Noguchi acted as if a large part of his firm's facilities were not being operated at the sufferance of the Railroad Board.[11] In a way, his attitude was warranted; the government, though owning the factory at Minamata, treated Nitchitsu as a private firm. Good management, effective capital accumulation, and government cooperation were necessary to reify Noguchi's dreams of developing Kagami.

Noguchi had several sources for investment capital in addition to aid from government institutions. Bonds were sold three times during World War I, but they were not a major source of funds until after 1926. Rather, as a joint-stock company, Nitchitsu relied heavily on its shareholders for capital. Noguchi frequently requested increases in authorized capital stock from his fellow stockholders. Each time capitalization was increased, the company's total worth increased. Because shares in Japanese corporations were issued by the company at a percentage of their par (face) value, shareholders' initial payments were lower than the par value of the stock. Whenever Nitchitsu needed additional capital, shareholders could be expected to pay additional installments toward the total par value. Paid-up capital, then, was often less than the authorized capitalization of the company; all shares were paid up to their par value when paid-up capital equaled total capitalization.

Purchasing shares of Nitchitsu stock was an excellent investment through most of the period of expansion before and during World War I. Stock-market prices for Nitchitsu shares were consistently at least 20 percent above par value (par value = 50 yen), and dividend payments averaged between 10 and 30 percent of par throughout and immediately after the war, jumping to an incredible 104 percent in 1920 before resuming more normal rates. (Taking a hypothetical worst-case scenario, assuming a buyer had purchased fully paid-up shares at their top 1920 stock market value of 230 yen, this would yield a still high dividend rate of 22.6 percent of purchase price. That same buyer would suffer serious losses, however, if he tried to sell his shares on the stock market the following year; see Tables 14 and 15.)

With such high return on investment, shareholders might be expected to pay requested installments readily. Indeed, they did, and also responded favorably to each additional stock offering.

The number of shareholders remained fairly low during the company's first three decades. Moreover, a small number of investors tended to control most shares, facilitating investment decisions (see Table 16).

In many companies, the domination of top management by the principal shareholders would have produced cautious investment decisions. Such managers would be more concerned with securing profits than with investment in new technology.[12] The case of Nitchitsu shows that the opposite tendency was also possible. Nitchitsu's investor-managers differed from those in a company like Tokyo Artificial Fertilizer in that they were also the company's leading scientists. Thus, their outlook bore a greater resemblance to that of the salaried managers in the modern American enterprise described by Alfred Chandler.[13] These American managers were more interested in preserving their firms—this was imperative for their career security—and thus were most concerned with their firms' long-term competitiveness. Both efficiency and improved technology could foster this. Scientist-entrepreneurs in Japan had similar views. To be

TABLE 14 Stock-Market Prices for Nitchitsu Shares, 1911–1927

Year	*High*	*Low*	*Year*	*High*	*Low*
1911	68.0	67.0	1919	205.0	125.0
1912	67.5	53.5	1920	230.0	75.0
1913	68.8	59.0	1921	75.0	61.0
1914	62.5	56.5	1922	76.5	40.0
1915	160.0	56.0	1923	86.8	71.0
1916	196.0	125.0	1924	97.5	78.0
1917	180.0	130.0	1925	131.2	109.5
1918	150.0	135.0	1926	136.5	111.0
			1927	126.8	93.0

Source: Suzuki Tsuneo, "Daiichiji taisenki," p. 153.

sure, they wanted dividends like any other investors, but they also had long-term goals. As scientists, they hoped their technological innovations would succeed and would become the basis for further technological advances. As in the case of DuPont in the United States, technical and managerial expertise coincided with capital control in Nitchitsu.

Indeed, Noguchi decided to augment his already active role in the company by taking over on-site management of operations in Kyūshū, the job abandoned several months earlier by Fujiyama. Moving himself to Kyūshū from the company's administrative headquarters in Ōsaka, Noguchi also managed daily operations at Minamata and accelerated planning for production of what were called the "three products" (carbide, calcium cyanamide, and ammonium sulfate) at nearby Kagami.[14] By expanding the research in cyanamide product quality which Fujiyama had conducted before his departure, he made further improvements in quality. In addition to expending time and effort on research, Noguchi worked at establishing, in 1913, another factory at Hiji in Ōita prefecture to manufacture carbide for use at Minamata.[15] As a result of this enormous burst of energy, by the beginning of the era of high profits during World War I, Noguchi had recovered from the destruction of a factory, the departure of a

TABLE 15 Nitchitsu Capital and Profits, 1908–1923

Semiannum	Capital[a] (¥1,000,000)	Paid-up[a] Capital (¥1,000,000)	Internal[b] Capital (¥1,000)	Bonds[c] (¥1,000,000)	Profits[d] (¥1,000)	Profit[e] Rate %	Dividend[f] Rate %	Dividends/Profits[g]
1908.1	1.0	0.64						
1908.2	1.0	0.64			25	7.8	10	
1909.1	1.0	0.82			42	10.2	10	
1909.2	1.0	1.00			53	10.4	10	
1910.1	1.0	1.00			74	14.9	10	
1910.2	2.0	1.25			68	10.9	10	
1911.1	2.0	1.50			67	8.9	10	
1911.2	2.0	1.75			79	9.0	8	
1912.1	4.0	2.00			123	12.3	10	
1912.2	4.0	2.00			163	16.4	10	
1913.1	4.0	2.50	65		198	11.7	10	81.3
1913.2	4.0	2.50	49		186	12.5	10	80.2
1914.1	4.0	3.00	-64		94	6.3	8	123.4
1914.2	4.0	3.20	199		370	13.7	10	69.6
1915.1	4.0	3.40	482		704	14.9	12	77.5
1915.2	4.0	3.80	601	1.0	354	18.7	15	75.7
1916.1	4.0	4.00	758		447	22.4	15	65.8

TABLE 15 *(continued)*

Semiannum	*Capital*[a] (¥1,000,000)	*Paid-up*[a] *Capital* (¥1,000,000)	*Internal*[b] *Capital* (¥1,000)	*Bonds*[c] (¥1,000,000)	*Profits*[d] (¥1,000)	*Profit*[e] *Rate %*	*Dividend*[f] *Rate %*	*Dividends/Profits*[g]
1916.2	10.0	5.50	772		654	23.8	20	79.0
1917.1	10.0	6.40	600	1.0	930	29.1	25	79.3
1917.2	10.0	7.00	977	1.0	1,150	32.9	25	74.6
1918.1	10.0	7.60	1,116		1,345	35.4	30	81.4
1918.2	10.0	7.60	1,376		1,684	44.3	30	67.7
1919.1	10.0	7.6	1,462		1,768	46.3	30	64.9
1919.2	10.0	7.6	1,707		1,830	48.2	30	62.3
1920.1	10.0	10.0			5,179	103.6	104	73.6
1920.2	22.0	13.0			1,519	23.4	20	82.3
1921.1	22.0	13.0			1,204	18.5	15	81.0
1921.2	22.0	13.0			1,186	18.2	15	82.2
1922.1	22.0	13.0			1,196	18.4	15	81.5
1922.2	22.0	13.0			1,199	18.4	15	81.3
1923.1	22.0	13.0			1,210	18.6	15	80.6
1923.2	22.0	13.0			1,196	18.4	15	81.6

TABLE 15 *(continued)*

Sources: a. Ōshio Takeshi, "Nitchitsu kontsuerun no seiritsu to kigyō kin'yū," *Keizai Kenkyū* 27:61–127 (March 1977), pp. 116–117.

b. Internal capital includes: reserves, balances forwarded, debt amortization; Suzuki Tsuneo, "Daiichiji taisenki Nitchitsu, Denka no tōshi to shikin chōtatsu," *Kurume Daigaku Shōgakubu seiritsu 30 shūnen kinen ronbunshu.* August, 1980. p. 147. Data for 1913–1919 only.

c. Ōshio Takeshi, "Nitchitsu kontsuerun," p. 109. These bonds were underwritten by Mitsubishi, Aichi, and Yamaguchi Banks. Interest rates were low: 7 percent for the first issue, thereafter 6 percent or lower. Most bonds were issued after 1926. The 7 percent interest rate is comparable with the 6–8 percent rate throughout the chemical industry.

d. 1980–1913: Shimotani Masahiro, "Nitchitsu kontsuerun," p. 69;
1913–1923: Suzuki Tsuneo, "Daiichiji taisen," p. 142.

e. Profit rates were calculated as a percentage of paid-up capital, adjusted to annual rates, a frequently used method in prewar Japan. Suzuki Tsuneo, p. 142.

f. Percentage of par value paid out as dividends. Suzuki Tsuneo, p. 142.

g. Dividends as percentage of profits. Suzuki Tsuneo, p. 142.

TABLE 16 Principal Shareholders in Nitchitsu, selected years

Name	*Number of Shares (%)*
1910	
Noguchi Jun	5,310 (26.6)
Fujiyama Tsuneichi	1,710 (8.6)
Viktor Hermann	1,439 (7.2)
Kosugi Tsuneyuemon	1,010 (5.1)
Tanaka Kiyo	980 (4.9)
Ichikawa Seiji	942 (4.7)
Kondō Renpei	750 (3.8)
Kondō Shigeya	750 (3.8)
Nakahashi Tokugorō	661 (3.3)
Yonei Genjirō	600 (3.0)
1913	
Noguchi Jun	16,696 (20.9)
Ichikawa Seiji	5,700 (7.1)
Viktor Hermann	5,356 (6.7)
Kushida Manzō	4,280 (5.4)
Kondō Renpei	2,500 (3.1)
Yonei Genjirō	2,400 (3.0)
Toyokawa Ryōhei	2,380 (3.0)
Kawakita Yoshio	2,174 (2.7)
Nakahashi Tokugorō	2,130 (2.5)
Kondō Shigeya	2,000 (2.5)
1916	
Noguchi Jun	40,617 (20.3)
Ichikawa Seiji	13,010 (6.5)
Kawakita Yoshio	11,890 (6.0)
Kushida Manzō	11,200 (5.6)
Aichi Bank	7,000 (3.5)
Toyokawa Ryōhei	6,990 (3.5)
Yonei Genjirō	6,514 (3.3)
Kondō Renpei	6,250 (3.1)
Nakahashi Tokugorō	5,024 (2.5)
Kondō Shigeya	5,000 (2.5)

TABLE 16 *(continued)*

1919		
Noguchi Jun	42,000	(21.0)
Ichikawa Seiji	12,980	(6.5)
Kushida Manzō	11,200	(5.6)
Kawakita Yoshio	8,640	(4.3)
Watanabe Yoshiro	6,800	(3.4)
Kondō Renpei	6,250	(3.1)
Katō Takeo	6,000	(3.0)
Aiichi Bank	5,900	(2.9)
Kondō Shigeya	5,000	(2.5)
Toyokawa Ryōhei	4,200	(2.1)

Sources: For 1910, 1913, and 1916, Ōshio Takeshi, "Nitchitsu kontsuerun no seiritsu to kigyō kin'yū," p. 123; for 1919, Suzuki Tsuneo, "Daiichiji taisenki Nitchitsu, Denka no tōshi to shikin chōtatsu,"p. 149.

highly talented collaborator, and one final setback—the loss of his respected teacher and advisor, Siemens's Viktor Hermann, when the latter was forced to leave Japan after his implication in the Siemens bribery scandal of 1914.[16]

Noguchi was ready to manufacture electrochemicals at Kagami by May 1914, even before electric facilities were completed on the Shirakawa in November. After a short period in which Nitchitsu bought electricity from Kumamoto Electric, Inc. (Kumamoto Denki KK), the coordinated production of ammonium sulfate and generation of electricity began.[17] The fertilizer production was particularly significant from a technological perspective; that is, Kagami promoted Japan's first real commercial manufacture of electrochemical ammonium sulfate. Noguchi's groundbreaking industrial production represented a blend of the technology of cyanamide, perfected at Minamata, with that of the conversion of cyanamide to ammonium sulfate, pioneered on a small scale at Hiejima.[18]

Successful at his large plant at Kagami—for its time, the 7,000-kilowatt hydroelectric capacity at Shirakawa was exceptional—Noguchi was encouraged to transfer Kagami's technology to Minamata after Minamata's reversion to company hands in 1915. His timing was again good. For the increased capital investment required by the expansion at Minamata, Noguchi was even better prepared financially than when planning Kagami three years earlier. World War I radically affected the availability and price of imported nitrogenous fertilizers. In June 1914, ammonium sulfate brought 122 yen per ton. This rose to 160 yen by August due to rises in marine insurance rates, dislocations in foreign exchange, scarcity of bottoms for transport, and farmers' typical mid-season demand. Bean cakes entered the market as a substitute in the fall, temporarily bringing the price of the competitive ammonium sulfate down to 120 yen in December. Thereafter, prices steadily rose to 190 yen for the extremely scarce British product in 1916, 230 yen in the spring of 1917, and 480 yen in August 1917. Prices fluctuated wildly from month to month in 1918, and averaged around 350 yen.[19]

Unstable and high prices were extremely frustrating to Japanese farmers but were most instrumental in helping infant Japanese fertilizer manufacturers gain a foothold in their own market. This had two benefits for manufacturers like Nitchitsu: Nitchitsu's large profits could be reinvested in major expansion of capital; and farmers' purchase of Nitchitsu's product accustomed them to domestic fertilizers and stimulated future demand.[20]

The company's extraordinary wartime prosperity (which continued during the years immediately following the war when European industries had not recovered to the point where exports were possible) made those connected with the firm rather prosperous. Shareholders earned enormous dividends, up to 104 percent of par value in the first half of 1920, white-collar employees received bonus shares in the company, and long-time blue-collar workers received gold watches.[21] Some sources note that all employees received bonus shares and that white-collar workers were also given cash bonuses that totaled 250,000 yen.[22] Though Nitchitsu was noteworthy among employers in being one of the first in Japan to institute the 8-hour work day, it would be remarkable if worker co-ownership had also been promoted at that time; bonus shares for workers seems inconsistent with contemporary company experiences. Given Nitchitsu's troubled labor relations after World War I, it is quite unlikely that more than a few blue-collar employees identified with the owners. Furthermore, records of a top-management meeting on 6 July 1918 indicate that employee bonuses were indeed more modest—3 months' salary to white-collar workers and 5 yen each to blue-collar workers—as profits had to be distributed among shareholders, public officials, and others in addition to employees.[23]

The healthy windfall profits Noguchi derived from wartime trade conditions were augmented by his carefully cultivated political contacts. Noguchi's sole interest was the affairs of Nitchitsu. Not only was he the largest single shareholder, but Nitchitsu was his creation, and he wanted to see it grow, prosper, and take the lead in introducing industrial technology. Here Noguchi had an effective ally in Nitchitsu President Nakahashi Tokugorō. Like Noguchi, he was a large shareholder, and so stood to gain from increased company profitability and higher dividend payments. Though President of Nitchitsu, Nakahashi had several other irons in the fire.

Nakahashi's government connections were ideally suited to assist Noguchi's ambitions. President of Nitchitsu until resigning in September 1918 to become Minister of Education, Nakahashi never allowed his duties in business to keep him far from the corridors of political power. He was, therefore, appropriately situated to help his firm when complications in production and transportation during World War I impeded European fertilizer deliveries in Japan and the rest of Asia. With 87 percent of the Japanese market in ammonium sulfate supplied by foreign sources before the war, import restrictions were bound to trigger increases in prices. Prices rose tremendously in Japan, but went up even more steeply in Southeast Asia. Japan had strict export controls on ammonium sulfate, which manufacturers wished to loosen to enter the lucrative Southeast Asian market. Large orders were pouring in from Java in 1917.[24] Nakahashi appealed to the government for permission to export and argued that, since Japanese farmers could not afford domestic ammonium sulfate at inflated prices anyway, the fertilizer might as well be sent overseas to earn foreign exchange. He downplayed the fact that resultant decreased domestic supply could further elevate fertilizer prices. In any case, his explanation persuaded the government to lift the ban on exports, opening the door to foreign sales.[25] Thus, Nitchitsu was passively aided by external circumstances beyond company control—prosperity brought about by the war—and actively aided by manipulation of ties linking the firm with policymakers in Tokyo, helping the company generate the support it needed to develop into a modern synthetic fertilizer company.

NITCHITU'S INCREASED SIZE AND RESULTING CONCERNS FOR MANAGEMENT

The Railroad Board sold the plant at Minamata back to Nitchitsu in 1915. Nitchitsu had benefited from the Board's control after 1912, but, by 1915, the company had regained financial stability. Reversion of the plant came at a good time. Soon converted to production of ammonium sulfate, Minamata had a projected capacity of 50,000 tons per year.[26] This conversion at Minamata and the almost simultaneous expansion of Kagami's facilities to produce 50,000 tons as well, set off an expansion of Nitchitsu that would continue unabated through World War II.

Nitchitsu's expansion occurred within an evolving company structure. The simple organization of management in the early Nitchitsu corresponded to the simple technology of production. As the technology developed in the following decades, the company itself became more complex. Indeed, it transformed itself from a simple company with a limited number of products which grew by horizontal development to a multifaceted, vertically integrated firm, usually called a "Konzern" (*kontsuerun*) or "new zaibatsu."[27]

Between 1912 and 1915 (the first year both Kagami and Minamata were effectively producing ammonium sulfate), Nitchitsu's organization became more sophisticated (see Table 17). In both 1912 and 1915, the organization was divided into three divisions (*bu*) under

Managing Director Noguchi Jun: General Affairs (*shomubu*); Production (*seihinbu*); and Temporary Construction (*rinji kōjibu*). Each division was subdivided into sections (*ka*), most of which were further subdivided into subsections (*kakari*). Within the General Affairs Division there was no Personnel Section and no differentiation between staff (such as accounting) and line (such as management) positions. As late as 1917, there were but 13 employees, including section heads, in the General Affairs Division.[28] There appears to be little modification in 1915 of the 1912 organization of General Affairs. The Production and Temporary Construction Divisions showed greater development; rather than having subsections (*kakari*) for each function at each plant (as in 1912), by 1915 functionally defined sections (*ka*) intervened between divisions and subsections. Priority went to function, not location of operations, by 1915. Management was becoming more specialized and professional.

In 1912, Minamata was the only significant operating production factory—the Ōsaka plant was tiny. By 1915, Minamata and Kagami were both producing fertilizer. They were similar because their technology did not inspire creative diversification in which new plants or workshops could be spun off to develop new products. For this reason, Nitchitsu was not yet a modern, multifaceted firm. Yet some important traits of the modern firm were emerging. For example, a layer of middle managers was created to coordinate the various functions involved in production, financing, and distribution. Management was increasingly hierarchical and professionalized. Plant heads were replaced because both plants were similar in function. Rather, section heads were under division heads and coordinated the activities of subsection heads.

Kagami had produced the important "three-product cycle;" this industrial process was replicated at other Nitchitsu facilities and remained in use until the synthetic process for ammonium sulfate was perfected in the 1920s. The "three-product cycle" recognized the chemical interrelationships of calcium carbide, calcium cyanamide, and ammonium sulfate. Although cement—one of the most important products in Japan before World War II—was produced in large quantities as a by-product of the "three-product cycle," the process

TABLE 17 Nitchitsu Management Organization

December 1912

Managing Director

- I. General Affairs Division
  - A. Accounting Section
    - 1. Accounting Subsection
    - 2. Treasurer Subsection
  - B. Management Section
    - 1. Sales Subsection
    - 2. Procurement Subsection
  - C. Archives Section
- II. Production Division
  - A. Minamata Plant
    - 1. Business Subsection
    - 2. Carbide Subsection
    - 3. Fertilizer Subsection
    - 4. Production Methods Subsection
    - 5. Electric Machinery Subsection
  - B. Sogi Electric
    - 1. Business Subsection
    - 2. Electric Subsection
    - 3. Water Subsection
    - 4. Electricity Transmission Subsection
    - 5. Power Distribution Subsection
- III. Temporary Construction Division
  - A. Ōme Plant
    - 1. Business Subsection
    - 2. Construction Subsection
    - 3. Limestone Subsection
  - B. Himekawa Plant
    - 1. Business Subsection
    - 2. Engineering Subsection
    - 3. Electricity Subsection

TABLE 17 *(continued)*

April 1915

Managing Director
- I. General Affairs Division
  - A. Accounting Section
    - 1. Accounting Subsection
    - 2. Treasurer Subsection
  - B. Management Section
    - 1. Sales Subsection
    - 2. Procurement Subsection
  - C. Archives Section
- II. Production Division
  - A. Business Section
    - 1. General Affairs Subsection
    - 2. Treasurer Subsection
    - 3. Accounting Subsection
    - 4. Purchasing Subsection
    - 5. Warehouse Subsection
    - 6. Packing Subsection
    - 7. Sales Subsection
    - 8. Maintenance Subsection
  - B. Production Section
    - 1. Carbide Subsection
    - 2. Calcium Cyanamide Subsection
    - 3. Ammonium Sulfate Subsection
    - 4. Cement Subsection
    - 5. Nitrogen Subsection
    - 6. Sulfuric Acid Subsection
    - 7. Carbolic Acid Subsection
    - 8. Analysis Subsection
  - C. Electric Section
    - 1. Generation Subsection
    - 2. Transmission Subsection
    - 3. Distribution Subsection

TABLE 17 *(continued)*

- III. Temporary Construction Division
  - A. Construction Section
  - B. Engineering Section
  - C. Electric Section
  - D. Business Section

Source: Ōshio Takeshi, "Nitchitsu Kontsuerun no kin'yū kōzō," insert between pages 114 and 115.

was primarily concerned with the three-stage method of fertilizer manufacture[29] (see Figure 3).

Ease in conversion recommended replication of the "three-product cycle" at Minamata. But carrying out the process on a scale projecting a 50,000-ton annual yield of ammonium sulfate demanded electricity in excess of that generated by Sogi Electric, the generating plant at Minamata. Noguchi therefore asked Nitchitsu's shareholders to authorize an increase in capitalization to 10 million yen. With the shareholders' approval, Nitchitsu's capitalization jumped from 4 to 10 million, with 1.5 million being paid up immediately; the increase also was financed by a 1-million-yen bond underwritten by three banks, Mitsubishi, Aichi and Kōnoike.[30]

What Nitchitsu got with its rapid expansion was a doubled output of the "three products" plus cement. On the other hand, expansion confronted Noguchi with additional concerns in finding skilled management, efficient labor, and plentiful energy. Noguchi ultimately solved these problems. The benefits of expansion outweighed even high cost.

Noguchi's first consideration was managers: not only how to find and retain them, but also how to structure his firm around them. Management was used in a way that was effective for the type of technology employed. Under Noguchi Jun, who was head of the Production Division in addition to being Managing Director, there were three section heads. One man was head of the Business Section responsible for both Minamata and Kagami, and he was located at Kagami. The head of the Electricity Section for both plants was also at Kagami, while the head of the Production Section was at Mina-

FIGURE 3 Three-Product Cycle

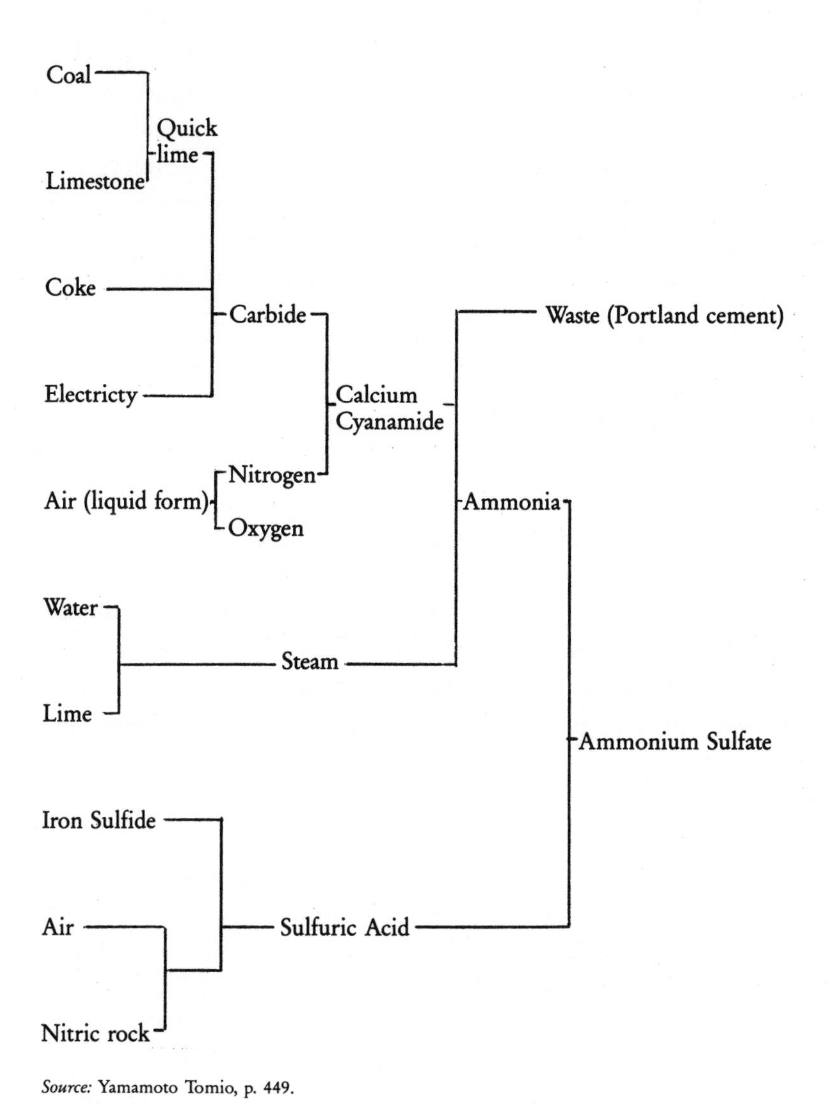

Source: Yamamoto Tomio, p. 449.

mata.[31] With managers coordinating various functions but sited at only one factory, the third level of management (subsection heads) received excellent on-the-job training. The middle managers directly under Noguchi could develop skills of company integration while encouraging cross-fertilization of ideas. Individual managers would view as their top priority company goals rather than goals of the plant where they were in residence. Effective managerial integration had important consequences during the next two decades when Nitchitsu created a far-flung empire of factories and separate subsidiaries. The sense of company solidarity was carried over into the later organization; top managers were shifted among various plants, but all had to receive their initial appointments and later promotions at headquarters.[32]

A second result of maintaining some physical distance between top and middle managers and some of their facilities was the transfer of much of the responsibility for daily operations to the third layer of management. Commuting by top managers often kept them from becoming intimately concerned with details of current affairs. This forced a large number of technically schooled employees to acquire skills of management. It also meant that Noguchi's operation would be structured along lines of production, as the third layer of management was divided into subsections concerned with discrete aspects of production, such as the carbide subsection, chemical-analysis subsection, and ammonium sulfate subsection. Similarities in structure were parallelled by similarities in plant size and output (see Table 18).

A structure like this created a pool of trained managers who, as they advanced through the ranks of company management, were able to take control, with relatively little difficulty, of Nitchitsu's transformation from a fertilizer company making just four products to a diversified chemical company. Training of management personnel during the 1910s was a step helpful to effective management after expansion. The first case of a middle manager's promotion to membership on the Board of Directors occurred early, right after the war, in 1919.[33] In time, Nitchitsu replaced many of the early directors who had been recommended because of their proven abilities at Mitsubishi with men who had worked their way through the ranks

TABLE 18 Productive Similarities of Kagami and Minamata, 1917

	Kagami		*Minamata*	
Electric Capacity (kilowatts)	Shirakawa	7,000	Sogi	6,000
	Naidaijingawa	4,000	Sendaigawa	14,000
	Midorigawa	4,700	Kurino	1,700
	Purchased from Outside Sources	10,000		
	Total	25,700	Total	21,700
	Kagami		*Minamata*	
Total Product (tons)	Carbide	45,000		40,000
	Calcium cyanamide	55,000		50,000
	Ammonium sulfate	50,000		40,000
	Cement (barrels)	250,000		250,000

Source: Yamamoto Tomio, p. 454.

Note: In 1917, Minamata was reorganized to permit increased production of ammonium sulfate.

at Nitchitsu itself. This pattern has an interesting parallel in the increasing sophistication of American business in the nineteenth century. Many American businesses, needing large amounts of outside capital, were dependent on financial institutions which often demanded representation on the board of the businesses they helped. In time, as those businesses grew and became more complex, the role of the financial institutions lessened because of their inability to monitor day-to-day decisions. "In many industries and sectors of the American economy, managerial capitalism soon replaced family or financial capitalism."[34] At Nitchitsu as well, talented middle managers rose to take over the decision-making Board positions held by men with Mitsubishi Bank connections.

To be sure, not all men of responsibility in Nitchitsu stayed on to accrue seniority and status; of the 17 top managers who joined the company during the Meiji period, 9 had already left by February 1912.[35] But this was a good record for that period of accelerated mobility for talented engineers and managers. Most of the 9 moved to establish their own firms or to accept government posts. Of the remaining 8, 6 died, after serving an average of 14.5 years in high positions within the company. The other 2 were Noguchi and Ichikawa, President and Vice-President of Nitchitsu during the 1930s and 1940s.

Mobility was possible not only for top management but also for other white-collar workers. One way to retain employees was good pay.[36] Many made over 65 yen per month in 1914, although starting salaries for technical employees in 1914 were 18 yen or 25 yen, depending on classification. These salaries rose dramatically each year from 1915 to 1918, making Nitchitsu a particularly generous employer.[37] Paying higher salaries was cost-effective for Nitchitsu. Loss of industrial secrets through employee mobility was more costly than higher pay. Nitchitsu management was so concerned about industrial security that the contract signed by technical workers stipulated that workers must not divulge company secrets.[38]

The possibility of occupational mobility also led Nitchitsu to reward perseverance and talent.[39] Most researchers had excellent training in Japan's technical universities or in the technical departments of comprehensive universities, but some entered with poorer preparation. Weak preparation was no barrier to future success at Nitchitsu, in contrast to most Japanese companies. After an employee entered the company, Noguchi downplayed the importance of academic training; rather, performance was to be rewarded. An emphasis on performance made employees feel that management took a continuing interest in their development. Many employees' written reminiscences of their years at Nitchitsu practically glow with nostalgia when describing the warm concern their bosses, including Noguchi, seemed to show for them.

Among those rewarded for diligence and loyalty were several young men from O.S.K. whom Nakahashi brought to Nitchitsu when he joined the company. These included Enonami Naosaburō, the first middle-management employee to move to the Board of Directors in 1919, and Hitotsubashi graduates Kaneda Eitarō and Ōgyū Den. Interestingly, none of these three was a scientist; this suggests that at least some white-collar employees were trained as professional managers, unlike the scientist-managers who were more typical at Nitchitsu. Enonami started his career in Nitchitsu in the sales department, and gradually was given greater responsibilities in general management. Kaneda was a general manager and Ōgyū an accountant who was promoted to the Board of Directors after the resignation in 1913 of the Mitsubishi accountant who had served as Auditor to the Board.[40] Other men who had worked their way through Nitchitsu's ranks in the following decades included electrical engineer Kubota Yutaka; technician Shimada Shikamitsu, in charge of construction of the plants at Minamata and later at Nobeoka and in Korea; and scientist Shiraishi Muneshirō, son of Shiraishi Naojirō, one of Nitchitsu's researchers during the 1910s (see Table 19).[41] Training and promotion of a wide group of people aided intra-firm mobility; mobility among sections, in turn, fostered flexible management capable of working openly and cooperatively with managers from other sections.[42]

Management fostered flexibility in other ways as well. Most important, researchers had unusual freedom to pursue their own ideas.[43] Moreover, Noguchi established a corporate culture encouraging study, which culminated in the establishment of the Noguchi Research Institute for pure scientific research in 1941.[44] Cooperation among sections in research and development, a noteworthy hallmark of later Nitchitsu operations, as well as mobility of management-level employees among production facilities—even between Nitchitsu's operations in Japan and Korea—had its roots in the mobility created during the period of rapid expansion during World War I.[45]

Rapid expansion produced the second of Noguchi's concerns—how to find, retain, and compensate labor. Blue-collar workers manifested problems ranging from the slight (the incompatibility of regional speech patterns among the diverse crew of workers, especially those with Kyūshū accents unintelligible to their northern bosses)[46] to the complex (finding and keeping appropriately trained workers). Manufacture of calcium cyanamide, the second of the three stages of production, demanded semi-skilled workers, and conversion of cyanamide to ammonium sulfate was at least as demanding.

TABLE 19 The Nitchitsu Board of Directors, selected years

	Before 1918	*1919*	*1929*	*1937*
President	Nakahashi Tokugorō	Toyokawa Ryōhei*	Toyokawa Ryōhei*	Noguchi Jun
Vice-President	–	–	–	Ichikawa Seiji
Managing Director	Noguchi Jun	Noguchi Jun	Noguchi Jun	Enonami Naosaburō
Director of Affairs	Ichikawa Seiji	Ichikawa Seiji	Ichikawa Seiji	Kanada Eitarō
Director	Shiraishi Naojirō*	Sengoku Mitsugu*	Watanabe Yoshirō*	Kirishima Zōichi*
	Watanabe Yoshirō*	Watanabe Yoshirō*	Kirishima Zōichi*	Shiraishi Muneshiro
	Toyokawa Ryōhei*	Enonami Naosaburō	Enonami Naosaburō	–
Auditor	Kagami Katsuichirō*	Kagami Katsuichurō*	Kagami Katsuichirō*	Hori Keijirō
	Hori Keijirō	Hori Keijirō	Hori Keijirō	Ōgyū Den
Counsel	Kondō Renpei	Kondō Renpei	–	–

Sources: Until 1918, Yamamoto Tomio, p. 439; 1919, Yamamoto, p. 454; 1929, Satō Kanji, *Hiryō mondai kenkyū*, p. 57; 1937, Yamamoto, p. 518.

Note: * Mitsubishi connection

Among Kyūshū farm hands—Nitchitsu's labor pool—mechanically and industrially trained workers were somewhat difficult to recruit. At the same time, the cyanamide process was far less efficient in its use of labor than the synthetic methods used later. Consuming at least as much electricity as some synthetic methods and up to 5 times as much as others, the cyanamide process demanded twice as much labor. Manufacture of 1 ton of ammonium sulfate by the cyanamide method required 23 workers, and by the synthetic method about 10 workers. Synthetic processes had few stages completed by manual labor, as opposed to the cyanamide, which had several.[47] It is, perhaps, partially because of the great investment in labor that, with the exception of the British, very few manufacturers in Europe used the cyanamide process to make ammonium sulfate, but instead proceeded early on to develop synthesis of ammonia, reserving cyanamide production for unadulterated use as a finished fertilizer product.[48]

Increased production of fertilizer meant that the number of Japanese workers in the fertilizer trades, regardless of their lack of preparation for industrial labor, grew rapidly during World War I. Many of them worked in Denka and Nitchitsu.[49] Kagami and Minamata each employed about 3,000 people by 1918, most of whom were involved in production rather than in electricity generation or general business affairs.[50] Japan's lower wage rates and higher profit margin due to wartime fertilizer insufficiency made the cyanamide method for ammonium sulfate temporarily more economical in Japan than in Europe. Ultimately, the method's inefficiency would allow it to be eclipsed by more efficient methods after the war. But, in the meantime, Nitchitsu found it profitable to run three shifts daily, despite the cost of labor, to prevent idling of expensive machinery.[51] This was particularly important after Nitchitsu agreed to let workers have an 8-hour work day.

Nitchitsu may have been the first Japanese company to institute the 8-hour work day.[52] Nevertheless, other conditions at Nitchitsu did not necessarily produce labor satisfaction. The excellent rapport often noted between Noguchi and his white-collar workers was absent in his relations with his blue-collar workers.[53] The Kagami plant, for instance, was the site of a major strike in 1918, owing to several possible factors. First, the work force at Kagami grew astronomically. In June 1914, there were 53 white-collar and 656 blue-collar employees. The total reached about 1,000 by December (63 white-collar, 903 blue-collar, plus 18 temporaries), and shot up to 3,000 by 1918.[54] This 4-fold growth in the work force occurred as output increased by a factor of 5. Thus, productivity per worker was expected to be 25 percent higher, though no technological innovations had been made. Second, the price of rice had skyrocketted during 1918, leading to strikes nationwide; Kagami was no exception. Workers demanded higher wages to keep pace with inflation and company assistance to purchase rice.[55] The strikers at Kagami really meant business. There was, at the time, no union to organize the workers. Workers simply signed a joint compact agreeing to strike. Some who signed went to work anyway, where they confronted violent opposition from their striking colleagues. Company housing was burnt, rocks were piled in front of offices, and, in the end, 46 workers went to trial for strike disturbances. All but 8 were convicted.[56]

Noguchi mediated personally in the strike, and many of the workers' grievances were addressed.[57] The problem of demanding excessive work was partially resolved by instituting three 8-hour shifts rather than two 10-hour shifts. Pay raises were instituted, but these were relatively small, averaging just 2 sen per day, to be granted every 6 months. More significant were rice subsidies. Cheap rice was made available. The market price for rice was 60 sen per 1.8 liter; Nitchitsu employees were entitled to 18 liters per month for each family member at 25 sen per 1.8 liter.[58]

Labor policies were viewed as generally more progressive at Nitchitsu than elsewhere, but the 8-hour shift was not without its problems. The day shift was convenient (8 a.m.– 4 p.m.) but, of course the swing (4 p.m.–midnight) and night (midnight–8 a.m.) shifts were not. Each worker changed shifts every 5 days. By 1935, that 5th day had become a holiday,[59] but at first it must have been exhausting for workers to alternate shifts without a break. Many of the rurally based factory workers were also farmers, and it was difficult for them to sustain their focus of responsibility in the village and to maintain their farms with an alternating work schedule. In general, however, work conditions at Nitchitsu were better than at many other plants, especially after management learned the lessons of the 1918 Kagami strike. This permitted Nitchitsu to hire easily and to retain valuable blue-collar labor.

Noguchi had a third major concern—raw materials and energy—in addition to the problems of labor and management accompanying expansion of Nitchitsu during World War I. As indicated in Figure 3, most of the raw materials for manufacture of the products in the "three-product cycle" were of domestic origin and thus available despite wartime slowdowns in imports. Nitric rock was the only imported material and, regardless of its apparent necessity for producing explosives in Europe, Japan suffered no decline in supply of this South American resource.[60] Raw-material *cost* was Noguchi's prime resources concern then, as availability was little affected by events abroad.

Finally, Noguchi had to consider the cost of electricity. The cost of electricity for manufacturers rose no faster than the manufacturing output price index during the war and throughout the period under consideration;[61] in any case, Nitchitsu generated most of its own requirements, permitting generation of electricity to remain a constant increment in overall electric costs. Each electrochemical plant was theoretically supplied with electricity by certain electricity-generating facilities, as noted in Table 18 above. Actually, the head of the Electricity Section of the Production Division determined where electricity should be transmitted over the interconnected electric lines between Kagami and Minamata. This gave additional flexibility to those in charge of production to respond to changing needs.[62]

The final cost of hydroelectricity was most strongly influenced by the cost of constructing new generating facilities. Initial costs of construction were high. A bond issue of 1 million yen in December 1916, underwritten by a group of banks including the Mitsubishi, Aichi, and Yamaguchi Banks, helped defray the cost of starting up the Naidaijingawa generating plant near Kagami that month. But even the new plant was not up to its task, and additional energy had to be supplied to Kagami by the Midori River and from outside purchases. As Noguchi was readying these facilities, he was also busy setting up the Sendaigawa generating station for use by Minamata. On line by May 1917, the plant's capital demands exceeded Nitchitsu's supply of funds, and management had to float another 1-million yen bond in November 1917.[63] Given the high rates of wartime profits, floating this bond was an astute decision. Interest rates were much lower than profits earned on Nitchitsu's products. Construction of the electric facilities permitted a 70-percent boost in output of ammonium sulfate at Nitchitsu between 1917 and 1918 and another 52-percent jump the following year.[64] Although foreign dumping during the early 1920s prevented a continuation of high wartime profits, thereby eroding Nitchitsu's fantastic profits of the preceding years, the company was able to maintain output, producing steadily for the next four years, and even raising its production again after 1924, the year Nitchitsu adopted the new technology of synthesis. Capital reserves accumulated during the 1910s, when Noguchi successfully managed the problems of expansion, would be used to fund costly innovations in the 1920s.

PREPARING FOR INNOVATION

Although Noguchi had the economic resources to be innovative during World War I, he waited several years before conducting expensive research and development of new basic science. Building Kagami and refitting Minamata for production of ammonium sulfate kept him quite busy. Profits were comfortable, and, facing little domestic competition and no foreign competition, Noguchi could afford to stay with tested though less efficient methods.[65] He was aware of the benefits of ammonia synthesis, as were a growing number of Japanese scientists. Some of these scientists even visited European pilot plants and laboratories, but few were able to enlist support from investors before hostilities impeded technology transfer. By contrast, some potential investors, like the consortium of firms discussed below which held German patents for synthesis, hesitated to produce ammonia commercially. This left the field open for Noguchi Jun and other scientist-entrepreneurs to inaugurate synthesis in Japan in the 1920s.

As early as 1914, Noguchi wrote an article that revealed his informed appreciation of European advances in methods of ammonia synthesis.[66] He clearly understood the chemical aspects of the process. He noted that, although the concepts involved were amenable to adaptation by Japanese chemists, the problems of large-scale commercial production of ammonia surpassed the industrial tools at hand. Requiring machinery capable of withstanding high pressure and high temperature, Noguchi wrote, ammonia synthesis would have to await advances in mechanical engineering. Certainly Japan was not alone in having a machinery industry that impeded industrial application of generally understood scientific principles. With the exception of Germany after 1911, other nations suffered the same problem.[67] Like scientists in those other nations, Japanese scientists had acquired, from their own research, foreign study, and reading articles like Noguchi's, an understanding of the basic principles of synthesis, permitting later industrial production. Research was sponsored by the government, universities, industries, and the military, and conditions for its application were created during the war and in the 1920s.

MATURATION OF THE TECHNOLOGY OF SYNTHESIS IN THE WEST AND JAPAN

Noguchi Jun perfected the cyanamide method for making ammonium sulfate during World War I. Though he was aware of the commercial applications of more advanced technology, he did not attempt even small-scale experimental synthesis of ammonia during the war. He preferred to accumulate capital reserves during this era of high sales and profits, which was possible even with his less advanced methods of production. Although Noguchi perceived no need to innovate until forced to upgrade productivity to compete in the saturated postwar market, Japanese scientists working in the West, academic scientists, and officials in government agencies continued research throughout the war.

Processes studied by Japanese during the 1910s had been pioneered in Germany, France, and Italy. As in Japan, chemists in these countries found that, although the chemical method of synthesizing ammonia from hydrogen and nitrogen was conceptually simple (ammonia is NH_3), industrial development was difficult. The state of chemical engineering had to be commensurate with that of chemistry. Ammonia synthesis was the first type of manufacture requiring high-pressure and high-temperature apparatus (available only in Germany at the turn of the century). With ammonia synthesis began the close industrial relationship between the chemical industry and chemical engineering. A brief description of the process of synthesis should indicate the necessity for advanced engineering technology.

Germans Fritz Haber and Karl Bosch collaborated on the world's first successful industrial production of synthetic ammonia. Haber, professor of physical chemistry at Karlsruhe, became interested in the question of making ammonia from two elements, nitrogen and hydrogen, common in the air. Badische Anilin und Soda Fabrik (BASF, later the central component of IG Farben) assisted his research, and, in 1908, as Haber neared success, BASF's Karl Bosch was assigned the project of making the reaction mechanically possible. Basically, the process called for combining 3 parts hydrogen with 1 part nitrogen. Obtaining the two gases was difficult and required high-pressure machinery.[68] Next, high-temperature machinery was needed to force the pure hydrogen and nitrogen to combine as ammonia.[69] Commercial production was predated by BASF's establishment in 1911 of a small test plant capable of producing 25 kilograms of the world's first synthetic ammonia. In September 1913, the laboratory expanded into the Oppau plant of BASF, making at first 36,000 tons of ammonium sulfate (8,700 tons of ammonia) and increasing to 250,000 tons of ammonium sulfate after the outbreak of World War I.

Germans were not alone in researching synthetic ammonia. Frenchman Otto Serpek began studying the chemical in 1909, and, although his process never achieved the success of the Haber-Bosch method, it attracted the attention of some Japanese chemists. Serpek's method, which resembled Haber's but differed in its simultaneous production of aluminum and ammonia, is more than a historical footnote mainly by virtue of its role in the development of Japan's ammonia industry. A third method developed during World War I was that of the General Chemical Company of Long Island. Differing little from the Haber process, the General Chemical method was also carefully investigated by the Japanese.

The first Japanese to stimulate his countrymen's interest in synthetic ammonia was Ōya Jun, an engineer in the Ministry of Communications. Ōya had been sent overseas in 1911 to report on ways Westerners used surplus electricity.[70] While in the United States, he attended an academic conference at which Badische's head of research discussed the development of Oppau in Germany. Ōya's enthusiastic reports of the conference encouraged scientists in Japan to begin studying methods of synthesis. It should be noted that Japanese research in synthesis, based on Ōya's 1911 observations, began the same year the first test plant was set up in Germany. The time lag for first use of technology was rapidly shrinking, indicating that the Japanese were catching up to the West in ability to innovate. Understanding the German technology of synthesis permitted them to use it as a basis for their own innovative studies.

In the years following, Ōya Jun reports, other Japanese had the opportunity to observe foreign methods at first hand. Shiraishi Genjirō of Japan Steel Pipe (Nippon Kōkan), sent to Europe to study railroads, returned via India where he investigated aluminum. Aluminum led him to France's Serpek method for ammonia.[71] Shiraishi reported his discovery to industrialist and investor Asano Sōichirō, who gathered his peers Shibusawa Eiichi and Tanaka Eihachirō to help sponsor a colloquium in 1913 on methods of producing ammonium sulfate. Takamine Jōkichi, recently returned from the United States, joined them in the Society for the Investigation of Ammonium Sulfate (Ryūan Chōsakai).

Among Society members there were differences of opinion regarding the best method of production. Takamine had little interest in the Serpek method, which was not yet commercially viable, and favored methods akin to the Haber-type he had observed at General Chemical in the United States. Shiobara Matasaku, another influential chemist, seconded Takamine's rejection of Serpek's method. They were joined by Tamaru Setsurō, former assistant to Haber at Badische, who enthusiastically supported his mentor's techniques over Serpek's.[72] Tamaru had introduced Suzuki Tatsuji (later head of Yokohama Technical College) to Haber in 1913, and the Yokohama man returned to Japan a convert to the Haber method.

Despite the support of scholars for the Haber process, the group of industrialists in the Society for the Investigation of Ammonia Research negotiated with Serpek for the French method. When World War I interrupted the negotiations, the most enthusiastic scholars began research on their own. Suzuki Tatsuji set up the Yokohama Chemical Research Laboratory (Yokohama Shami Kenkyūjo) in 1915, and expanded on the technology he had observed in Germany.[73] Takamine Jōkichi made use of his friendship with General Chemical's William H. Nicols to study that firm's methods. Banker Watanabe Katsusaburō retained dye chemist Ishisaka Gojirō and a staff of 6 chemists and 30 technicians in a special nitrogen-research laboratory in Tokyo during the war.[74] Clearly, a large number of scientists were being exposed to methods of synthesis. The resulting diffusion of technology advanced Japan's industrial technological level during the 1920s when production of synthetic fertilizers finally began.

Suzuki, Tamaru, Takamine, Shiobara, and Ishisaka worked on the chemical aspects of ammonia synthesis; but no one in Japan could manufacture machinery capable of withstanding the application of high pressure. Several years earlier in Germany, Badische had found only Krupp, the arms maker, capable of making the high-pressure machinery needed in the production of ammonium sulfate. Japanese researchers were unable to use their own munitions manufacturers, who were too busy filling munitions orders during World War I, in a similar way, that is, to make specialized machinery.[75] Powerful zaibatsu possessed the financial means to develop a heavy-machinery industry for chemical engineering, but they saw little need for investment at that time. The government, on the other hand, did recognize the strategic importance of synthetic processes and invested heavily in nitrogen research. The Japanese Navy was developing high-pressure machinery that could eventually be used by the ammonia industry, but it would not be ready until after the end of the war. In sum, there were people in Japan who understood both the basic science of synthesis and the need for its commercial application. But the technology was still inadequate, and those with capital remained unwilling to invest in the research and development necessary to advance that technology to world levels.

Indeed, government agencies in Japan and in the West had far more incentive than private firms to invest in research during the war. Most of the belligerents faced a similar problem of lack of nitrogen; this was particularly acute, since smokeless powder, a universally used gunpowder, required nitric acid. Nitric acid could be obtained either by processing Chilean sodium nitrate—imports of which were eliminated in most European countries in August 1914—or by oxidizing ammonia manufactured by the synthetic or cyanamide method.

In Germany, the government sponsored primarily the synthetic method after a brief attempt to develop the calcium cyanamide method.[76] Fritz Haber and another scientist, Emil Fischer, persuaded the military and the Raw Materials Board, from whom any producer would have to obtain permission to acquire the raw materials for manufacture, that oxidation was possible. Badische was authorized to oxidize ammonia in the spring of 1915. But the Haber-Bosch method of ammonia synthesis was still in its infancy, and most of the ammonia continued to come from old sources, like by-product production from coke works. Using the by-product for nitric acid meant that German farmers lost their best source of by-product ammonium sulfate. In war, food is no less a strategic commodity than munitions, so the government ordered the substitution for ammonium sulfate of calcium cyanamide, a substance highly unpopular with German farmers. Relief was in sight, however, since the Oppau plant, which had begun full-scale industrial production in 1913, was working up to capacity by 1915. Persuaded that synthesis was the better technology, the government threw its support to Badische's methods, extending a large loan to the company in April

1916 following its order, three months earlier, that the company build a second plant. In Germany, government intervention was, therefore, crucial in increasing the quantity of nitrogenous chemicals produced by the most advanced technological processes.

In the United States, the National Defense Act of 1916 gave President Woodrow Wilson authority to advance government research and development of nitrogen plants.[77] General Chemical, where Takamine Jōkichi had studied a method of synthesis similar to Haber's, was given the nod to build a plant at Sheffield, Alabama (called the Number One Plant), which opened in September 1918 and closed two months later. The Number Two Plant used the cyanamide method. The government started from scratch with this project and was producing carbide within ten months of beginning construction at Muscle Shoals, Tennessee. Despite the massive effort at mobilization for construction of the Number Two Plant, it also closed with the termination of hostilities, and was later sold to the Tennessee Valley Authority. Two other plants were begun but were not completed by the end of the war. In toto, the American government spent $107 million on nitrogen research and development.

The British government was similarly interested in fostering production of ammonia.[78] Shortages of the chemical were at first alleviated by increasing output of by-product ammonia through replacement of beehive ovens with recovery furnaces, but this measure soon proved insufficient. In June 1916, the Ministry of Munitions formed the British Nitrogen Products Committee, charged with investigating new ways of producing nitrogen. It established research facilities, pilot plants, and subcommittees to study electricity and related matters. The British government was able to turn its advanced projects over to the Brunner, Mond Company, which developed into one of the world's largest ammonia interests in the postwar years.

Official concern with the nitrogen problem also caused the Japanese government to seek ways of expanding production through elevation of technological levels. Encouragement of research took two principal forms: establishment of the Institute of Physical Chemistry (Rikagaku Kenkyūjo) in 1917, whose stated purpose included fostering good relations between chemists in business and academia; and formation of the Special Nitrogen Research Laboratory (SNRL, Rinji Chisso Kenkyūjo) of the Tokyo Industrial Experimental Laboratory (TIEL) in 1918, which gave particular attention to equipment design and catalytic agents for synthesis.[79]

Potential shortages of imports during the war lent a sense of urgency to the scientists' demands for research laboratories such as the Institute and SNRL. But, even before the war, scientists recognized the need for encouragement of research and commercial development of strategic chemicals. In 1913, Takamatsu Toyokichi and Takamine Jōkichi, encouraged by entrepreneur Shibusawa Eiichi, gathered a group of leading Japanese scientists to petition the Presidents of the House of Peers and the House of Representatives to call for government establishment of a research institution.[80] In October 1914, the group received official endorsement of their plan for national support of scientific research in a document signed by Takamatsu, the Prime Minister, and the Ministers of Education, Finance, and Agriculture and Commerce.[81] The outbreak of war persuaded the documents' authors to intensify efforts to mobilize government assistance to chemical research, and produced a series of deliberations lasting from 24 November to 1 December 1914. Sponsored by the Ministry of Agriculture and Commerce, these meetings assembled members of a group called the Chemical Industry Investigation Commission, principally prominent academic chemists.[82] Both the Institute of Physical Chemistry and the SNRL of TIEL originated in these deliberations.

The first report of the Commission's conferences began with a statement similar to those enunciated by European governments, that World War I was essentially a "chemical war."[83] Specific proposals followed in three major areas: soda products, coal-tar products, and electrochemicals. Members were divided into subcommittees for purposes of investigating and reporting, and, showing unusual speed, produced a complete document on 3 December 1914. Members of the Commission included those who had first petitioned the government for research support in 1913 and a dozen other leading scientists. The head of the Commission, Vice-Minister of Agriculture and

Commerce Ueyama Mitsunoshin and Minister of Agriculture and Commerce Oka Minoru were the group's only non-academic members.[84] Active research chemists with intimate knowledge of Japan's resources and requirements in the areas of strategic and commercially necessary research, then, were free to establish policy with the imprimatur of a government agency able to implement it, the Ministry of Agriculture and Commerce. In succeeding months, subcommittees of the scholars' group reported in greater detail the various requirements stipulated in the original reports, further enhancing the relationship between scholarly and government interests.

The first important result of the Commission's reports—legislation for establishment of government-run dye works—preceded recommendations for fertilizer.[85] In the area of nitrogenous fertilizer, the first development was organization of the Institute of Physical Chemistry in 1917. The Institute was not the only laboratory spawned by increased interest in research during the war. Facilities set up by the Imperial Universities at Kyoto and Tōhoku, the City of Ōsaka, the Ministry of Communications, and dozens of private enterprises all predated the Institute by several months.[86] But the Institute was the first highly sophisticated laboratory established specifically as a national facility bringing together the expertise of scientists from throughout the country without regard to institutional affiliation. Created with 2 million yen from government funds, 1 million yen from imperial coffers, and 2,187,000 yen from private investors, the Institute began operating in March 1917. The first Director of the Institute was Kikuchi Dairoku, and his staff included Assistant Director Sakurai Jōji, Physics Department Head Nagaoka Hantarō, and Chemistry Department Head Ikeda Kikunae.[87] Despite the Institute's apparently strong beginning, its contributions to research were minimal until its eventual reorganization under Director Ōkōchi Masatoshi in 1921. Laboratories established by local governments and universities were more productive during the war.

Japan's participation in World War I on the side of the Allies permitted its acquisition, as spoils of war, of the German Haber-Bosch patent. Wishing to use the method in Japan, the government made it available to interested developers. The Japanese Navy, with its need for ammonia for conversion to explosives, was one such interested party. But the Navy's leadership failed to press for exploitation of the patent, figuring that others could develop it and spare the Navy the expense.[88] A consortium of business interests, including Mitsui, Mitsubishi, and Sumitomo, also expressed interest, but similarly lagged in developing the method. To make use of the patent, the government was forced to carry through another of the proposals of the Chemical Industry Investigation Commission, establishment of the Special Nitrogen Research Laboratory.

The SNRL was set up in March 1918 under the directorship of Kodera Fusajirō. A man with a broad-ranging background in government, teaching, and research, Kodera had also studied with Haber in 1908.[89] He was a good choice for the post, since his training included research in both areas necessary for successful ammonia research—high-pressure chemical engineering and chemistry. Under Kodera were Shibata Katsutarō, head of research in catalysts, and Yokoyama Buichi, head of research in high-pressure machinery.[90] Though able scientists, Shibata and Yokoyama initially faced setbacks in their attempts to solve technical problems in their respective areas. Yokoyama had read about appropriate machinery in the German *Zeitschrift fuer Elektrochemie* but found no manufacturer in Japan capable of making machinery able to withstand 200 atmospheres pressure. The Navy had experimented with high-pressure machinery, but their results initially failed to meet Yokoyama's standards.[91]

Discouraged by what he saw in Japan, Yokoyama traveled abroad in search of technology and machinery. He wished to observe the German Badische plant, but was denied admission in 1920. Yokoyama and an assistant then journeyed to Paris where, through the intercession of representatives of the Kōbe firm Suzuki Shōten, they were able to inspect a tiny plant for synthesizing ammonia by the French Claude method. As Suzuki was in the process of negotiating the purchase of Claude machinery, the two SNRL men were treated to an extensive inspection tour lasting five days. On the way home, in October 1921, they purchased high-pressure machinery from Hydrendt, Messerschmidt, and other German and American firms.[92] Although the trip contributed significantly to the government researchers' understanding of high-pressure technology and ultimately permitted the researchers' modification and improvement of techniques, it also indicated the government's backwardness as compared with private investors in chemicals. Suzuki Shōten, a private company, was a step ahead of the official government researchers in acquiring foreign technology, and therefore contributed significantly to diffusion of that technology in Japan in the years after World War I.

Catalyst research under Shibata Katsutarō at the SNRL also hit some snags. He and his assistants experimented with several types of potential catalysts, all of which either forced the chemical reaction to take place with explosive force or else produced fumes. A reference to use of pure iron in a report issued by the Fixed Nitrogen Laboratory in the United States put the temporarily derailed experiments back on track and permitted Shibata to patent his method (hereafter known as the TIEL method) in October 1922.[93] Though making slow progress, Yokoyama and Shibata eventually succeeded in producing ammonia by a synthetic method sufficiently modified from any European or American method to justify its own patent. The process would, within a few years, be used by a new firm, Shōwa Fertilizer, one of Japan's largest during the 1930s. Not only would the new firm use a process that was Japanese, but it would also use domestically produced high-pressure machinery. Chemical engineering did continue to lag, but was forced to develop by advances in chemistry and the chemical industry. Japanese chemists were able to conduct research at international levels by the 1920s; chemical engineering and industrial machinery would approach international levels during the next decade.

LICENSE ACQUISITION AND ADAPTATION: NITCHITSU AND ITS COMPETITORS

Noguchi Jun was Japan's first successful producer of ammonia by synthesis, but, though the Japanese TIEL method was nearing perfection in the early 1920s, Noguchi preferred using an imported process. His diligence in acquiring an Italian license was matched by his hard work in adapting the license for successful manufacture. Though the most successful, Noguchi was not the only early developer of synthesis of ammonia; his enthusiasm for innovation was shared by leaders in Suzuki Shōten, a firm involved in trading, manufacturing, and finance. As a trading company, Suzuki was known for aggressive expansion of its business opportunities. A desire to exploit good opportunities in manufacturing impelled the company's acquisition of license rights to a process for ammonia synthesis in the early 1920s.

A 1921 article in the London *Times* about Georges Claude's research in high-pressure ammonia synthesis in France sparked the interest of Isobe Fusanobu, a researcher at Suzuki Shōten.[94] At a cost of approximately 2 million yen, Isobe acquired the Claude license, which he used to establish Suzuki's Claude Nitrogen Industries Incorporated (Kurōdo-shiki Chisso Kōgyō KK) in April 1922. By October of the following year, Isobe's small plant at Hikoshima (5 tons daily capacity) appeared to be nearing success in producing ammonia at the ultra-high pressure of 1,000 atmospheres. The promise of success at the pilot plant, despite numerous problems with explosions and unsatisfactory equipment, persuaded Suzuki's management to proceed with plans to manufacture ammonium sulfate on a commercial scale.[95] In 1926, one year before its bankruptcy, Suzuki began building a large plant on the site of the pilot plant at Hikoshima, to be called First Nitrogen Industries (Daiichi Chisso Kōgyō KK). When the parent company failed in 1927, First Nitrogen had not yet begun producing, but the experience accumulated by the plant's researchers was not lost. Its equipment and technology were sold to Mitsui, and its employees were transferred to several companies.

While Suzuki was struggling with the explosions and faulty equipment inherent in installing a process not yet perfected even in its country of origin, Nitchitsu advanced the Casale method of ammonia synthesis, named for its innovator, Luigi Casale. Executives at Nitchitsu were well aware of the need to find cheaper methods of fertilizer production. Britain's wartime export embargo of ammonium sulfate was lifted in 1919, releasing a flood of imports into Japan which had immediate repercussions for Nitchitsu. In September 1920, Nitchitsu received only 235 yen per ton for expensively produced ammonium sulfate, despite the firm's contract with its distributor to sell the fertilizer at 385 yen per ton.[96] Noguchi did not wait to see the price continue plummeting. By February 1921, Noguchi had arrived in Italy where he decided to buy production rights for Luigi Casale's method of making synthetic ammonia. Noguchi's trip to Europe actually had other motives as well: Not only did he wish to take the vacation he felt he deserved to view the revolutionary political and economic changes sweeping Europe in the wake of World War I, but Nitchitsu's license for the Frank-Caro (cyanamide) method had to be renewed before its imminent expiration.[97] Thus, Noguchi was at an important threshold. He had to decide whether his European junket should involve buying new technology or renegotiating the purchase of old methods.

While in Italy, Noguchi toured the tiny laboratory and pilot plant run by Casale. Producing only a quarter ton of ammonia daily, the facility was hardly impressive, but Noguchi, author of articles lauding the benefits of synthesis of ammonia, was able to see merit in Casale's laboratory.[98] Noguchi visited Casale's laboratory with one of his managers, Kusu Kannosuke, who, along with another manager, Taniguchi Takaichi, was accompanying Noguchi on his European tour. After an inspection by Noguchi and Kusu, a manager at the Italian laboratory allowed the latter to see a typewritten request from Suzuki Shōten's New York office to purchase licensing rights to the Casale method. Neither Noguchi nor Kusu nibbled at the manager's bait intended to stimulate their interest in the method, but instead feigned a lack of interest. Other Japanese firms were less successful in their negotiations. Both Kaneko Naokichi of Suzuki Shōten and entrepreneur Kuhara Fusanosuke were also interested and sent technicians to inspect the Casale laboratory. But the facilities were deceptively small, and these technicians failed to see the potential merit of the project.[99]

Following their meeting with the Italian manager, Noguchi and Kusu boarded a train leaving the area, apparently to go to renew their cyanamide license. Then Noguchi decided that, although the other scientists seemed to have lost interest in the Casale method, haste was important if he wished to secure the license. Noguchi returned to

Casale's house, determined to see him. The Italian scientist was neither at home nor at the laboratory, and Noguchi discovered that he had gone to Rome. While Mrs. Casale phoned her husband, Noguchi hurried to Rome where he impressed Luigi with his interest in and ability to use the process.[100] Casale demanded a steep price for the privilege, about 1 million yen, which Noguchi, though the top managing officer in Nitchitsu, was unwilling to authorize alone. Requesting additional time to consult with his colleagues in Japan, Noguchi paid a deposit of 10 percent, or 100,000 yen, for a two-week extension of the period of decision.[101]

Noguchi raced through the winter snow to Bern, Switzerland, where, after consulting with a man named Itamoto (no personal name given in the reference) at Bern University, he composed an explanatory telegram to his colleagues at headquarters in Ōsaka.[102] His message spoke with alarm of the dangers of another company's prior acquisition of the license. His colleagues at Nitchitsu were completely surprised by the telegram. Enonami and Ichikawa, his closest associates on the Board, decided to call a meeting of the entire Board of Directors to respond to Noguchi's note. The price for the license was high, and few at Nitchitsu knew the extent of modification of facilities necessary for the Casale method. The Board, therefore, wired Noguchi to continue to negotiate for time, to pay the deposit only, and to return home to explain what he had observed in Italy. The members may not have shared Noguchi's ability to make rapid decisions, but they were equally unwilling to abandon innovative projects with profitable potential. Casale accepted the deposit. After making a detour to study the rayon industry in Germany and Italy, Noguchi returned to Japan.[103] The day after he landed at Yokohama, he met with his top management.[104]

All this traveling and meeting with his managers lasted several months. Casale's granting such a long extension appears somewhat illogical in view of his earlier offer of a shorter two-week extension. Perhaps Noguchi's enthusiasm and apparent understanding of the complex process impressed the Italian. Perhaps Casale's laboratory needed the large deposit. More likely, Noguchi had succeeded in sufficiently stalling the process of negotiation to allow the initial interest of Kuhara and Suzuki to dissipate; Suzuki, in particular, may have been either nearing completion of its negotiations with Claude or else using the Casale option as a bargaining tool in its dealing with Claude. Suzuki may have had only a weak interest in the Casale method all along. In either case, Casale's decision was astute, since Noguchi represented the best chance for selling the license to a Japanese concern at that time.

Noguchi spent the months after his return to Japan in early summer inspecting prospective plant sites. Taniguchi Takaichi's preliminary investigations persuaded Noguchi to check the area around Nobeoka in Miyazaki prefecture in northern Kyūshū. Accompanied by local dignitaries, Noguchi took a motor tour of the area, deciding that Tsunetomimura appeared adequate from several perspectives. Not least important was the assistance Noguchi received from Kenseikai Party politicians from Kyūshū who persuaded the Hara Cabinet (Seiyūkai) to overturn Communications Minister Noda Utarō's decision to encourage transmission of electricity from Miyazaki prefecture.[105] This resulted in retention of more electricity in the prefecture. That a greater supply of electricity would be available for his planned chemical factory helped persuade Noguchi that this site was ideal.

Having selected a site, Noguchi and four of his colleagues were ready to return to Europe to negotiate with Casale. Enonami Naosaburō went along in his managerial capacity; Taniguchi Takaichi and Shimada Shikamitsu accompanied him as skilled technicians; and Ōhata Goichirō as a special negotiator.[106] The quintet first stopped in New York, where they successfully negotiated with American Cyanamide, the American holder of the Italian Casale license, to relinquish its rights. Next stop was London, where they assessed their company's ability to pay for the Casale license. The steady decline of the lira made collecting the necessary capital (10 million lira) rather easier than anticipated. Mitsubishi Bank connections led the Nitchitsu group to an Italian bank in London which handled their exchange of funds to lira.[107] From London, the five Japanese went to Germany, postponing the trip to Italy while they examined the rayon industry in Berlin and invited Shiraishi Naojirō's son,

Muneshiro, a student in Germany, to join them. In Rome, a "happy" Casale sold the rights to his process plus some machinery to Noguchi and his colleagues on 12 December 1921. As December 1921 was a time of social and political turmoil in Rome, the Japanese did not tarry but took off for home within two weeks of signing the purchase agreement.[108]

An unexpected yet important development of the overseas trip by the group from Nitchitsu was the acquisition of the services of the talented Shiraishi Muneshiro. Shiraishi had entered Nitchitsu in 1915 but left in 1918 when Noguchi told him to postpone a trip to the United States until his seniors had had their opportunity to go. He did not wish to wait, so found another way to get to Europe. A recipient of a scholarship from the Iwasaki family to study in Germany, he was bound, at least temporarily, to the Mitsubishi company. This was no new relationship. The Shiraishi family had enjoyed a history of relations with Mitsubishi. Muneshiro's father Naojirō had advanced through the ranks there before moving to Nitchitsu. The son Muneshiro enjoyed himself traveling with the Nitchitsu group from Germany to Italy but was deeply concerned about the propriety of his leaving Mitsubishi after having received such a generous scholarship. Noguchi reminded him that, being over 30, he should be committing himself to his life's work.[109] Although several members of his Board of Directors had Mitsubishi connections, Noguchi was increasingly distancing himself from the Mitsubishi influence; his rehiring one of the company's brightest scientists away from Mitsubishi helped to underscore this.

Indeed, Noguchi was establishing himself as the dominant figure within Nitchitsu. His strong position in decision making within the company was augmented by acquisition of the Casale license. It was during the 1920s that he began to play the role of the "one man," the powerful leader of a multi-faceted chemical empire. This was a sobriquet which characterized Noguchi during the last two decades of his life.[110] One noteworthy instance of his growing dominance occurred within weeks of his return to Japan from Europe with the license. He called a meeting of the Board of Directors in late March 1922 for the ostensible purpose of discussing the advisability of plans to establish a plant at Nobeoka.[111] Noguchi had already made up his mind on that issue, and was merely getting the Board to ratify his decision. Many of the members thought that production should begin at Kagami, but Noguchi countered with ready explanations of Kagami's inadequacy. Though unable to make unilateral decisions, Noguchi was clearly the dominant member of the Board of Directors.

A few days later, on 2 April 1922, Noguchi took a group of managers from Kagami to the Nobeoka site, where they worked assiduously, "forgetting to eat," until the plant was set to begin operation in September of the following year.[112] He brought the men with whom he had worked most closely in recent years: Iwabashi Yū (Production Division head and head of the Minamata plant), Yanagiya Sayū (Electricity Division head), Shimada Shikamitsu (Temporary Construction Division head), Taniguchi Takaichi (head of the Nobeoka plant) and Shiraishi Muneshiro (assistant head of Nobeoka and head of the reorganized Research Division). Casale himself arrived in Japan the day of the great Kantō Earthquake, 1 September 1923, and was temporarily unable to proceed on his journey to Kyūshū. But, within three weeks, the Italian inventor threw the switch that commenced production at Nobeoka, thereby starting Japan's first synthesis of ammonia on 20 September 1923.[113]

Just as when he was planning production of cyanamide, Noguchi had to solve problems in management of personnel and resources in his synthesis operations. As before, electricity was a crucial problem. And, again, he found ready sources (see Table 20). Beginning with Gokasegawa Electric—a company founded in 1920 by Nagamine Yōichi, Diet member from Miyazaki, and jointly managed by Nagamine's family and Nitchitsu executives—Noguchi's Nobeoka plant was able to draw on numerous sources of hydroelectricity, the major requirement for manufacture of synthetic ammonia.[114]

With his success at creating new facilities for generating electricity at Nobeoka, Noguchi had solved an important problem in synthesizing ammonia. Less than two years after Nobeoka began production, Noguchi began to contemplate expansion of the plant facilities at Minamata. As Minamata was converted to the synthesis operation, it developed a symbiotic relationship with Nobeoka reminiscent of its

TABLE 20 Hydroelectricity at Nobeoka

Source	Amount	Date of First Generation
Gokasegawa	12,000 kw(a)	August 1925
Mamihara	4,500 kw	February 1926
Sensogawa #2 Plant	3,200 kw	November 1926(b)
Sensogawa #1 Plant	1,800 kw	January 1927(b)
Ichinosegawa	9,800 kw	July 1927
Takenogawa	8,000 kw(c)	September 1927
Tōchi		March 1928

Sources: Yoshioka, p. 141; Yamamoto, p. 456.; and Kondō, p. 138.

Notes: (a) Yamamoto notes initial generation at Gokasegawa as 4,500 kw.
(b) Yamamoto, official company historian, notes the initial generation of electricity in the above months. Yoshioka and Kondō each offer two other sets of dates for Sensogawa plants.
(c) Generative capacity is not given for the Takenogawa and Tōchi facilities; total capacity is estimated by Yamamoto at 40,000 kw, however, permitting an estimate of approximately 8,000 kw capacity for the two plants together.

relationship with Kagami in the 1910s. Before and after World War I, Minamata and Kagami had shared management structures, allowing the plants to share top production managers, permitting high intrafirm mobility, and inducing a sense of company loyalty rather than plant identification in most employees. If structural similarity was the basis for successful integration, the construction of Nobeoka should have temporarily undermined this integration, since the plant differed from its two predecessors in several respects. First, it was located at some distance from the other two plants, which were both in Kumamoto prefecture, diminishing the potential for sharing functional operations. And, second, because of differences in the technology of production, management structure in the plants also differed. By 1923, Nitchitsu company organization required that heads of each plant report directly to the managing director, just as heads of divisions did, because each plant used different processes.

Company organization had sustained several significant changes since 1915. By November 1923, Nitchitsu had grown to the extent that management felt compelled to undertake a formal study of company organization. This resulted in new company regulations stipulating that separate divisions be created to deal with separate problems rather than having a few generalist divisions handling a wide variety of duties.[115] Thus, in the area of business affairs an Accounting Division and a Subsidiary Division were added in 1923. The Subsidiary Division, which would later play an important role in the company, played a relatively minor role at that time because Nitchitsu defined subsidiaries not as companies controlled by the subsidiary division but rather as companies in which Nitchitsu held stock. These included 6 power companies in Kyūshū, 3 other power companies, and a mining company. The Subsidiary Division did not even have its own head and staff; the Accounting Division head also served as head of the Subsidiary Division. By contrast, the new Accounting Division filled an immediate need. The tremendous increase in records and other paperwork during and after the period of rapid acquisition, construction, and installation of the synthetic method of ammonia manufacture convinced Nitchitsu's managers to include an Accounting Division in the new regulations on company structure.

New divisions were added in other areas as well. Electricity was elevated to a division, rather than being a section under the former Production Division and former Temporary Construction Division. Research was also elevated to divisional status; research had become far more sophisticated and the need to coordinate diverse types of research more critical after the importation of the synthetic method.

The financial side of management had grown substantially by 1923, but the weight of management was still on the production side. More divisions were responsible primarily for production than for management. This emphasis was also reflected in company finances. That is, most expenses until 1926 were associated with setting up synthesis of ammonia at Nobeoka and Minamata.[116] The major method of capital accumulation at that time was building equity by calling in payments on shares already issued and by issuing more shares. After 1926, as Nitchitsu expanded rapidly in Korea, both these patterns changed. In the later period, company expenses were more often associated with acquisition of shares in subsidiaries, and selling bonds was as important as selling shares as a method of obtaining funds.[117] But, in the first half of the 1920s, Nitchitsu was primarily a chemical company whose shareholders supported its goals and whose financial emphasis was on funding production. It was not yet a "Konzern," a term to be discussed in Chapter 5.

To manage this production company, Noguchi was able to find the right persons for the right jobs. Management at Nobeoka consisted of the same men responsible for the plant's creation, who had previously been Noguchi's closest colleagues at Kagami and Minamata. Some of the division heads and heads of sections in production were engineers and managers who had worked their way through middle-management ranks as technicians. Shimada had been responsible for construction of Kagami and repeated his role at Nobeoka. Iwahashi Yū, a cyanamide specialist, had joined Nitchitsu in its early days in Minamata, and stayed with Nitchitsu even after cyanamide production was terminated at Minamata and transferred to a subsidiary in Niigata.[118] Taniguchi Takaichi, an engineer, demonstrated his ability while accompanying Noguchi on his European trips in 1921. Nagasato Takao was introduced to Nitchitsu by his father, an early (1906–1907) executive of the company.[119] Ōshima Eikichi was brought to the company under somewhat different circumstances. His father had been head engineer of the coal company where Ichikawa Seiji was first employed after graduation from Tokyo Imperial University; the son Eikichi, a popular young assistant professor of chemistry at Tokyo University, attracted the attention of Noguchi and Ichikawa as a magnet for bright young graduates.[120] The first of these young scientists Ōshima attracted was Kudō Konogi, head of the ammonia-synthesis section at Nobeoka and later an important developer of a Nitchitsu subsidiary, Korean Coal.[121] Indeed, Ōshima continued to bring in new talent even in later years. Most of the new scientists and engineers at Nobeoka and Minamata were in their twenties; their average age was 28.[122] The engineers at Nobeoka formed the core of skilled employees transferred to Minamata and Korea, and almost half the technicians at Nobeoka served as training personnel for their successors in other Nitchitsu plants.[123] Sharing of technical information later helped enhance a sense of interdependence among the company's factories (see Table 21).

Another method of inducing shared interests between Minamata

Table 21 Nitchitsu Management Organization, 1923

Managing Director (Noguchi Jun)

Business Manager (Ichikawa Seiji)

- I. General Affairs Division (Enonami Naosaburō)
  - A. Management Section
  - B. Archives Section
  - C. Procurement Section
- II. Accounting Division (Ōgyu Den)
  - A. Accounting Section
  - B. Treasurer
- III. Electricity Division (Yanagiya Sayū)
  - 1. Shirakawa Subsection
  - 2. Midorigawa Subsection
  - 3. Sendaigawa Subsection
  - 4. Kurino Subsection
  - 5. Sogi Subsection
  - 6. Sendaigawa Subsection
  - 7. Sensogawa #1 Subsection
  - 8. Sensogawa #2 Subsection
  - 9. Ichinosegawa Subsection
  - 10. Tōchi Subsection
  - 11. Takenogawa Subsection
  - 12. Kagami Subsection
  - 13. Minamata Subsection
  - 14. Nobeoka Subsection
- IV. Research Division (Shiraishi Muneshiro)
- V. Temporary Construction Division (Shimada Shikamitsu)
  - A. Business Section
  - B. Procurement Section
- VI. Subsidiaries Division (Ōgyu Den)
- VII. Engineering Division (Hirano Asakichi)
  - A. Business Section
    - 1. Accounting Subsection
    - 2. General Affairs Subsection

TABLE 21 *(continued)*

B. Production
    1. Design Subsection
    2. Research Subsection

VIII. Minamata Plant (Iwahashi Yū)
  A. Business Section
    1. General Affairs Subsection
    2. Treasurer Subsection
    3. Accounting Subsection
    4. Purchasing Subsection
    5. Storage Subsection
    6. Packing Subsection
    7. Transport Subsection
    8. Inspection Subsection
    9. Insurance Subsection
  B. Production Section
    1. Carbon Subsection
    2. Nitrogen Subsection
    3. Cyanamide Subsection
    4. Sulfuric Acid Subsection
    5. Ammonium Sulfate Subsection
    6. Cement Subsection
    7. Analysis Subsection
    8. Construction Subsection
    9. Electricity Transmission Subsection
    10. Carbide Subsection

IX. Kagami Plant (Yanagiya Sayū)
  A. Business (same subdivisions as at Minamata)
  B. Production Section (same as Minamata, minus the carbon subsection).

X. Nobeoka (Taniguchi Takaichi)
  Assistant Head (Shiraishi Muneshiro)
  A. Business Section
    1. General Affairs Subsection
    2. Accounting Subsection

TABLE 21 *(continued)*

3. Storage Subsection
4. Purchasing Subsection
5. Shipping Subsection

B. Production Section
1. Nitrogen Subsection
2. Oxygen Subsection
3. Synthesis (Kudō Konogi) Subsection
4. Sulfuric Acid Subsection
5. Ammonium Sulfate Subsection
6. Pharmaceuticals Subsection
7. Analysis Subsection
8. Construction Subsection
9. Electrolysis Subsection
10. Electricity Generation Subsection

Sources: Yoshioka Kiichi, pp. 142–143; Yamamoto, p. 460; Ōshio Takeshi, "Nitchitsu Kontsuerun no kin'yū kōzō," Diagram #1, between pp. 114 and 115.

and Nobeoka was by making the functions of the plants similar. This was the method developed several years earlier for Minamata and Kagami. Although the emphasis on identical operations was short-lived and was soon replaced by emphasis on complementarity of production in the firm's plants, Noguchi temporarily saw copying of tested methods as a certain route to successful expansion. He brought in scientists Kudō Konogi and Hashimoto Hikoshichi from Nobeoka to oversee adaptation of the Minamata plant for synthesis of ammonia.[124] Conversion was carried out with little trouble. Noguchi had hoped to use Japanese-made machinery and regretted that the still low level of technology in the Japanese machinery industry forced him to buy compressors and other heavy equipment from Italy and elsewhere. But he wished to expedite production at Minamata, so he had no choice. In the event, Kudō and his staff managed to begin production of ammonia on 25 February 1925, and the plant attained its initially planned output of 20 tons, 3 times that of Nobeoka, by April 1927.[125]

The success of ammonia synthesis at Minamata highlighted the inefficiency of the cyanamide process for ammonium sulfate. Within months of Minamata's conversion, plans to terminate the latter production methods had been carried out at Nitchitsu. Some facilities for production of calcium carbide were rebuilt at Minamata later, in 1932, for use in the manufacture of organic synthetics (another branch of the chemical industry), but the fertilizer sections of Minamata were purged of the older "three-product cycle" of making nitrogenous fertilizers. Minamata plant head Iwahashi Yū, accompanied by a group of young scientists, technicians, and workers, thereupon established the Shin'etsu Electric Company, with machinery and capital from Minamata, as a subsidiary of Nitchitsu. In September 1926, Shin'etsu Electric was incorporated as Shin'etsu Nitrogenous Fertilizer, Inc. (Shin'etsu Chisso Hiryō KK) and received much of the calcium cyanamide machinery formerly installed at Kagami. Kagami, forced to close by its uneconomical production methods and the insufficiency of electricity as increased production of synthetic ammonia at Nobeoka and Minamata drained its supplies, donated some of its machinery to Shin'etsu. In October 1927, it sold the rest to Dai Nihon Fertilizer, which had recently begun diversifying.[126] The next year, Shin'etsu could no longer sell calcium cyanamide competitively and gradually switched to other chemicals.[127]

Sales of machinery to Dai Nihon as well as use of skilled engineers already trained at Nobeoka helped offset the expenses of converting the plant at Minamata. Maintenance of a high profit rate throughout the period of conversion, combined with a 10-million-yen bond issue underwritten by the consortium of banks that had previously supported Nitchitsu—Mitsubishi, Yamaguchi, and Aichi Banks—supported the effort.[128] Nitchitsu's capital needs were very high at that time, as the company was also in the process of constructing the first of its large subsidiaries in Korea.

Access to capital was a requisite for expansion, but financial costs were not Noguchi's only concern. As at his other plants, there were problems in managing resources and labor. The most important technological consideration in synthesis was finding adequate sources of electricity. Minamata was ideally situated for easy solution of this problem. It had long generated its own hydroelectricity for manufacturing cyanamide and also used electricity generated at Kagami. Another consideration in expansion of output was the possible need for extra labor. Nitchitsu's management recruited large numbers of farmers living in the neighborhood. Improvement in the schedule of work shifts at the plant permitted the day-shift workers to leave the factory at 3:15 p.m.;[129] this may have helped encourage farmers to take blue-collar jobs at the plant, despite low remunerative benefits compared with those given engineers and higher-ranking employees.[130] In the end, Noguchi solved the problems of energy, capital, and labor at Minamata with relatively little effort and so could begin his major undertaking in Korea at the same time. In the next decade, the Korean plants were to become the heart of Noguchi's fertilizer production and the third largest facility in the world for manufacturing synthetic ammonia.

Also noteworthy during the early 1920s was Noguchi's adventurous development of the rayon industry in Japan, using surplus synthetic ammonia as the major raw material in the fabric. Rayon was the first of an ever-widening circle of closely related products made by subsidiaries of Nitchitsu in Japan and Korea. The products' technological relationships permitted productive complementarity of these subsidiaries.[131]

A few years later, engineer Murayama Kyūzō of the Ministry of Agriculture and Commerce was invited by Noguchi to join the staff at Minamata, where he developed an internationally patented method of oxidizing ammonia to produce nitric acid.[132] This discovery set Nitchitsu on the road to becoming one of Japan's largest producers of explosives, leading in turn to large-scale production of other components of explosives. Another scientist, Hashimoto Hikoshichi, also made important discoveries in carbide-based organic synthetics, the production of which was later found to induce the dread Minamata Disease, mercury poisoning. This change accelerated the pace of research and development in a wider range of technologies and allowed Nitchitsu to become, during World War II, one of Japan's largest suppliers of nitric acid and undiluted sulfuric acid for explosives and methanol for airplane fuel.[133]

Noguchi's pioneering development of synthetic ammonia within Japan was also significant. Clearly, he recognized the potential for exploiting the market for chemical fertilizers by undercutting his competitors' prices through technologically based increases in productivity. He was willing to gamble on untested processes, guided by a bold entrepreneurial spirit and a solid scientific training. He was able to anticipate spin-offs and interrelated uses of new technology. Thus, although ammonia was initially synthesized to fill a domestic need for fertilizer, and fertilizer continued to consume much of the company's ammonia, the development of new production technologies for rayon, nitric acid, and synthetic plastics initiated a fundamental structural change in Nitchitsu's character from "fertilizer company" to diversified "chemical company." Nitchitsu was Japan's prototypical technology-driven chemical company.

FUJIYAMA AND ELECTROCHEMICALS

Fujiyama Tsuneichi's investment in electrochemicals offers an informative contrast to Noguchi's. He moved to Hokkaidō and founded a chemical company in 1912. But his investment was more cautious and generally less successful than Noguchi's. Unlike Noguchi, Fujiyama chose to stay with the old cyanamide technology, and it was not until after he stepped down as Managing Director that his company, Denka, began to diversify and innovate. As long as there was little competition, this strategy worked, but, as competition increased in the 1920s, Fujiyama's company fell behind. International and even national levels of technology surpassed his. His company had the same technology as Noguchi's in 1912, but Noguchi used it as a springboard for further advances. Fujiyama did not. "Yet, not to innovate is to die."[134] By the mid-1920s, when Fujiyama was replaced as managing director, Denka began vigorous innovation.

The story of Fujiyama's experience is illuminating. Soon after his departure from Nitchitsu, Fujiyama Tsuneichi was invited by an old college friend to the office of Mitsui Mining's coal mines in Kyūshū. There he was introduced to Makita Tamaki, head of the mining facility. Makita expressed great interest in Fujiyama's request for help in establishing a facility for producing carbide at Ōji Paper, the Mitsui-affiliated company at Tomakomai (in Hokkaidō) which produced surplus electricity it wished to sell.[135] Fujiyama had taken 50,000 yen with him when he left Nitchitsu to use as seed capital. This was a considerable sum for an individual in those days. In addition to Makita, Fujiwara Ginjirō, then managing director of Ōji Paper, and Dan Takuma listened to Fujiyama's request. All three later rose to the highest positions in Mitsui. From Fujiwara, Fujiyama received permission to use surplus electricity generated by the paper plant, and, from Mitsui itself, he received a matching grant of 50,000 yen for starting production as well as for continuing operational expenses.

Fujiyama established the Hokkai Carbide Factory in May 1912 on land neighboring the paper plant.[136] Though supported by Mitsui, production remained for the time being in the hands of the four men. The new business soon encountered problems. Hokkai Carbide's plans for manufacture of the "three products" (calcium carbide, calcium cyanamide, and ammonium sulfate) were based on methods similar to those used by its competitor, Nitchitsu. Fujiyama had perfected a continous operation furnace before his departure from Kyūshū, but patent considerations prevented his using it at Hokkai Carbide.[137] He was legally forced to redesign the furnace at great cost. The expenses involved in the furnace studies can be estimated by comparison with previous research. To be sure, research costs were undoubtedly lower the second time Fujiyama studied furnace efficiency, but furnace studies had cost Nitchitsu 140,000 yen when Fujiyama first investigated the apparatus.[138] Despite the likely high costs, he continued to modify the Frank-Caro process, and his changes were put into operation within a year. Production of calcium cyanamide began at Hokkai Carbide in April 1913, one month after he received a patent for his process. Fujiyama's Hokkai process raised quality significantly from the level it had attained before he left Nitchitsu; he was able to produce a cyanamide with 18 percent nitrogen content.[139] By October of the following year, Hokkai Carbide was turning some of the calcium cyanamide into ammonium sulfate.

Setbacks remained relatively minor, and the small plant was aided by the onset of World War I. Rising fertilizer prices enticed an increasing number of investors to the field, and, in 1915, a group of 22 influential businessmen signed a declaration of support for incorporating Hokkai Carbide as a separate company, Electrochemical Industries, Inc. (Denki Kagaku Kōgyō KK, hereafter referred to as Denka).[140] Following the declaration, Denka was incorporated on 1 April 1915, capitalized at 5 million yen (of which 1.26 million yen was paid up). One hundred thousand shares were issued to 267 investors; Mitsui Hachirōemon held 49,000 of them. Fujiyama held 2,535 himself, some of his family members held other lots of one thousand, and managers at Denka as well as influential venture capitalists like Shibusawa Eiichi held most of the rest. Most of the group of 22 influential men were among the largest shareholders.[141] Despite its legal status as an independent corporation, Denka was clearly a Mitsui-related company. Furthermore, Fujiyama's role as an investor in the reorganized Denka was far smaller than it had been in the predecessor company, Hokkai Carbide. In this respect Denka differed from Nitchitsu, which had a weaker connection with Mitsubishi Bank. Top Denka management and technicians were from Mitsui, including Umagoe Kyōhei, a septuagenarian former beer maker, who became President of Denka. The company was an integral part of Mitsui as Ōji Paper was.

Almost immediately, it became apparent that Tomakomai could sell as much fertilizer as it was able to produce. But Tomakomai was increasingly seen as having some problems as a production site. First, the supply of electricity was becoming less certain. Just as war-induced import restrictions had made domestically produced nitrogenous fertilizers more popular, they also increased demand for paper. Ōji Paper, still supplying its surplus electricity to Denka, simply needed it for paper manufacture and wished to decrease transmission to Denka. At the same time, Hokkaidō farmers' demand for chemical fertilizers had failed to keep pace with that of other Japanese farmers. As a result, company leaders concluded that Tomakomai might not be the best place for Denka's factory and moved most production facilities to a new major plant at Ōmuta near Mitsui's Miike mines in Fukuoka prefecture.[142] Tomakomai continued producing some fertilizer until May 1922, when all facilities there reverted to Ōji Paper.[143]

Denka's first move toward expansion occurred within a year of the company's incorporation. In March 1916, Denka embarked on the Japanese chemical industry's first foreign investment. Making use of the high-grade coal available on the continent, Denka began a cooperative effort with the South Manchurian Railway Company. This operation yielded, during its three and a half years of existence (from November 1916, when carbide was first produced, until June 1920), 19,000 tons of carbide, 21,000 tons of calcium cyanamide, and 13,000 tons of ammonium sulfate. Eventually, production in China stopped for three reasons: The price of Manchurian coal was too high relative to the price of chemical fertilizer after wartime prices fell; electric supply was insufficient; and the Chinese government attached an exceptionally high tax on export of the chemicals to Japan, making their production unprofitable.[144] What is significant about the venture is that it was the first instance of a chemical company's moving overseas. Chemicals were not primarily an extractive industry like coal or iron-ore mining, ventures frequently undertaken by developed countries in developing countries. The chemical industry's move to the continent was not a natural development.

Even before the Chinese operation got off the ground, Denka was planning its major facility at Ōmuta in Fukuoka. Ōmuta was totally integrated with Mitsui's nearby mining operations. The factory took advantage of surplus electricity from the mines (6,000 kilowatts) and from Kumamoto Electric (5,000 kilowatts). High-grade coal came from Miike, as did sulfuric acid.[145] Ōmuta seemed a natural and easy site for Mitsui's expansion of Denka. Company planners recognized that organization was as important as access to resources in making the new factory productive. Aware of potential problems at Ōmuta resulting from ineffective organization, they brought in a specialist, engineer Hibi Masaji, to organize the facility.[146] A year of organizing the plant and tentative production of each of the "three products" resulted in good productivity by 1917. That year, 6,000 tons of carbide, 5,600 tons of calcium cyanamide, and 3,000 tons of ammonium sulfate were manufactured.[147] While still below the totals achieved at

Nitchitsu's Minamata and Kagami, and even at its own Tomakomai, Denka's Ōmuta plant made a respectable contribution to Japan's supplies in these important chemicals.

Denka's plants, as part of the larger Mitsui conglomerate producing numerous products requiring energy, took advantage of any surplus energy Mitsui's other close-by factories and mines could provide. By 1918, energy dependence on other plants seemed impractical, however, and plans for construction of a new fertilizer plant at Tagawa were accompanied by plans for generation of electricity as well. Denka's chief competitor, Nitchitsu, had always constructed fertilizer plants with plans for self-generation of electricity. Officials of Denka apparently saw the wisdom of self-generation and, starting with Ōmuta in 1919, began construction of electric facilities at their already established fertilizer plants.[148]

Despite such attempts to economize and institute effective operation, Denka was plagued by unwise policy decisions after the war. Indeed, by the middle of the 1920s, Fujiyama, as managing director, had led the company to the brink of disaster. He made two major blunders. First, he planned to build another facility at Ōmi just as the Japanese market for nitrogenous fertilizer was collapsing due to foreign dumping, demonstrating a decided lack of foresight. Furthermore, he doggedly pursued an outmoded process, rejected in Europe, for making calcium cyanamide.

The Ōmi plant had actually been carefully planned. It was located on the site of Nitchitsu's flood disaster in Niigata in 1911; the area contained not only several facilities for generation of electricity but also large deposits of limestone amounting to one-tenth of Japan's total supply. These deposits had earlier attracted Nitchitsu, too. In 1921, Denka acquired and merged two local electricity firms and beefed up their facilities for generation. The company forged ahead. By December 1921, carbide was being produced, followed in July 1922 by calcium cyanamide, and in August by ammonium sulfate. Denka's total sales from both its fertilizer plants at Ōmi and Ōmuta (the other facilities were sold to other investors in Mitsui and the South Manchurian Railway) doubled in 1923 over the 1922 level, but plunged precipitously in 1924.[149] Although sales recovered the following year, Denka's method was ultimately less productive, and its continued ability to sell was not assured. Denka's response to the European dumping that produced the low prices and poor sales of expensive domestic product was to continue the "three-product cycle." Nitchitsu, by contrast, responded with technological innovation. Denka's eventual recovery from problematic sales in the 1920s required the firing of Fujiyama and rationalization under a new head of production, his successor, Fujiwara Ginjirō.

Fujiyama's other poorly timed effort was his attempt to use the Soederberg-arc method of producing calcium cyanamide at Denka's plants.[150] An article about the arc method developed by Norwegian C. W. Soederberg which appeared in the American journal *Chemical and Mechanical Engineering* caught Fujiyama's eye in July 1922. Immediately contacting Mitsui's London office to serve as agent, Fujiyama purchased the license for the process from the Norwegian.[151] The process was sophisticated, using large amounts of electricity and magnets to shape the alternating current which oxidized air. But the method never reached its commercial potential even in northern Europe where it was developed and where electricity, its chief requirement, was cheap.[152] Hibi Masaji attempted to use the process successfully at Denka, but the mechanical technology remained inadequate. Finally, in 1929, Norwegian engineer Knut Hylland traveled to Japan to start the process, years after it had been generally abandoned in Europe.[153] Although effective production of cyanamide by the arc process eventually enabled Denka to make high-quality fertilizer, its cost remained high. Fujiyama would have better served his company had he invested, as had other makers of fertilizer, in technology for synthetic ammonium sulfate rather than anxiously developing overly expensive though elegant methods of perpetuating the "three-product cycle."

Denka's preoccupation with maintaining the "three-product cycle" had several results. First, it cost the company money. Ammonium sulfate and calcium cyanamide competed for the same market, especially in the first half of the 1920s, when most of the calcium cyanamide produced was still being used as the basis for ammonium sulfate. In competition with ammonium sulfate, the price of calcium

TABLE 22 Price of Calcium Cyanamide as a Percentage of Price of Ammonium Sulfate, 1924–1930

Year	*Change in Ammonium Sulfate Price from Previous Year*	*Change in Cyanamide Price from Previous Year*	*Price of Cyanamide/ Price of Ammonium Sulfate (%)*
1924	-13.7	-12.8	97
1925	9.3	1.9	91
1926	-15.2	-13.1	93
1927	-14.3	-20.3	86
1928	-3.6	-7.9	82
1929*	-5.4	4.5	91
1930	-19.8	-26.5	94

Source: Calculated from data in Table 4, Hashimoto Jurō, "1920 nendai noryūan shijō," p. 58.

Note: * indicates year in which cyanamide price did not fall in conjunction with that of ammonium sulfate.

cyanamide could not be established independently. Therefore, cyanamide's price fluctuated with that of ammonium sulfate, and dropped when improvements in production technology and heavy imports of foreign ammonium sulfate drove down its price (see Table 22). Lower prices meant lower profits available to cyanamide producers.

Another result of Denka's adherence to the "three-product cycle" was structural. Expansion of the company meant construction of additional production facilities, all making essentially the same products and using the same processes. While it was a convenient way to expand production facilities when a known market existed, establishing nearly identical plants was not conducive to intra-firm technology transfer and augmentation.

By contrast, Nitchitsu began to move away from emphasis on the horizontal expansion it fostered during World War I. Nitchitsu's management encouraged expansion into new areas after production by synthetic methods began in the early 1920s. They established plants that made various types of intermediate products and encouraged transfer of ideas from technicians who studied one type of product to those who studied others. Thus, the store of knowledge available among scientists at Nitchitsu was quite sufficient for innovation or successful imitation. Denka, on the other hand, expanded only horizontally, cloning its facilities without noteworthy creativity until the company's management finally decided to diversify after Fujiwara replaced Fujiyama. Limitation to one principal method of production would, in itself, tend to stifle creativity, and the "three-product cycle" was particularly limited in its application. Adoption of newer methods would have permitted Denka to develop advanced spin-off technology several years before it actually did and to invest in a wider variety of processes. In turn, Denka might have developed a more complex and efficiently integrated structure of diversified plants, as Nitchitsu did. The foremost cause of Denka's stagnation during the 1920s was its inability to advance technologically.

FOUR

Nitchitsu's Diversification and Search for Resources: 1924–1933

Access to his own sources of electricity had permitted Noguchi Jun to invest in advanced technology. Early on, these sources were cheap and readily available. By the mid-1920s, concerns about possible rising costs of electricity and about tightening regulation of the electric industry compelled him to seek more certain sources of electricity. The colonies offered a hospitable environment for businessmen like Noguchi. Although controls were tighter in Korea and Manchuria, chemical companies took advantage of their special relationship with colonial administrators who could use their authority and discretion to offer the companies cheap electricity.[1]

The decade following the first large-scale commercial production of synthetic ammonia in the mid-1920s saw several new developments deriving from changes in technology and resource availability. New manufacturers using a variety of production methods entered the market. These new investors, who represented the second and third waves of investment in electrochemicals, will be considered in Chapter 6. In addition, this period was marked by the first major industrial investment in Japanese colonial enterprise; in particular, the lion's share of Japanese investment in Korea was concentrated in the chemical and electrical industries and undertaken by Noguchi Jun.

Active industrial expansion at home and in the empire is an important factor in the history of the industry and, in particular, the history of Nitchitsu. Also critical to an understanding of the company's history is an assessment of its response to domestic and international developments initiated outside the company boardroom. An important exogenous factor affecting decisions about marketing, sales, and production was the politicizing of the fertilizer issue. Administering a heavily agrarian country with a largely agrarian electorate, Japanese governments were required to make policy for agriculture. Both these issues—Noguchi's expansion to Korea and the politicization of the fertilizer industry—will be considered in this chapter.

Changes in access to resources inspired complicated strategic decisions. In the case of Nitchitsu, the search for hydroelectricity led to the company's expansion in Korea, its increasing independence of Mitsubishi as a source of funds, and the reorganization of its management structure, since access to large quantities of electricity permitted Nitchitsu to produce a wider variety of products. The first decade of the Shōwa period (1926–1989) was, thus, an era of successful diversification of manufacturing processes of chemical products within Nitchitsu; the next decade was an era of diversification into industries other than chemical manufacturing, particularly marketing and finance as well as manufacture of non-chemical products. While Nitchitsu was a fertilizer company until the mid-1920s and became a diversified chemical company during the next ten years, it became a large vertically integrated multi-unit conglomerate during the mid-1930s.

A pioneer in developing synthetic ammonia, Noguchi Jun also preceded other producers in expanding his company's output, number of plants, and types of products. He was aware of the financial risks involved in simultaneously expanding production of ammonia, inaugurating new types of products such as rayon and explosives, and beginning construction in Korea of the world's second largest hydroelectric facility and third largest fertilizer firm. But, risky though diversification was, not diversifying would have been more destructive to Nitchitsu. The early 1920s was a time of intense international competition in fertilizers, and producing new products as well as old products more cheaply was the only way to remain competitive. Not all investors and potential investors in electrochemicals responded as Noguchi did. Luck, careful planning, and fortuitous technological breakthroughs permitted continued solvency.

INTERNATIONAL COMPETITION AND THE EASTERN NITROGEN ASSOCIATION

The end of World War I signaled the revival of the fertilizer industry throughout the industrialized world. Within a few years, German output of ammonia increased by 43 percent, American more than doubled, and Italian almost tripled.[2] Most of the ammonia was used for ammonium sulfate, and all that fertilizer had to find markets. Japanese trading companies were quick to see profits in importing fertilizer; following the lead of Shima Trading, Inc. of Ōsaka, merchants began to import synthetic ammonium sulfate in ever-increasing amounts after 1923. Prices began to fall when English manufacturers, responding to Japanese farmers' preference for the purer synthetic "white" ammonium sulfate produced in Germany over the "red" by-product version exported by England, drastically cut prices.[3] Japanese companies like Nitchitsu thus faced their strongest competition from cheap imported fertilizer.

Despite the acknowledged value in import substitution, low import prices were actually a major disincentive for new companies to start up production of ammonium sulfate. Nowhere is this clearer than in the response to low prices of the Eastern Nitrogen Association, a group of eight firms which had originally come together in 1921 to acquire and develop the Haber method. In the years between 1921 and 1923, foreign price competition had made exploitation of the highly expensive Haber license completely impractical. Therefore, rather than using the license for commercial production, the eight members of the Eastern Nitrogen Association chartered themselves as a nitrogen-importing association and negotiated a deal with the German nitrogen syndicate in June 1923. As license holders, the Association members were to receive an ad valorem royalty of between 2 and 3 percent on imported ammonia made by the licensed Haber method. (This netted the members of the Association 2,888,000 yen in royalties between 1924 and 1928.)[4] In addition, the right to import German ammonium sulfate was to be transferred in 1924 from H. Ahrens of Kobe to Mitsui, Mitsubishi, and Suzuki.[5] With a strong vested interest in maintaining and increasing imports, members of the Association, and in particular those who made additional profits from importing, had little incentive to manufacture ammonium sulfate until investment conditions improved at the end of the 1920s.

To be sure, the royalty helped raise the price of fertilizer, making Nitchitsu, Denka, and other manufacturers a bit more competitive. But the effects of the deal struck between the Association and the German ammonia syndicate were generally negative. First, farmers had to pay the equivalent of an import fee, which was then channeled into a small number of private companies. This "quasi duty" eventually became the object of political controversy. Second, Association members were making money without producing, while new producers of synthetic ammonium sulfate like Nitchitsu were being undersold by foreign competitors. In such a business environment, diversification and colonial investment, along with technological breakthroughs, appeared the most promising means for Nitchitsu to remain competitive.

NITCHITSU'S EXPANSION: DIVERSIFICATION AT HOME

Diversification was both a response and a stimulus to a changed business environment. In this respect, Nitchitsu resembled successful chemical companies in other countries. DuPont in the United States, for instance, began extensive diversification in the World War I era in response to various governments' needs for munitions and other chemicals. After the war, it was apparent that diversification had made the American company, which had been organized according to various functions (like purchasing, manufacturing, and sales), structurally unwieldy. This, in turn, stimulated change; it led to DuPont's reorganization along product lines.[6] In the long run, the structural changes, which had by no means been universally applauded by DuPont management, as well as sales of the products of diversification themselves, helped the company through the rocky years of the Depression. Likewise, American Cyanamid and Union Carbide diversified actively after studying processes related to their original products and manufacturing methods. A fairly common pattern for all these firms was to invest in ammonia manufacture, hydroelectricity generation, and rayon synthesis. A notable exception in the United States was Allied Chemical, a large synthetic ammonia manufacturer that resisted diversification; Allied suffered great losses in the Depression when farmers were unable to buy fertilizers.[7]

When DuPont entered rayon manufacture in 1923 (the name "rayon" for what had been called artificial silk was coined the following year) the market was still wide open. By the end of the decade, both American and foreign (Italian and Japanese) competitors threatened DuPont's position. This encouraged DuPont, an innovative company, to do intensive research in new synthetic fibres and to diversify production accordingly. Most significantly, each type of diversification was related in some way to products or processes already available within the company. Investment attempts in areas related only remotely, if at all, to products and processes used within the company were rarely as successful as diversification in closely related areas.

The same was true for Nitchitsu. Diversification appeared to be a valuable strategy for surviving excess competition in the fertilizer industry in the 1920s. As we shall see in Chapter 5, Nitchitsu, like DuPont, was ultimately forced to change its company structure as diversification proceeded, especially after Nitchitsu's center of gravity moved to Korea. And, in later years, attempts to manufacture products not closely related to goods already produced by Nitchitsu were unprofitable and made sense only as responses to the particular environment of World War II, when government contracts were widely available.

Rayon, on the other hand, was closely related to ammonia production. As Noguchi's company matured, and as he faced increasing competition in fertilizer sales, diversification into rayon manufacture appeared to be a wise move. Noguchi was interested in producing synthetic fabrics for other reasons as well: He believed there was a large potential market in Japan for rayon; and he enjoyed developing new industries for their own sake.[8] He was hardly the first entrepreneur in Japan to be interested in rayon manufacture; Suzuki Shōten, another early ammonia producer, had established Teijin (Imperial Rayon) during World War I. But the field was underexploited, and Noguchi recognized that.

Always seeking new investments, Noguchi had taken advantage of his time in Italy negotiating with Casale to travel to Rome to observe rayon manufacture. Unfortunately, his colleagues on the Board of Directors at Nitchitsu did not share his enthusiasm for the artificial fibre. They claimed it was foolhardy for Japan, the world's leading producer of raw silk, to attempt to make a synthetic silk.[9] Undaunted, Noguchi took Uehata Goichirō, manager of Asahi Rayon (Asahi Jinken), on a second tour of inspection of rayon plants in Europe. They found Berlin's Glanzstoff Company impressive, and decided to buy the rights to its patent. Unsupported by the rest of the Nitchitsu Board, Noguchi exclaimed, "O.K. I'll do it myself!"[10] and put up his own funds to buy the Glanzstoff license. Around the same time, Japan Cotton Co. (Nippon Menkasha) President Kita Matazō, who was seeking European technology for rayon, met Noguchi, and the two joined forces to develop the license at Kita's Asahi

Rayon plant at Ōtsu. Noguchi and Kita began to renovate the plant for the new process in May 1922, and the first rayon appeared two years later. Sales were brisk, so Noguchi decided to expand capacity. In 1926, Nobeoka was selected as a good location for a new plant, but construction was temporarily halted in 1927 due to financial difficulties and policy disagreements among management.[11]

As it happened, the delay helped Noguchi because it gave him time to study better production methods. In 1928, he traveled to Germany to investigate and possibly purchase a license for a new method, the Bemberg, to upgrade rayon technology at Nobeoka. At the same time, Noguchi saw acquisition of a new license as a way to strike out on his own, without his rayon partner Kita. In fact, when Kita asked him about his new license, Noguchi snapped, "I didn't go to Germany on behalf of Asahi Rayon!"[12] True to his word, in 1929 Noguchi independently established Japan Bemberg Silk (Nippon Bemberg Kenshi) to develop his recently acquired license (capitalization: 10,000,000 yen). In one sense, he did not entirely cut himself off from his original rayon firm; the majority of the technicians at Asahi left to join him at his new company. And, by 1933, the old and new firms apparently patched up their differences and merged to form Asahi Bemberg Silk, Inc. (Asahi Bemberg Kenshi KK).[13]

Asahi Bemberg was one of the few consistently profitable subsidiaries of Nitchitsu. It differed from most other Nitchitsu subsidiaries in that it had an unusually large number of shareholders (although Nitchitsu did hold 70 percent of the shares) and in that foreign institutions were among the biggest investors in 1933. (Foreigners had withdrawn from the company by the time of its merger in 1943 with Japan Explosives (Nippon Kayaku.)[14] Noguchi's successful diversification into production of rayon was a significant step toward acquisition of new technology and expansion of the company's profitability.

Minamata and Nobeoka were the scene of advanced experimentation in two additional areas of chemistry during the 1920s. Hashimoto Hikoshichi—Mayor of Minamata when the tragedy of industrial mercury poisoning came to light in the 1960s—discovered new processes at Minamata using the culpable mercury catalysts in the manufacture of acetic acid, the basis for plastics.[15] Nobeoka was also the site of advanced research in explosives. Noguchi became interested in explosives after he began to use large quantities in the course of construction in Korea. His two suppliers–Tsuno Seiichi, President of Chōsen 'Mite, Inc. (Chōsen Maito KK), and Kido Shōzō, President of Kido Trading (Kido Shōji)–urged Noguchi to produce his own explosives. Noguchi had considered producing the explosive carlite because he already had plenty of the electricity required in its manufacture, but the carlite Nitchitsu made emitted fumes and was inappropriate for the uses most needed by Noguchi, particularly mining and road building. So he began to consider producing dynamite.[16]

A bit earlier, in 1925, Murayama Kyūzō, a researcher affiliated with the Temporary Nitrogen Research Laboratory of the Ministry of Agriculture and Commerce, accepted an invitation to establish a pilot plant and research laboratory at Nobeoka. The following year, Murayama succeeded in synthesizing nitric acid and, by June 1928, had set up a small pilot plant for its production. His discovery was recognized internationally as a major advance in chemistry, and permitted Noguchi to contemplate production of explosives seriously, a move that took advantage of the new technology and, his contemporaries would add, permitted the chauvinistic Noguchi to demonstrate his dedication to national security.[17]

Noguchi brought up dynamite production at the December 1929 Board meeting. Several members were reluctant to sanction production for a variety of reasons, including concern about danger. Noguchi argued convincingly that explosives would be an important industry in the future and that Nitchitsu possessed the necessary raw materials within the firm, thereby placing the company in a particularly competitive position. The Board eventually agreed.

Having surmounted the first hurdle–his own Board–Noguchi had to face a few more.[18] First, no one at Nobeoka had any experience with explosives. The market for industrial explosives was dominated by Japan Explosives (Nihon Kayaku KK) and some smaller firms. The market for munitions was dominated by the military arsenals.[19] Through businessman Kido Shōzō, Noguchi was introduced to Miyamoto Shōji, an explosives technician at Yamasaki Explosives (Yamasaki Kayaku), whose expertise in explosives was just what Nitchitsu needed.[20] Second, Nitchitsu needed government approval to begin production. Explosives production was restricted by law, and many industrialists, military men, and even some in the Home Ministry opposed Noguchi's petition to begin production. Miyamoto Shōji used his connections in the Home Ministry to persuade the Police Bureau of the Home Ministry to issue Nitchitsu permission to produce explosives. In December 1930, Noguchi established Nitchitsu Explosives (Nitchitsu Kayaku), capitalized at 1 million yen. Then a third problem impeded Noguchi's start-up in explosives: finding a proper site for construction. Local landlords opposed selling Noguchi land, so he sat down and negotiated personally with the influential families in the village. Soon the persuasive Noguchi had access to land at a reasonable price.[21]

Other hurdles followed. Fourth, Miyamoto had little experience with setting up a plant specifically for dynamite, and the Army and Navy, which did, were not handing over their blueprints. As a result, the process of construction, which eventually succeeded, was more difficult that it should have been.[22]

Gathering a talented work force in a new field, Noguchi's fifth challenge, was not easy either. Noguchi succeeded in attracting a wide variety of employees from government offices (particularly the military) and from other companies involved in explosives. Some were recruited by their university professors. One student who had planned to do research in the Navy was told by his professor that Nitchitsu showed promise of becoming an important producer. After entering the company, the student was immediately won over by the freedom to do research in areas that had been denied him at the university. This stress on research must have been attractive to many young scholars.[23]

Sixth, Noguchi had to obtain technologically sophisticated machinery not generally available in Japan. With the help of Mitsubishi Trading (Mitsubishi Shōji) he was able to import machinery from several foreign manufacturers.[24] And seventh, he had to make sure he had a steady supply of the glycerine necessary for production of nitroglycerine; this he got when his own plant at Hungnam, Korea, began to produce glycerine in October 1931. Nitchitsu Explosives would become, like Asahi Bemberg, one of Nitchitsu's more profitable subsidiaries; the two firms merged in 1943 and produced munitions under Navy directives.[25]

Nitchitsu thus began its successful diversification in Kyūshū during the late 1920s. By then, fertilizer output at the new plant at Hungnam, Korea, was about to come on line, and was projected to be so high (400,000 tons of ammonium sulfate were anticipated), that fertilizer was abandoned at Nobeoka in favor of rayon and nitric acid. In the end, the scale of Nitchitsu's diversification in Japan was dwarfed by the company's expansion in Korea.

NOGUCHI'S ENTRANCE INTO KOREA

One of Noguchi Jun's admirers wrote in 1943 that the "American continent existed before Columbus was born, but modern America was created by Columbus's discovery; modern Korea was created by Noguchi's discovery."[26] This exaggeration holds within it more than a grain of truth. Working in concert with administrators in the office of General Ugaki Kazushige, Governor General of Korea from the eve of the Manchurian Incident (1931) until 1936, Noguchi Jun developed most of Korea's hydroelectric and chemical plants. As late as 1942, after other corporations had had the opportunity to follow Noguchi into Korea, capital invested by Nitchitsu continued to account for 27 percent of all capital invested in Korea.[27] Small wonder that Noguchi earned the sobriquet "Entrepreneurial King of the Peninsula" (*hantō no jigyō-ō*).

Noguchi cooperated with government authorities wherever Japan's military held sway throughout East Asia during the 1930s and 1940s, but Korea was the heart of his operations. Although the period after 1937, when Japan's war with China broke out, was marked by increasing restrictions of Nitchitsu's managerial freedom, his company's early development abroad was characterized by profitable government-business cooperation.

Indeed, the logic of Noguchi's entire endeavor in the 1920s—which, as we have seen, was a form of technology-intensive industrialization for import substitution—demanded aggressive exploitation of colonial opportunities. Otherwise, he would have been powerless to compete with underpriced imported products sold at home at that time. Almost two decades of risk taking and development of technology might have been lost had Noguchi not aggressively used the opportunities his environment offered him. From 1920 on, that environment, shaped by the Japanese government's colonial policies, included the chance to invest in Korea, a Korea controlled by Japan. Japan's colonial control permitted Noguchi to obtain rights to develop river-based hydroelectric systems, to market his products without worrying about protective tariffs, and to take advantage of a huge market not limited to Korea or even Japan. Expansion to Korea was necessary for Noguchi's enterprise. And because Noguchi's enterprise played such a large role in Japan's importation, diffusion, and innovation of technology in the process of import substitution, and in its contribution to Japan's Gross National Product, it can be identified with Japan's twentieth-century developmental story. It is no exaggeration to say that colonial investment itself played a crucial role in that story. How and why did Noguchi become involved in Korea?

Noguchi's interest in Korea antedates his investment there by several decades. His initial contact with Korea occurred around 1907, when he was busily building his first hydroelectric plant at Sogi Electric. A boyhood friend, Ogura Seinosuke, solicited his technical advice on matters pertaining to generation of electricity at a facility Ogura proposed to build in Korea. Little mention of Korea in the next seven years appears in accounts of Noguchi's life, but, in 1914 or 1915, he apparently made a trip to visit Ogura there.[28] Whenever Noguchi became bored with his life in Hiroshima during the next few years, he would "dream of" Korea. It is difficult to say why, in the absence of economic motives, Noguchi seemed drawn to the colony; reaching out for Japan's geographic frontiers could be said to resemble his aspirations to discover the frontiers of science.

But Noguchi failed to implement his dreams immediately. He continued to be preoccupied with his work in Japan; moreover, the Corporation Law (1911) promulgated by Korea's Japanese colonial administrators severely limited investment opportunities in Korea. Concerned that Japanese investors would take advantage of cheap Korean labor and start-up costs to invest capital in industries competitive with still-developing Japanese industries, the Government General of Korea regulated investment in non-agricultural areas.[29] The Governor General had extraordinary administrative independence from Tokyo and could regulate investment by fiat.[30] Until the abolition of the Corporation Law in 1920, most Japanese investors saw little reason to seek permission to set up factories. The Korean infrastructure remained underdeveloped, and, without adequate sources of energy and transportation, industrial investment appeared relatively uneconomical.

The Japanese Rice Riots after World War I, however, illuminated the necessity to stimulate Korean agricultural production of cheap rice to export to Japan. The Japanese authorities in Korea announced a "Program to Increase Rice Production" in 1920, which called for reclaiming 800,000 hectares of paddy land, irrigating that land, and encouraging farmers to use fertilizer-sensitive seeds.[31] This, in turn, encouraged production of chemical fertilizers. The Governor General's development policies continued to evolve, and, by the 1930s, they urged development under the slogan of "Industrialization" (*kōgyōka*).[32]

Development required cheap energy. During the first decades of colonial administration in Korea, until 1930, the Government General laid the groundwork for eventual electrification of rivers. Although most energy was initially thermoelectricity, as it was in most other developing countries, the Japanese recognized the enormous potential of harnessing Korea's seasonally abundant flow of water. The first investigations of potential rivers for electrification were conducted by the Government General between 1911 and 1914 and indicated that electrification of Korean rivers could yield at least 57,000 kilowatts of power. Later investigations between 1922 and 1929 revealed a massive underestimation; the estimate was revised dramatically upward to 2.25 million kilowatts.[33] Construction of hydroelectric plants took time, and, until 1930, coal-fired thermoelectric plants continued to supply most of Korea's needs (see Table 23).

TABLE 23 Development of Electric Stations in Korea, 1910–1938, selected years (in 1,000 kilowatts)

Year	*Water*	*Coal*	*Other*	*Total*
1910	–	1.7	–	1.7
1917	0.1	6.5	1.4	8.0
1923	3.5	18.6	3.3	25.4
1929	13.4	34.5	–	47.9
1931	109.4	53.4	–	162.8
1938	522.3	145.8	–	668.1

Source: Andrew Grajdanzev, *Modern Korea*, p. 134.

Noguchi's subsequent investments changed the balance of Korean energy generation in favor of hydroelectricity. When his projects began operation after 1930, the electric power available to Korean industry soared. As quid pro quo for permission to develop water rights, in most cases Noguchi had to promise public transmission of one-half to two-thirds of the power generated at each facility. This is not to say that the average colonial subject benefited from increased availability of electricity. On the contrary, in the 1930s, only one in ten Koreans enjoyed electricity at home compared to 90 percent of all Japanese.[34] Nevertheless, the availability of electricity helped justify Noguchi's petition for rights to various rivers.

Noguchi's company acquired those rights with some difficulty. The question of the potential destruction of nature by damming of rivers, commonly heard today, was not part of the 1920s debate, but the question of who had the right to do the damming was. Mitsubishi Gōshi Kaisha, the parent company of Mitsubishi Bank, Noguchi's longstanding supporting institution, was actively considering developing two important rivers in Korea, the Pujon and the Changjin, in cooperation with Germany's I. G. Farben. Although the multinational effort came to naught in the adverse economic conditions of the mid-1920s, for a time both Mitsubishi and Nitchitsu expressed interest in the same rivers.[35]

Noguchi, when informed of their potential, worked to acquire the rights to electrify both rivers. While the Government General was conducting major surveys of hydroelectric potential, Noguchi's help was solicited by two intrepid engineers who had discovered opportunities for advancement in the colonial economy, Morita Kazuo and Kubota Yutaka.

Noguchi had known both men for several years, but in separate contexts. Morita Kazuo had been Noguchi's classmate at Tokyo Imperial University, although the two had had little in common, and Morita did not feel particularly close to Noguchi. Morita had been a close friend of Ichikawa Seiji at the University, however, which probably helped bring Noguchi and Morita together in 1924.[36] Kubota Yutaka was much younger, having graduated from Tokyo Imperial University in 1914. After a five-year stint in the Home Ministry's Civil Engineering Office, Kubota left public life to join a firm and, eventually, to establish his own engineering consulting firm in a small office rented from Mitsubishi in the Marunouchi business section of Tokyo.[37] His former superior in the Home Ministry decided to run for elective office, so Kubota joined his campaign. While fund-raising, he met with and was impressed by Noguchi, who made quite generous political contributions.[38] After the campaign, Kubota journeyed to Seoul, where he acquired several hundred topographical maps of Korea (scale=5,000:1) prepared by the Army Ministry's Land Survey Division.[39]

Back in his Marunouchi office, Kubota was visited by Morita. Morita had left his job at Hayakawa Electric in 1923 and traveled to Korea at the invitation of his friend Fukushima Michitada, Editor-in-Chief of the newspaper *Keijō Nippō*.[40] While in Korea, he became fascinated by the prospect of electrifying Korean rivers. On his return to Japan, he dropped in on Kubota to find out if the younger man had any information about Korea. Indeed he did—the stack of excellent topographical maps. Morita poured over them for a fortnight and then returned them to Kubota. He was sure that the deep, swift rivers of the northern part of Korea, where the mountains ranged from 1,000 to 2,000 meters high, could be profitably harnessed. The two engineers decided to work the Pujon and Changjin Rivers and drew up tentative plans.[41] Morita went back to Korea and discussed these plans with Fukushima who, as a journalist, had extensive contacts with Koreans and Japanese in Korea. He thought the idea of electrifying the Pujon would be a good start—the Changjin could come later—and added that Morita should consider transmitting to Seoul some of the power generated because, he stated, there surely was insufficient demand for electricity in the north![42] The men planned to apply to the Governor General for permission to develop the river, but they found they had competition. Later they discovered that Mitsubishi had also applied for the rights to develop the Pujon and the Changjin.[43]

Undaunted, the 54-year old Morita journeyed, in the late autumn of 1924, into the undeveloped wilderness to study the Pujon, braving the extraordinarily cold winter climate of the Pujon Highlands on horseback. (The average annual temperature was 1.8° C, and January temperature averaged -19.8° C. November must have been quite cold.) Enduring the rigors of the trip paid off; he found Kubota's topographical maps to have been surprisingly accurate, permitting him to pinpoint a dam site. He raced back to Seoul where he joined Fukushima and Kubota in forming a corporation to apply to the Government General for the water rights.[44] With what they believed to be a careful plan, even showing such detail as the proposed dam site, the group expected to be well received. But Governor General Saitō Makoto and Political Affairs Officer Shimooka Chūji had another application in their office. Wishing to select the best applicant, they requested additional information of Morita, Kubota, and Fukushima, including a more detailed explanation of their development plans and, more important, of their corporation's expected customers.[45] After all, the Pujon project promised to be more than 4 times as large as the largest generating plant operating in Japan and to generate 8 times the total number of kilowatts currently generated throughout Korea.

Kubota and Morita considered possible industrial uses for that huge output of electricity. Kubota hit on electrochemicals, and the two recalled the two pioneers of electrochemistry, Noguchi and Fujiyama. Kubota preferred Noguchi, remembering his generosity in the 1924 elections and being impressed by his use of political connections. Morita was not initially pleased with the prospect of working with Noguchi, but recognized that, unlike Fujiyama, when Noguchi tackled a project, he completed it with dispatch.[46] Noguchi showed his eagerness—and his speed—by replying to Morita's explanatory letter with a telegram inviting Morita to Osaka to meet with him and with Morita's friend Ichikawa. The meeting went well, particularly when Morita satisfied the only major concern Noguchi had, a concern about who would direct the civil engineering of the monumental project, by mentioning "a man named Kubota who has opened a small civil-engineering office in Marunouchi."[47] Noguchi had great respect for Kubota, despite the fact that, at 34, he had not had much experience for such a large job.[48]

Responding, "I'm coming right away!" Noguchi saw his "dream" of going to Korea as near to fruition.[49] As 1924 drew to a close, Morita took Noguchi and a party of three others (including Shiraishi, Kaneda, and Shigematsu Jūji) to the proposed dam site. Noguchi was impressed.[50] Noguchi's joining the project of Fukushima, Morita, and Kubota made the undertaking essentially a Nitchitsu project. It remained for the Governor General to accept Nitchitsu's claim in preference to Mitsubishi's initial expectation that they could develop the Pujon and find consumers for the power generated.[51]

Political Affairs Officer Shimooka expressed some confidence in Noguchi, but would legally be unable to reject Mitsubishi's prior claim without sufficient reason. A fire in the offices of the Government General destroyed the legal documents Mitsubishi had filed, but the responsible staff would not ignore the zaibatsu's claim.[52] Other reasons would have to be advanced to permit Nitchitsu's entry into Korea. Shimooka thought hard and concluded that he simply disliked zaibatsu. He declared that "zaibatsu are absolutely out of the question. A new person like Noguchi would be better."[53] He felt that colonial policy would best be served by developing the industry as quickly as possible, and that Nitchitsu would get to work more quickly.[54] Furthermore, Noguchi was perceived as able to build a large complex capable of serving as a defensive bulwark against the Soviet Union to the north of Korea.[55] From the company's point of view, the tensions between Japan and the Soviet Union cast a cloud over Noguchi's plans, but the benefits of developing in Korea outweighed these costs. While Noguchi spent some time persuading the Iwasaki family, who personally owned shares in Nitchitsu, that Nitchitsu's development of the Pujon made good business sense, some of his political friends who were also friends of Shimooka lobbied on his behalf.[56] Nitchitsu may also have appeared attractive to Shimooka because of Kubota's excellent preparation. Kubota was on contract as a consultant to Nitchitsu at the time; he had not yet joined the company. But his studies of the requirements for construction were extensive. He calculated the size of the reservoir required by using such factors as average rainfall over a dozen years. Unfortunately, Shimooka found errors in calculation in Kubota's work, so approval was not immediate. In the end, however, Shimooka was impressed by Kubota's efforts and conceded that, after all, any company could make errors in calculation.[57]

Shimooka's approval may have been postponed for another reason as well. It was rumored that Kubota had stolen his plans from Mitsubishi. So he and Morita marched in to see Mitsubishi Gōshi manager Okamura and requested a statement attesting to Kubota's originality. Soon thereafter, in June 1925, Nitchitsu was given the right to develop the Pujon and Mitsubishi the right to develop the Changjin.[58] Both Mitsubishi and Nichitsu were pleased with the decision. Mitsubishi bankers, as Noguchi's creditors, approved of his acquisition of the Pujon because it was a good investment for a company trying to remain solvent. And the Iwasaki family, who personally held much stock in Nitchitsu, were happy to see their investments well handled.

The move to Korea helped Noguchi in several ways. Most important was the need for additional electricity. Even as Nobeoka and Minamata in Kyūshū began production of ammonium sulfate, they were unable to produce as much electricity as they needed for substitution of fertilizer imports. Noguchi had been searching for other rivers in Japan to electrify when Morita approached him. Rivers not controlled by the five major Japanese electric firms were fast disappearing in Japan.[59]

A second reason for Noguchi's move to Korea was that the cost of electricity was far lower than in Japan. Because costs of electricity figured largely in the final costs of ammonia made by the Casale method, Noguchi had to find a cheap source of power. Within Japan itself, the cheapest electricity averaged at least 1 sen (100 sen=1 yen) per kilowatt hour. Morita and Kubota had calculated that electricity generated at Pujon would cost just 4 rin (100 rin=1 sen) and even less at Changjin, where charges would be just 2.5 rin for internal use within the Nitchitsu group and 4.5 rin for outside customers. The savings would be substantial.[60] And construction costs were very low, even for Korea, at the sites developed by Nitchitsu. Construction costs per kilowatt in 1943 at the Pujon were 4.24 sen, at the Changjin 2.71 sen, and at the Yalu 5.66 sen, compared to 6.14 sen for already completed plants in Korea and 123.7 sen for plants under construction.[61]

Third, taxes were lower in Korea. A savings of 7 to 8 yen per ton could be realized as a result of these tax savings.[62] To encourage Japanese investment, the Korean Government General permitted corporations to subtract repayment of loans from profits to lower their taxable profits.[63] Noguchi took advantage of other measures to save on taxes in Korea. Until the first half of 1937, taxes paid by Chōsen Chisso were minimal; income taxes were practically avoided for four years until 1933. The company applied to the Government General on several occasions to have its taxes reduced. These petitions contained a complicated set of calculations of taxable and tax-free income and expenses. Certain fertilizer products were tax-free whereas hardened oil or coal were taxable. Costs for production or interest payments or other expenses were prorated and subtracted according to this formula from the profits for each category—taxable or tax-free. Using this method, Chōsen Chisso, which earned most of its profits from fertilizer, was able to use creative accounting to reduce its tax burden substantially.[64] In effect, these tax measures amounted to a government subsidy.

Fourth, labor was generally less expensive in Korea. The existence of an increasingly repressive colonial government guaranteed that labor, although restive in the 1930s, would continue to be under-compensated.[65] In 1935, Chōsen Chisso employed 582 white-collar and 7,116 blue-collar workers.[66] (This number does not include employees at other subsidiaries based in Korea.) Most of the former were Japanese and were well compensated. In 1938, white-collar workers had the following starting salaries: 75 yen per month for university graduates, 70 yen per month for technical-school (*senmon gakkō*) graduates, and 50 yen for middle-school graduates, with a 30 percent allowance added in all categories for working in Korea.[67] Blue-collar labor was not so fortunate. Workers with great seniority could make more than white-collar employees, but raises were determined by subsection heads, and Japanese workers were usually paid more than Korean.[68] Because of discrimination, Korean workers averaged just 57 percent of Japanese employees' wages. Even at the end of the war, when labor was in short supply, Japanese male workers averaged (with bonuses and overtime) 7.33 yen per day; Korean males, 3.58 yen; Japanese women workers, 2.66 yen; and Korean women, 1.46 yen. The percentage of Koreans employed by Noguchi increased dramatically after 1940, since Japanese employees were being drafted.[69] By the end of the war, labor was in extremely short supply; children, prisoners of war, convicts, and forced laborers of various types were coerced into working for Nitchitsu.[70]

Fifth, Korea itself was an ideal market for locally produced fertilizers. The Government General's policies of encouraging increased agricultural production for export to Japan promised increasing demand for fertilizers. Although the Government General would retreat from the agricultural-encouragement program in the mid-1930s, the future looked bright to Noguchi from the vantage point of the 1920s.[71] In fact, Noguchi wrote that Korean demand (see Table 24) was a major reason for his move to Korea.[72]

Sixth, European producers had resumed dumping in Japan in the late 1920s. While Japanese trading firms reaped a windfall importing fertilizer, domestic producers had to struggle to sell at competitive prices. Eventually, Japanese firms formed cartels in response to European dumping, but that would take five years. In the meantime, firms that could cut costs might remain solvent.

And, seventh, Noguchi's personal views about the role of his com-

TABLE 24 Rising Fertilizer Demand in Korea, 1915–1938, selected years (1,000 tons)

Year	*Ammonium Sulfate*	*Other*
1915	0.06	0.47
1925	12.50	8.80
1932	164.60	67.80
1938	311.50	201.10

Source: Sung Hwan Ban, Pal Yong Moon, and Dwight Perkins, *Rural Development*, p. 100.

pany and about his own mission impelled his interest in Korean development. Analyses written before postwar biographers saw reason to deemphasize the benefits of imperialism show no embarrassment in commending Noguchi's patriotism. Not only was he the "Entrepreneurial King of the Peninsula," but he also had the appropriate "spirit" (*seishin*) of selfless dedication to develop Japan's frontier—without interest, it was claimed, in personal gain.[73] His product, fertilizer, was seen as uniquely patriotic because its manufacture permitted a sure supply of the raw materials for explosives and required large-scale electrification, which also had strategic benefits.[74]

In addition to patriotism, Noguchi held other beliefs that encouraged his move to Korea. His company goal of growth demanded diversification—that is, production of numerous interrelated products. A firm using a high level of technology requires constant improvement of technology to remain competitive. Making new and diverse products usually advances a company's store of practical knowledge. But diversification requires cheap materials and byproducts of already implemented processes. It was the cheapness of Korean electricity with its potential to give Noguchi the flexibility necessary for experimentation with new products that permitted extraordinary growth of the Noguchi empire in the 1930s.[75] As an entrepreneur whose actions have been universally characterized as based on intelligent planning followed by bold decisions, his ability to diversify in Korea was surely not simply fortuitous.

THE FOUNDING OF CHŌSEN CHISSO HIRYŌ KK

Noguchi threw himself into his new development project. By August 1925, he was in Korea, had acquired a horse, and had ridden into the mountains to investigate the Pujon.[76] Kubota, still working on contract to Noguchi, was also busy in the summer of 1925. In a four-month period, he and his staff of 5 or 6 completed the construction plans for the Pujon project, a job that usually took one to two years. When the planning was completed in October, Kubota realized how much he had enjoyed working on the Pujon project and told Noguchi that he did not wish to abandon it unfinished. So Noguchi brought Kubota in as a top manager, a position Noguchi refused to allow Kubota to leave to resume work as a private consultant when the project was completed.[77]

After a hectic summer and fall, on 27 January 1926, Chōsen Hydroelectric (Chōsen Suiden), called "East Asia's Largest Plant," was founded.[78] Capitalized at 20 million yen, the firm had many familiar faces from Nitchitsu serving on its Board of Directors. Noguchi Jun was President, as he was in most Nitchitsu subsidiaries. His Managing Director, second in command, was Morita Kazuo. While many of the other members stayed in Osaka, Morita continued to live in Korea to oversee the operation. Other new members included Fukushima Michitada of the *Keijō Nippō*, and Shigematsu Jūji, the Ōita prefecture politician who had been instrumental in persuading Shimooka to transfer the Pujon water rights to Noguchi.[79] Most other directors were long-time managers at Nitchitsu (Shimura Gentarō and Kirishima Shōichi were Mitsubishi men). Later in the spring of 1926, after plans for building the hydroelectric plant in four stages had been drawn up, work began. By April or May, Noguchi and his close colleagues Morita and Kubota prepared for large-scale construction, including such expensive items as a railroad, 12,000 meters of water pipes, and dams. They planned to begin generating electricity by 1928, but serious typhoon damage set the opening date back by one year.[80] Estimated total costs of the project were 55 million yen.[81]

Such large costs were justified by the extraordinary output Nitchitsu management expected of the electrical generating plant. The potential capacity of the Pujon project was re-estimated upward to 200,000 kilowatts, which could allow Nitchitsu to produce 3 times as much fertilizer as all the other Japanese plants taken together. These estimates were not to be attained, however. Stage 1 of the Pujon plant, which began generating in July 1929, had a capacity of 65,000 kilowatts; stages 2 and 3, finished in November 1930, were to generate 40,000 and 18,000 kilowatts, respectively; and stage 4, completed in 1932, was expected to put out 12,000 kilowatts.[82] Although these capacities indicated a total capacity of over 130,000 kilowatts—still far short of the 200,000 originally anticipated—the Pujon project actually failed to generate more than 80,000 to 90,000 kilowatts, necessitating later development of additional energy sources.[83]

Despite its shortcomings, the Pujon project was a monumental advance in the Japanese electrical generating industry. The scale of the project was far larger than other Japanese plants. The 87-meter dam formed a reservoir of 9 square miles, and the waterfalls dropped 707 meters (stage 1), 216 meters (stage 2), 94 meters (stage 3), and 41 meters (stage 4).[84] Construction was difficult, and the technology developed was advanced, including an ill-fated cable car called the "Inkline."[85] Stage 1 required machinery imported from Poland and Germany, but stages 2, 3, and 4 used sophisticated machinery designed and built in Japan. Noguchi's large-scale use of Japanese-made machinery helped the Japanese chemical and electrical-machinery industries to come of age.[86] In addition to fostering technological advances in machinery, construction of Chōsen Hydroelectric at Pujon encouraged progress in cement making and civil engineering, and helped train scores of young technicians and engineers sent out from Nitchitsu's Kyūshū facilities.[87]

Additional scores of young engineers were sent to Korea to construct the fertilizer works at Hungnam, which were to use the electricity from the Pujon River. Arriving in cold northern Korea from sunny southern Kyūshū, the engineers shivered through the winters attempting to design and build the enormous plant and the city for 45,000 workers (70,000 residents during the 1930s; by the end of World War II, the population reached 180,000) on the site of a tiny fishing village of perhaps 40 houses.[88] Founded on 2 May 1927 (capitalization 10 million yen), Chōsen Nitrogenous Fertilizers (Chōsen Chisso Hiryō), the official title of the Hungnam complex, was ready for operation by the end of 1929, that is, within two years and eight months. The extraordinary speed of this construction, despite a year's delay because of the typhoon, depended on the experience of men with whom Noguchi had worked for several years in Kyūshū. He chose them because he felt speed was necessary to remain competitive in the increasingly adverse market conditions for ammonium sulfate in the late 1920s. In fact, he was so eager to start production that he paid local Korean landowners three times the market price for the factory site land.[89]

With himself as President of Chōsen Chisso, Noguchi brought two trusted scientists from Kyūshū–Taniguchi Takaishi and Shiraishi Muneshiro–to share responsibilities as Managing Directors. (Taniguchi was unable to continue in his position for long because of illness.) Three hundred others, including leading managers from Kyūshū, joined them as the chemical firm's first employees. The Board of Directors also had men whose ties to Noguchi antedated the Korean venture. Most served on the Board of Chōsen Hydroelectric as well, with the notable exceptions of two men who had expedited the creation of Chōsen Hydroelectric–Fukushima and Shigematsu–who were accorded Board status only in the electric firm. But the electric firm did not last long as an independent entity. On 9 November 1929, the two companies merged as Chōsen Chisso KK, capitalized at 30,000,000 yen; this was raised to 60,000,000 yen on 20 October 1932.[90]

A short time after Hungnam's poured concrete structures hardened in November 1929, the plant began operating. The first ammonium sulfate, phospheric acid, and carbolic acid were produced in January 1930, just as world market prices for the fertilizers were approaching their nadir due to overproduction (production rose to 128 percent of demand)[91] (see Tables 25 and 26).

All Japanese manufacturers had to respond to worsening market conditions, and Nitchitsu's response was to invest at Hungnam, where resources, especially electricity, were plentiful and cheap.

TABLE 25 Average Ammonium Sulfate Prices in Japan, 1920–1942 (yen per ton)

Year	*Domestic*[a] *Price*	*Average*[a] *Import Price*	*German*[b] *Price*	*Import Market Share*[c]
1920	287 (305.1)	274 (274)	220	47
1921	165 (177.6)	138 (157)	145	45
1922	171 (189.6)	136 (177)	181	50
1923	183 (216.8)	171 (191)	162	58
1924	176 (184.3)	158 (169)	169	61
1925	188 (197.3)	163 (185)	170	61
1926	157 (172.0)	151 (156)	136	66
1927	127 (141.9)	131 (134)	125	58
1928	127 (135.2)	131 (130)	123	55
1929	127 (126.4)	127 (123)	109	62
1930	90 (88.5)	97 (87)	76	45
1931	71.7	71		
1932	71.5	72		
1933	93.9	95		
1934	93.9	95		
1935	111.2	105		
1936	98.4	–		
1937	99.7	–		
1939	102.9	–		
1940	102.9	–		
1941	102.4	–		
1942	102.2	–		

Domestic Price Fluctuations by Month[d]

	1930	*1931*	*1932*	*1933*
January	109.9	73.1	74.4	101.6
February	101.9	73.3	73.9	99.5
March	97.1	75.7	70.7	93.6
April	94.9	87.7	62.7	95.7
May	92.0	84.3	60.5	91.7
June	92.8	80.0	56.3	99.2

TABLE 25 *(continued)*

	1930	*1931*	*1932*	*1933*
July	92.0	76.0	56.8	92.0
August	86.7	65.9	58.7	89.9
September	81.9	61.3	70.4	89.3
October	73.9	60.5	72.3	89.1
November	69.9	60.0	98.9	
December	69.1	63.2	102.1	

Sources: (a) 1920–1930 calculated from figures in Hashimoto Jūrō, "1920 nendai no ryūan shijo," p. 51; 1930–1942 and calculations in parentheses by Shimotani Masahiro, "Nitchitsu kontsuerun," p. 83; Shimotani Masahiro, *Kagaku kōgyō shiron*, p. 150.
(b) Nihon Ryūan Kōgyō Kyōkai, p. 90.
(c) Suzuki Tsuneo, p. 74.
(d) Shimotani Masahiro, *Kagaku kōgyō shiron*, p. 150.

Note: German prices are included because German imports accounted for more than half the fertilizer imported throughout most of the 1920s.

Hungnam was built up, while fertilizer production was decreased at Nitchitsu facilities in Japan to permit concentration on smaller batches of more profitable commercial chemicals (see Tables 27 and 28).

The problem of foreign dumping worsened in the late 1920s, however, and shifting production to Korea, while helpful, was not enough to fight the competition. Moreover, the responses of other Japanese fertilizer manufacturers frequently exacerbated the problems Nitchitsu encountered. Importers took advantage of the overvalued yen while Japan was on the gold standard to release large quantities of suddenly cheaper European fertilizer. This marketing tactic hurt manufacturers. Noguchi turned to other fertilizer producers in response. But going outside his own company to strike an agreement with other fertilizer producers meant that some aspects of company planning would be beyond Noguchi's control. Undesirable as relinquishing his freedom may have been, suffering the effects of excessive competition alone, outside the group of chemical-industry businessmen, was even less desirable. So Noguchi chose to place limits on his freedom to make plans. In his circumscription of managerial decision-making ability Noguchi showed he could respond

TABLE 26 Prices of Imported Ammonium Sulfate in Japan, 1928–1933
(yen/ton; parentheses contain prices set by exporting country, where available and different from Kanagawa market price)

Month	*1928*	*1929*	*1930*	*1931*[a]	*1932*[b]	*1933*[b]
January	129.76 (131.0)	127.61	106.05 (98.50)	73	74.4	101.6
February	132.97	130[a]	96.91	73	73.9	99.5
March	133.85 (131.0)	131[a]	95.00	76	70.7	93.6
April	133.50 (128.0)	129.57	91.00[a]	84	62.7	95.7
May	136.97 (131.5)	126.63 (126.0)	89.00[a]	84	60.5	91.7
June	126.60 (122.5)	124.35 (126.0)	89.00[a]	78	56.3	99.2
July	121.02	122.56 (116.50)	89.00[a]	72	56.8	92.0
August	n.a.	115.05	89.00[a]	64	58.7	89.9
September	124.5	113.18 (108.5)	85.00[a]	60	70.4	89.3
October	122.74 (123.5)	109.62 (109.5)	74.00[a]	60	72.3	89.1
November	n.a. (125.5)	110.00 (110.50)	70.00[a]	60	98.9	
December	n.a. (127.75)	107.29 (106.50)	69.00[a]	63	102.1	

Sources: Figures without superscripts are from Satō Kanji, *Hiryō mondai kenkyū*, pp. 127–128; figures marked (a) are from Watanabe Tokuji, *Gendai Nihon sangyō hattatsushi*, p. 274; and those marked (b) are from Shimotani Masahiro, *Nihon kagaku shiron*, p. 150. Those for 1929 and 1930 are from two sources—Satō and Watanabe.

TABLE 27 Synthetic Ammonium Sulfate Output at Nitchitsu Plants, 1923–1935, selected years (in 1,000 tons)

	1923	*1924*	*1927*	*1930*	*1932*	*1933*	*1934*	*1935*
Minamata	–	–	60*	n.a.	99.1	60.2	40.8	49.0
Nobeoka	12.5*	25*	25	n.a.	–	25.6	13.3	9.5
Chōsen Chisso	–	–	–	120.0	225.0	244.2	307.0	341.4
Fertilizer used in Korea					165	184	213	274

Sources: 1923, 1924, 1927, Yamamoto Tomio, p. 461; pp. 463–464. 1930, Nihon Ryūan Kōgyō Kyōkai, pp. 136–137. 1932–1935, Shimotani Masahiro, "Nitchitsu kontsuerun," p. 73. Fertilizer used in Korea in Watanabe Tokuji, *Gendai Nihon sangyō hattatsushi*, p. 398.

Note: Asterisk (*) denotes capacity and not necessarily output.

TABLE 28 Fixed Capital in the Noguchi Empire, 1941

Location	*Amount*	*Percentage of total Investment*
Japan	170 million yen	17
Korea	659 million yen	66
Manchuria	138 million yen	14
China	29 million yen	3

Source: Kobayashi Hideo, p. 144.

flexibly to market forces and such exogenous conditions as the politicization of the fertilizer market in the late 1920s. Viewed at the time as a setback and a challenge for Noguchi, the politicization of the fertilizer market radically changed the environment in which Noguchi operated. His strategies were shaped by his perceptions of that environment, which were themselves affected by his values and company goals. In the end, Noguchi's responses to the political challenges helped shape his investments in technology and ultimately strengthened his company.

FERTILIZERS IN INTERNATIONAL AND NATIONAL POLITICS

Nitchitsu and other Japanese producers of nitrogenous fertilizers found themselves facing a formidable challenge from European manufacturers in the late 1920s. While world output of ammonium sulfate rose to 2.11 million tons in 1929, demand reached only 1.87 million tons. A large part of the world's supply was controlled by two national syndicates, the British Sulphate of Ammonia Federation, organized around Britain's Imperial Chemical Industries, and the Deutsche Stickstoff Syndikat, organized around IG Farben. Not only were these two syndicates sponsored by their respective governments, but, until 1930, they competed aggressively with each other for the Japanese fertilizer market.[92] Lacking protective tariffs for fertilizers since 1899, Japanese makers sought protection in two ways: through organization and through government assistance.[93] Both these responses had domestic political ramifications. The main players in the complicated political situation were the Seiyūkai and Minseitō Cabinets, farmers' organizations, agricultural cooperatives (which were called *sangyō kumiai* but were actually agrarian cooperatives), fertilizer importers like Mitsui and Mitsubishi, and Japanese manufacturers—Denka, Dai Nihon, Shōwa Fertilizer, and especially Nitchitsu.[94]

Government-level discussions about marketing fertilizer in Japan began in the mid-1920s. Members of the Diet petitioned the government of Tanaka Giichi to set up the Fertilizer Investigation Committee (Hiryō Chōsa Iinkai) in June 1927, made up of 6 representatives from the Agriculture and Forestry Ministry, 4 from Commerce and Industry, 3 from Tokyo Imperial University, 5 from the Diet, and the Presidents of Dai Nihon, Nitchitsu, and Denka.[95] Ultimately, the committee broke into two factions, with the Commerce and Industry people urging support of fertilizer manufacturers and the Agriculture and Forestry people of the farmers. Each group had different priorities regarding fertilizer.[96]

A year earlier, in November 1926, the Imperial Agricultural Association (Teikoku Nōkai, primarily a landlord group, founded in

1910), called for government assistance to farmers in purchasing fertilizer. The Association recognized that, although imported fertilizer was currently cheaper than domestically produced fertilizer, market domination by a handful of importers left farmers at the merchants' mercy. Therefore, the Association recommended that government agencies either manufacture fertilizer themselves or else subsidize producers in order to create a stable and cheap domestic source of fertilizer.[97]

The Imperial Agricultural Association's efforts were unsuccessful. The major fertilizer importers, Mitsui, Mitsubishi, and Suzuki Shōten, were all considering the eventual production of fertilizer and therefore did not wish to have any potential competitors receiving government subsidies. In addition, these firms felt threatened by changes in the market in the 1920s, since direct marketing was gradually replacing marketing by trading firms.[98] After 1923, Brunner, Mond (later the major part of Britain's ICI) marketed its own fertilizer in Japan, although it continued to use designated regional merchants, many of whom had ties to Mitsui or Mitsubishi. Fertilizer produced by the German company BASF was marketed exclusively through Mitsui and Mitsubishi until 1929, but then the German trading firm H. Ahrens re-established direct sales. Mitsui and Mitsubishi sales of fertilizer from Germany and England had accounted for as much as 87 percent of all fertilizer imports in 1928.[99] Japanese importers therefore worked hard to block the efforts of the Imperial Agricultural Association because they saw their profits already eroding from the trend toward direct sales. The Imperial Agricultural Association and many other less privileged farmers came to regard importers as their political opponents. This opposition was compounded by the fact that regional dealers frequently were creditors to farmers, thereby limiting farmers' ability to purchase from alternate suppliers.

Domestic manufacturers became increasingly antagonistic to importers as well, although much domestic fertilizer was also marketed through dealerships run by the importers.[100] Nitchitsu relied on Mitsubishi and Mitsui to sell its fertilizer from 1919 to 1932. At the same time, farmers were the political opponents of manufacturers whenever the latter sought government support for prices and marketing. Each of the three interest groups was, therefore, alternately hostile to one or both of the others.

Misperceptions among these antagonists abounded. Farmers felt that fertilizer prices, always the largest single item in farm budgets, were unfairly high during agriculture's economic doldrums in the 1920s. Rice prices plummeted in 1920 and failed to recover adequately throughout the decade.[101] But ammonium sulfate prices fell even farther, dropping to one quarter their post-World War I value by 1930.[102] Better technology, exploitation of colonial labor, and cheaper electric rates softened the blow of lowered prices for Nitchitsu.[103] These lower prices helped farmers somewhat, though many continued to believe that fertilizer costs were too high.

Like the farmers, manufacturers misinterpreted the cause of their marketing problems. Manufacturers blamed the foreign companies for dumping. Clearly, the British and German syndicates were locked in a battle for the Japanese market and were willing to cut prices to maintain market share. Noguchi Jun was particularly incensed at foreign fertilizer manufacturers and once proposed to buy up whole shiploads of British ammonium sulfate before they reached Japanese ports. He proposed to turn these ships around, sail them back to England and, he said, dump their cargoes on the British market.[104]

Of course, foreigners were an easy target because their lower prices did affect the Japanese market. What many manufacturers failed to see, however, was that the importing and marketing policies of Japanese distributors were as important as the actions of foreign exporters in establishing the price and quantity of imports. A good example is the experience of American exporters. Although the Americans sold ammonium sulfate in Japan for less than the Germans or British in the late 1920s and early 1930s, they rapidly lost the 59 percent of market share they enjoyed in 1923; by 1924, American exporters supplied just 20 percent of the Japanese market, and in 1928 just 10 percent.[105] Japanese importers, who largely determined access to the fertilizer market, were part of the problem for domestic manufacturers. Domestic manufacturers had to come up with more creative solutions to their marketing problems. Eventually, they set up marketing cartels, made agreements with foreign companies, and used farmers' cooperatives to regain the power to affect the market.

The way toward use of farmers' cooperatives was paved by legislative action. The government was not of one mind on the issue of government involvement in fertilizer marketing and distribution. While the Commerce and Industry Ministry opposed any government involvement, the Ministry of Agriculture and Forestry proposed, in 1929, a number of measures for better distribution of fertilizer based on ideas generated by the Fertilizer Investigation Committee during the previous two years. The fertilizer problem had thus become a political issue by August 1929, when the Seiyūkai Government was replaced by Hamaguchi Yūkō's Minseitō Cabinet. Although formally in opposition to the fertilizer-control bill supported in the preceeding months by the Seiyūkai, the Minseitō Government found themselves politically forced to do something for the farmers. But they changed the thrust of the Seiyūkai bill. Instead of supporting government intervention in fertilizer marketing, as the previous government had done, Hamaguchi's cabinet proposed to have cooperatives buy and distribute fertilizers.[106]

In August 1930, the Ministry of Agriculture and Forestry announced Fertilizer Distribution Improvement Regulations which gave clout to the ineffectual cooperatives that had been established in 1923.[107] The regulations provided 4,083,065 yen in government subsidies for private cooperatives to buy fertilizers directly from manufacturers and mix them in proportions appropriate to specific areas and crops. Cooperatives could buy mixing machinery with loans from the Production Cooperative Bank (Sangyō Kumiai Ginkō).[108] Working in cooperatives gave farmers an education in fertilizer sales and distribution. When concerned about continued erosion of fertilizer prices between 1930 and 1932, manufacturers could turn to direct negotiations with the increasingly sophisticated farmers in the cooperatives. By circumventing fertilizer distributors, domestic manufacturers could sell their products at prices more competitive with those of foreign producers. The cooperative's cost dropped steadily (see Table 29). Farmers could thus afford more fertilizer, thereby increasing producers' profits.

TABLE 29 The Role of Cooperatives in Fertilizer Distribution, 1923–1930
(tons; 1,000 yen)

Year	*All Types of Fertilizer*		*Ammonium Sulfate*	
	Amount	*Value*	*Amount*	*Value*
1923	15,568	1,463	1,497	245
1924	26,610	2,590	3,757	614
1925	31,659	3,166	5,290	794
1926	36,343	3,122	6,234	898
1927	46,399	3,638	6,767	911
1928	64,912	5,358	7,614	924
1929	116,769	8,418	7,423	758
1930	214,432	10,365	24,512	1,667

Source: Hashimoto Jūrō, "1920 nendai no ryūan shijō," p. 62.

distribution but not in importation. Producers and importers approached the problem of excess competition from imports in different ways. The most noteworthy example of this difference was the disagreement between the trading companies and producers like Noguchi Jun over the Fujiwara-Bosch Agreement of 1930. The disagreement originated in their divergent reactions to the formation of the International Nitrogen Cartel (the Convention de l'Industrie de l'Azote, hereafter I.N.C.). Initially, some of the largest Japanese producers banded together against the I.N.C., but, within months, those with close ties to importers, as Denka had with Mitsui, changed their approach. This led to Noguchi's counter proposal to the cartel.

Discussions among European manufacturers and governments about the necessity of international cartels arose in a climate of frustration. The Great Depression had begun, world fertilizer supply far exceeded demand, and German and English manufacturers were exhausting each other in competition. International cartels, in which participating nations agreed on market shares permissible for each participant and in which floor prices were rigidly fixed, were frequently discussed in 1929 and 1930.[109]

The first international organization for restraint of fertilizer trade was the International Agreement on Cyanamide Sales. Signatories to the May 1930 document included Germany, France, Italy, Belgium, Czechoslovakia, Yugoslavia, Rumania, Sweden, Switzerland, Norway, and Poland; 92 percent of world production was covered, but the signatories pointedly did not include Japan.[110] The following August, the group met again in Berlin and, with a few important additional members present, signed the International Convention for Nitrogen. Great Britain, the moving force behind the nitrogen convention, was allotted a large percentage of all possible exports.[111]

As under the cyanamide agreement, nations self-sufficient in fertilizer were allowed to prevent entry of imported fertilizer. Any demand a country might have that its own producers could not fill could be supplied by foreign manufacturers in a ratio pre-established in London. The I.N.C. members produced 80 percent of the world's nitrogen (96 percent of Europe's), and figures would be higher had two producers not been excluded: the United States, because the Sherman Anti-Trust Act (1890) forbade international agreements in restraint of trade; and Japan, because, in a world divided into colonies, trusteeships, and other forms of imperialism, it was the only major newly industrializing country that could import freely from nations other than a mother country. The I.N.C. had an immediate impact on the Japanese market. Germany and England ended their rivalry, and, because of substantial savings realized through elimination of advertising and other market-securing techniques, the price of ammonium sulfate dumped in Japan continued its downward slide.[112]

In reaction to the threat of a European cartel, three of Japan's four largest producers of ammonium sulfate formed the Nitrogen Deliberative Association (Chisso Kyōgikai) in March 1930.[113] The Association members, Nitchitsu, Denka, and Dai Nihon, together with four other non-members, requested that the Finance Ministry apply the Unfair Dumping Law of 1910 (*Futō renbai hō*) against the international cartel. Fujiwara Ginjirō, President of Denka, represented the producers' group as spokesman.[114] Noguchi was also outspoken. He argued before the Minseitō Cabinet that the government's failure to tax imports sufficiently deprived it of customs revenues. Moreover, he stated, the fertilizer industry had never received any government help except during World War I, when the government relaxed fertilizer export controls.[115]

As might be expected, the members of both the Hamaguchi Cabinet and the later Wakatsuki Cabinet split in their support of the manufacturers. Bureaucrats from the Commerce and Industry Ministry and other financially related government men sided with the manufacturers, while the Agriculture and Forestry people, especially Ishiguro Tadayuki, head of the Agricultural Affairs Bureau, strongly objected to any support of manufacturers that could raise prices for farmers.[116] These differences of opinion within the government led manufacturers to negotiate directly with foreign companies.[117] Cabinet debates continued until 1931, when even Ishiguro was persuaded that farmers' interests would best be served by a viable fertilizer industry that was domestically based.[118] By the time the Minseitō Government had accepted in principle the idea of import restrictions, it was replaced by the Seiyūkai Cabinet of Inukai Ki; the Cabinet's lifting of the gold standard in December 1931 did more to aid the Japanese fertilizer industry than any restrictions could have done.[119] As the yen was devalued, an effective trade barrier was established, and Japanese fertilizer was freed of import competition.

While the producers in the Nitrogen Deliberative Association were still negotiating with the government, Fujiwara Ginjirō, President of Denka, began secret negotiations with Herman Bosch, President of the Japan office of the German trading company H. Ahrens. Noguchi was a party to these negotiations when they began in November 1930. Fujiwara's plans called for reductions of German imports from 380,000 tons in 1929 to 200,000 tons in 1930.[120] Furthermore, the proposed agreement provided:[121]

1. That Japanese ammonium sulfate producers be absolutely prohibited from exporting.

2. That Japanese producers form an importing cartel, and agree to take 200,000 tons of imports during the first year of the agreement, to be followed by scheduled reductions of 50,000 tons each year for the next four years until imports were phased out.

3. That the import price be set at 85 yen per ton.

4. That Japanese producers be prohibited from building new plants or expanding capacity.

5. That all imported ammonium sulfate be sold through Mitsui or Mitsubishi.

6. And that the agreement be in effect for five years.

After Fujiwara negotiated his agreement with Herman Bosch, the Minseitō Government decided that Japan's compelling national interest in preserving an ammonia industry demanded modification of the agreement. The Commerce and Industry Ministry also proceeded to expedite the stalled implementation of the Unfair Dumping Law by proposing a protective tariff.[122]

Several scholars have stated that Fujiwara Ginjirō's terms were extremely injurious to Japanese makers of fertilizer.[123] Not only did the negotiated selling price undermine their profitability, but restrictions on expansion were detrimental to firms in the process of building at the time of the agreement. Nitchitsu, Showa Fertilizer, Sumitomo Fertilizer, and Mitsui Mining would have to abandon investments in partially constructed plants which, under the terms of the agreement, constituted prohibited additional capacity.[124] Nitchitsu's Hungnam facility, Shōwa Fertilizer's Kawasaki plant, and Sumitomo's Niihama plant were almost ready—Hungnam was to begin producing in 1930 and the other two in the early spring of 1931—but Mitsui Mining's plant at Miike would not be ready for another year, and its loss would presumably be less costly. Manufacturers who sold their product through Mitsui would be pressured into acceptance of the agreement, according to this scenario.[125] Was the Fujiwara-Bosch Agreement an attempt by zaibatsu to dominate fertilizer marketing?

Even if Fujiwara was attempting to redefine the market, most of the provisions of the Agreement were, on closer examination, not particularly injurious to other companies. In 1929, 0.79 percent of Japan's fertilizer production was exported, and this grew to 5.6 percent in 1930.[126] These figures appear to suggest rapid growth in production which might require an expansion of foreign markets. But that interpretation would be misleading; Japanese manufacturers were not able to supply their own market, let alone foreign markets.

Imports continued to surpass 300,000 tons in 1930. The establishment of the price of imported fertilizer at 85 yen, as stipulated in the third clause, hurt some fertilizer companies, including Fujiwara's own inefficient Denka, although Denka was in the processs of implementing technological improvements which soon permitted cheaper production.[127] Noguchi's costs were somewhat lower, so the impact of the agreement was less adverse.

Unlike the other clauses, the one stipulating restrictions on plant expansion would have had a serious impact on the firms under construction. The requirement was vigorously opposed even by firms interested in disposing of inefficient excess capacity. Furthermore, managers within Sumitomo, Mitsubishi, and Mitsui were involved in negotiations to obtain licenses for ammonia at that time.[128] The fifth stipulation would mainly harm the European fertilizer distributors in Japan. Rather than stressing the ostensible, if exaggerated, problems of the Fujiwara-Bosch Agreement, a more interesting approach focuses on the strategies Noguchi employed to deal with the conditions of his business environment.

Noguchi parted company with Fujiwara over his interpretation of how the fertilizer market should be controlled. Noguchi decided to call the German in to renegotiate the agreement. The Noguchi-Bosch proposal was recast so that it could benefit Nitchitsu's business strategy. Most significantly, Noguchi eliminated the clause ending plant construction. Moreover, Noguchi's draft rejected restrictions on Japanese exports, eliminated sales of American fertilizer in Japan, and prohibited direct sales to Japanese consumers by foreign firms.[129]

The Fujiwara-Bosch Agreement, revised in April 1931 after consideration of the Noguchi-Bosch proposals, stipulated:[130]

1. That English and German manufacturers be required to cease direct marketing. All imports were to be handled by Mitsui and Mitsubishi, including imports from the United States, which had formerly been contracted only to Mitsubishi.

2. That the agreement be in effect for three years from 1 July 1931.

3. That British and German imports total 100,000 tons during the first year, 80,000 tons the second year, and 50,000 tons the third year, unless Japanese-made supply failed to keep pace with demand.

4. And that, although exports of Japanese-made ammonia were not recognized by this agreement, future exports could be negotiated in good faith with the Germans and British.

The April 1931 agreement gave Japanese manufacturers almost everything they wanted. But it had become almost irrelevant by the time of its completion. Japan's trade relations with the Europeans were already changing. At the time of the formation of the International Convention for Nitrogen in 1930, Germany and England dominated the world market. By the end of 1931, several European countries, including France, Holland, Belgium, and Poland, had become self-sufficient in fertilizer and were beginning to erect their own stiff tariff walls against imports. In retaliation, Germany also imposed tariffs. The international cartels simply could not stay together. Despite the beneficial effects of the 1932 devaluation of the yen and rapidly changing international conditions, Japanese fertilizer manufacturers, led by Noguchi Jun, decided they should make even greater efforts to stabilize the fertilizer market. Attempting to further structure their environment, the manufacturers formed a distribution cartel, the Ammonium Sulfate Distribution Association (Ryūan Haikyū Kumiai).[131]

The ASDA grew from a group of company presidents who had gained political sophistication in joint lobbying efforts with the Hamaguchi and Wakatsuki Cabinets. The leading member of this group, Noguchi Jun, had years of political experience. As a single actor–the "one-man" character he had also displayed in his work with Nitchitsu–he had had, however, little opportunity to work with his peers in other companies. But environmental change forced him to adjust, and joint efforts were one possible strategy. As his industry gained international status, it confronted problems of the international market necessitating industry-wide cooperation. Within Japan, the industry had become economically important as a high-technology industry, as a pivotal maker of intermediate products in the agricultural sector (fertilizers), and as a manufacturer of strategic products (explosives). When a young man, Noguchi had used friends of influence to further his plans; by the 1930s, he was himself a figure to be reckoned with and an intimate of the most influential leaders in

Japan and Korea. Noguchi developed his role as an industry leader, and the other ASDA member companies were also leading figures in later industrial organizations and national policy-making. But ASDA policy affected prices and supply only as long as the crisis it was addressing continued; in the end, companies preferred acting as individuals planning for individual companies.

ASDA was a response to particular conditions. It was formally organized in October 1932. With the exception of companies making ammonium sulfate as a by-product of coke manufacture or other processes, the Association brought together all producers—certainly a more successful organizing effort than had been achieved by the ad hoc lobbying group. There were six original members—Nitchitsu, Dai Nihon, Denka, Miike Chisso, Shōwa Fertilizer, and Sumitomo Fertilizer—and two additional later members—Chōsen Chisso and Asahi Bemberg. As the last two were both Noguchi's companies, he had extraordinary influence within the ASDA. Four other companies which were founded later—Ube Chisso, Yahagi Industries, Tōyō Kōatsu, and Manchurian Chemical Industries—did not become members but cooperated with the Association.[132] At the time of ASDA's inauguration, 95 percent of all ammonium sulfate produced was made by Association members.

The ASDA had two major stated purposes: maintenance of market prices for fertilizer and control of the quantity reaching the marketplace.[133] The latter goal resembled the goal of the earlier superphosphate cartel in limiting access to the market of surplus product. But domestic manufacturers were not yet capable in 1932 of meeting Japanese demand. Imports could have easily filled the gap between supply and demand if the exchange rate had not made them prohibitively expensive after the beginning of 1932. Although farmers preferred ammonium sulfate to other nitrogenous fertilizers, they could find cheaper substitutes if ammonium sulfate eventually became too expensive. Thus, ASDA had to balance their own need for higher prices with farmers' refusal to pay them.

ASDA played a role in the gradual rise of fertilizer prices in 1932 and 1933, together with market forces and yen devaluation. Prices steadily increased, from 56.2 yen per ton in June 1932 to 72.2 yen at the time of ASDA's founding. The Association's first price fixing was at 76 yen, and, by the spring of 1933, market forces had raised the price to over 90 yen per ton. In May, its mission accomplished, the Association ceased to function.

Throughout the early 1930s, production costs for ammonium sulfate did not rise and, in fact, in some cases like that of Chōsen Chisso, dropped. Until production costs began to rise gradually in the mid-1930s, manufacturers reaped a windfall. Their profits strengthened in mid-decade as farmers, recovering from the Depression, began to get higher prices for their crops and, in turn, could spend more on fertilizers.[134] Their success in controlling prices notwithstanding, manufacturers took one additional step and negotiated from a position of strength with the International Convention for Nitrogen. They signed three separate agreements, in January 1934, February 1935, and November 1936.[135] Basically these provided:[136]

1. That Japan would accept imports in the following amounts: 150,000 tons (under the 1934 agreement); 60,000–125,000 tons (under the 1936 agreement).

2. That Japan could export up to 60,000 tons to South America, North America, China, Hong Kong, Southeast Asia, and Manchuria.

3. ASDA members would have priority in handling, setting prices, and distribution of imports.

By the 1930s, the ammonium sulfate industry, led by Noguchi Jun, had come of age in Japan. Against a backdrop of politicization of the fertilizer market, Noguchi decided that expansion and further diversification of production would serve the company well.

GROWTH AND STRUCTURAL CHANGE, 1930–1933

When Hungnam came on stream, Asia's largest and the world's third largest chemical facility was inaugurated. This tremendous scale gave Noguchi the flexibility to install production facilities for other types of chemicals. Combining these different products led to yet additional products. Spin-off technology abounded in the following years as formal intra-firm mechanisms for data sharing were pioneered by Noguchi. Indeed, by 1935, the firm had greater sales volumes in products other than fertilizers (see Figure 4) and, by 1939, fertilizers, while dominating Nitchitsu's profits, accounted for only 40 percent of sales.[137] An equally important development beginning with Noguchi's expansion into Korea was the gradual transformation of Nitchitsu's structure from a purely manufacturing company to a holding company, and the growth of structure permitting highly effective control of subsidiaries' financing and production.

The transformation from a manufacturing company made Nitchitsu into what is often called a "new zaibatsu" or a "Konzern" (*kontsuerun*). This is not to say that parts of Nitchitsu did not continue to manufacture; indeed, they were, singly, the largest Japanese producers in many of their respective fields. Much of Japan's ammonium sulfate came from one Nitchitsu subsidiary corporation alone, Chōsen Chisso. But the main company office itself increasingly resembled a holding company built of its subsidiaries' securities. By the early 1930s, fixed capital accounted for only 10 to 15 percent of the assets of the parent company, and after 1933—as the next chapter will show—the pace of Nitchitsu's transformation to a Konzern accelerated.

Throughout its history, but especially before the mid-1930s, Nitchitsu and its subsidiaries remained dedicated to chemicals. Indeed, the extent to which the securities column surpasses the fixed-assets column in Table 30 is most significant as an indicator of the expansion of the subsidiaries—separate corporations were created rather than branch plants—while Nitchitsu's manufacturing operations in Kyūshū grew more slowly. Electrical generation was developed in conjunction with electrochemicals. Mining operations, coming on stream in the mid-1930s, extracted those minerals closely tied to electrochemicals and munitions—coal and magnesium. And the railroad to Hungnam was built in the late 1920s because the infrastructure of the area was so underdeveloped that it prevented productive use of the chemical plant and hindered transportation of construction supplies to the work site. Spin-off chemical products engendered new subsidiaries which were all technologically interrelated. Technological dependence of one subsidiary on another tended to encourage profitable cooperation.

Figure 4 Relative Position of Sales of Fertilizers and Other Products, Nitchitsu, 1905–1935

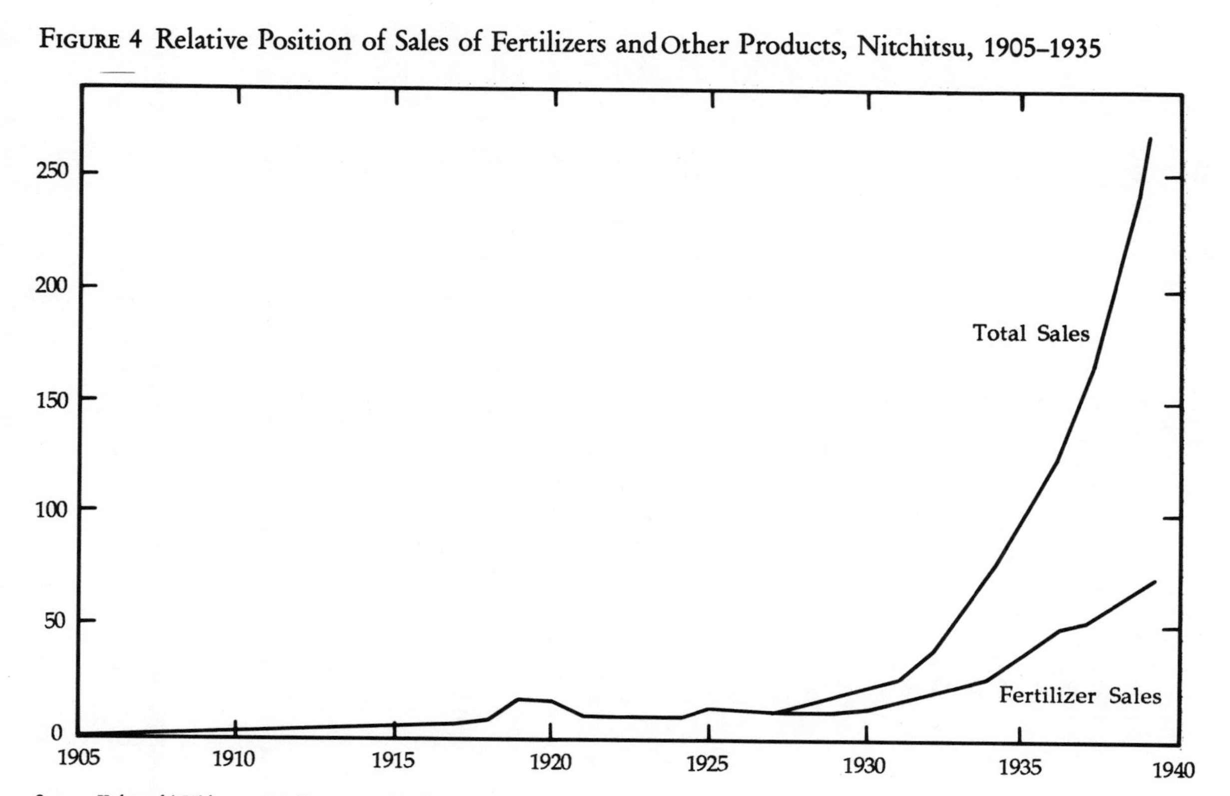

Sources: Kobayashi Hideo, p. 162; Yamamoto Tomio, p. 45.

TABLE 30 Nitchitsu Assets, 1918–1940
(%)

Semiannum	*Negotiable Securities and Loans Outstanding*	*Fixed Assets*
1918.1	21.0	79.0
1920.2	26.5	73.5
1923.1	30.6	69.4
1925.2	44.6	55.4
1926.1	50.8	49.2
1926.2	35.1	64.9
1927.1	44.7	55.3
1927.2	52.1	47.9
1928.1	47.2	52.8
1928.2	53.3	46.7
1929.1	62.7	37.3
1929.2	58.8	41.2
1930.1	65.7	34.3
1930.2	71.1	28.9
1931.1	80.9	19.1
1931.2	82.3	17.7
1932.1	82.9	17.7
1932.2	90.1	9.9
1933.1	90.0	10.0
1933.2	88.8	11.2
1934.1	86.6	10.4
1934.2	90.0	10.0
1935.1	90.3	9.7
1935.2	90.6	9.4
1936.1	91.5	8.5
1936.2	92.4	7.6
1937.1	90.3	9.7
1937.2	90.1	9.9
1938.1	88.0	12.0
1938.2	88.1	11.9
1939.1	88.4	11.6
1939.2	87.7	12.2
1940.1	86.0	14.0
1940.2	87.0	13.0

Source: Shimotani Masahiro, "Nitchitsu kontsuerun," p. 61.

Management staffs at Nitchitsu subsidiaries were also interrelated. Noguchi had developed a system of management transfer and sharing during the 1910s when initially expanding his operation from Minamata to Kagami in Kyūshū. He extended this system to all his subsidiaries.[138] Not only were people from Nitchitsu members of the boards of directors of all subsidiaries, as might be expected, but staff and their associates involved in daily operations at the subsidiaries were also trained and promoted through Nitchitsu. Technicians were often moved from subsidiary to subsidiary. Only production workers could be employed from local labor pools, and advancement of workers to white-collar status had to go through the main office. Shared experience fostered company loyalty, close ties, and cooperation among management and personnel. In addition, the company sponsored twelve scientific conferences between 1931 and 1943, effectively promoting cross-fertilization of ideas as well as cameraderie.[139]

The organization of Nitchitsu also underwent change in 1926, within months of the founding of Chōsen Hydroelectric. The General Affairs Division, which had housed the Business Section and Procurement Section, divided, and these two sections became divisions. In 1934, Procurement had grown so quickly to accommodate the needs of the rapidly blossoming subsidiaries that it, too, subdivided into two sections. The number of staff members in Procurement alone doubled in eight years; in General Affairs, it increased 700 percent in twelve years; in Accounting, it rose 400 percent in the same amount of time; and, in Business, it went up 1400 percent.[140]

The growth of new subsidiaries, especially Chōsen Chisso, required great expansion of the parent company's management structure. Each function of each subsidiary was overseen by the appropriate division of Nitchitsu. Thus, the Accounting Division handled financial decisions for Chōsen Chisso, and the Procurement Division obtained and shipped to Chōsen Chisso all the production supplies the Korea-based subsidiary needed. In some ways, because of the extensive control by Nitchitsu, subsidiaries might be viewed as little more than glorified branch plants of headquarters. But, controls by Nitchitsu notwithstanding, there were important benefits in having a structure made of several corporate entities rather than of branch plants, even in cases where subsidiaries were wholly owned by Nitchitsu. Tax considerations were one such benefit. Subsidiaries could apply for tax relief as new corporations. Another benefit was that, as independent corporations, each subsidiary had to balance its own books and could be bailed out of difficulties only by formally obtaining loans either from Nitchitsu or from outside bankers.

Nitchitsu itself was greatly aided by the subsidiary-parent company relationship. In those cases where it controlled a majority or all of the shares of its subsidiaries (Nitchitsu counted as subsidiaries some cases where its control was as low as 30 percent),[141] it could transfer shares as easily as it could transfer loans and other liquid funds among subsidiaries. This gave Nitchitsu tremendous flexibility to finance expansion within the Nitchitsu conglomerate. Nitchitsu was able to use one established subsidiary company's reserves—which were usually placed "on deposit" with the parent company, Nitchitsu—to "buy" shares of a subsidiary company in the process of formation. Those shares could later be transferred back to the older company, whose reserves had been used, if conditions warranted. For example, Chōsen Chisso came to hold majority shares of several companies that had been Nitchitsu subsidiaries because Nitchitsu had earlier used Chōsen Chisso "deposits" to build those subsidiaries. When Chōsen Chisso's deposits with Nitchitsu became excessive—Chōsen Chisso was quite profitable—Nitchitsu balanced the books by giving shares of Nitchitsu subsidiaries to Chōsen Chisso. Thus, independently incorporated subsidiaries, with their own shares, gave Nitchitsu accountants a great deal of flexibility.

Independent subsidiaries held an additional benefit for Nitchitsu. Small matters like official reports or calculations necessary for daily operations were made by managers at the subsidiary level. Larger decisions were made by top Nitchitsu management, to be sure—Noguchi earned his nickname "one man" because *he* made final decisions after hearing discussion by his leading staff and advisors[142]—but most daily decisions were made by men with a degree of autonomy they themselves found unusual. Top management at Nitchitsu was relieved of daily decision making in the far-flung enterprise, while managers in the subsidiaries received invaluable managerial experience.[143]

Because of their size, Chōsen Hydroelectric and Chōsen Chisso were unusually important subsidiaries. With the founding of these two companies, subsidiaries gained a new significance within Nitchitsu. The Subsidiary Division had still appeared unimportant in 1930; the Accounting Division controlled all the finances of subsidiaries and the two divisions shared one head, the head of Accounting. But, in May 1932, the Subsidiary Division was deemed sufficiently significant to have its own head with no other responsibilities.[144] The importance of subsidiaries in finance, production, transportation, and research continued to grow in the 1930s.

No subsidiary contributed more to Nitchitsu than Chōsen Chisso. And most of the profits derived at Hungnam came from manufacture of fertilizer and fats and oils. In the second half of 1931, 99.57 percent of Chōsen Chisso's income came from fertilizer.[145] This percentage continually declined thereafter as more spin-off products were developed, but always remained around two-thirds. Nevertheless, management faced serious problems at Hungnam in the months after production began.

The Pujon hydroelectrification project was alternately beset by floods and droughts. The latter, occurring in 1930, wrought the most serious damage. The reservoir filled to only 40 percent of its capacity, allowing only half the planned amount of electricity to be generated. Noguchi and Kubota were nervous the entire summer of 1930, a period in which the sentence "One millimeter (of rain) is worth 10,000 yen" was commonly heard in Hungnam.[146] Furthermore, the first ammonium sulfate produced came out pink, indicative of faulty production.[147] Thus, Nitchitsu had to respond to a number of problems: European dumping, insufficient generative capacity, and, of course, the Great Depression. Small wonder that employees' salaries were frozen for four years from 1928 through 1931.[148]

Under such conditions, and especially with the beginning of the Depression, Kushida Manzō, the banker at Mitsubishi with whom Noguchi had negotiated earlier loans, was becoming "cranky" (*tsumuji*) about increasing the level of financing of Nitchitsu.[149] Within

Nitchitsu management meetings themselves, men with Mitsubishi connections tended to be more cautious about investments during the troubled Depression era, while those more indebted to Nitchitsu for their positions tended to be more supportive of investment as a strategy to overcome challenges.[150]

Thus, two mutually contradictory approaches to the problems of the early 1930s gained support within Nitchitsu management: financial caution, on the one hand; and investment in increasing electrical generative capacity, on the other. In the end, those supporting investment prevailed and Noguchi proceeded to build one of the world's largest electrical plants on the Changjin River. Some at Nitchitsu believed the victory for the pro-expansion side to have been Pyrrhic, as Mitsubishi Bank, Noguchi's supporter for the previous quarter century, withdrew its support. Nevertheless, Noguchi's enterprise went on to successful growth independent of Mitsubishi and to cultivation of important ties with politicians and military men in the Korean colony, even as Mitsubishi began to unravel most, but not all, of its ties with Noguchi. These ties were decreasingly relevant, in the event, because Noguchi's plans for expansion served the interests of the Japanese colonial authorities. At the same time, Noguchi's growth and survival required Korean investment; and his company was particularly advantaged by the support of the Japanese authorities in Korea.

NOGUCHI'S ESTRANGEMENT FROM MITSUBISHI

Noguchi's biographers cast the Mitsubishi bankers in a rather unfavorable light when discussing their reluctance to back Noguchi's venture in the 1930s. They emphasize the bankers' unadventurous spirit compared to that of the entrepreneur. They imply that Mitsubishi bankers probably wished to limit Noguchi's production of fertilizers while garnering profits from importing European fertilizers. Some writers believe that Mitsubishi bankers resented Noguchi's acquisition of the rights to develop the Changjin River in 1933, which had originally been given to Mitsubishi when Noguchi got the Pujon.

Some note that Mitsubishi bankers may have been aware of Mitsubishi's own recent work with electrochemicals, which competed with Noguchi's. Finally, some show Mitsubishi bankers to be concerned about technological problems facing Noguchi in his Korean venture. Indeed, Noguchi's Casale process had been used on a commercial scale only in Nobeoka and Minamata before 1930. Results there were good, but what if Noguchi failed in Korea? In fact, he did have some initial problems with technology, particularly with pink fertilizer. Several reasons, then, are given for Mitsubishi Bank's lopping off financial support for Nitchitsu.

Mitsubishi Bank did discontinue its support, but that happened later (1933) and the reasons were quite different. To understand the decisions made by Nitchitsu management which lost them much of their Mitsubishi support but gained them important friends in the colonial government, we must consider both the ways in which Nitchitsu managers went about promoting their companies' needs and the importance to these managers of influential friends. As we have seen, Chōsen Chisso faced a number of problems in the early 1930s, and one solution would have been to develop sufficient electricity to permit increased fertilizer output. Kubota, among others, was interested in continuing construction of the Pujon hydroelectric project to complete stage 4. This had been delayed due to insufficient funds for development and to uncertainty in the fertilizer market, particularly (in the view of those Mitsubishi bankers concerned with Nitchitsu) in a colonial environment.[151] But Kubota forged ahead with plans for stage 4, deciding to implement "creative financing." He figured the expensive machinery could be paid for in installments over an 18-month period. Even the bold Noguchi was reluctant to implement Kubota's suggestion and said that "Mitsubishi would never agree to deferred payments."[152] Kubota countered with a financial analysis, saying that another factor in the decision to invest—labor—would be extremely cheap during the Depression. Noguchi was eventually won over, but Mitsubishi bankers were still not convinced. Because of the Depression, they were reluctant to spend money and sent around inspectors to see what work was being done at Pujon.[153]

Around the same time, Noguchi indicated some tension in his relationship with Mitsubishi managers through his flippant behavior at his annual reading of the Nitchitsu company report.[154] The last time Noguchi invited Mitsubishi men to a reading of the annual report was 1931. The President of Mitsubishi Trust was in attendance and brought Noguchi's attention to some confusing figures Noguchi had written on the blackboard. The banker wondered how Nitchitsu could repay its loans if shareholders earned 10-percent dividends at a time when ammonium sulfate brought just 60 yen per ton. Noguchi tried to use wit to divert attention from his blackboard error—that is, he had omitted loan repayments from the calculations on the board—but the Mitsubishi bankers were suspicious of his motives, believing he was trying to avoid complete disclosure of company finances. This anecdote is indicative of the differences in attitude between Noguchi and members of Mitsubishi; Noguchi often treated questions about finances rather lightly, while Mitsubishi bankers were understandably more conservative.[155] Such differences in manner did not suffice to sever the ties between Nitchitsu and Mitsubishi, but they did cause some concern to Noguchi in 1931 and 1932.

This concern was reflected in Noguchi's hesitance to agree to Kubota's plans to develop stage 4 of the Pujon and, soon thereafter, the more promising Changjin River. Noguchi was uncharacteristically slow to agree to these projects because they might require loans from Mitsubishi. When he finally did agree to develop stage 4 of the Pujon, he asked to discuss the project with Kubota. Kubota's enthusiasm for development was apparently unbounded. He tried to talk Noguchi into developing the Changjin River simultaneously, because, he said, with construction costs of 2.5 sen per kilowatt of capacity, it was cheaper to dam than the Pujon, where construction costs were 4 sen per kilowatt.[156] Noguchi expressed some reservations. He was concerned about Mitsubishi's response to Chōsen Chisso's bid to develop the Changjin. After all, Shimooka had given Mitsubishi the rights. Noguchi suggested developing the Hochon River instead. Kubota pushed for the Changjin and begged Noguchi to go to the proposed dam site. He knew Noguchi would be convinced if he saw the Changjin. He was right. Noguchi visited the

Changjin and, on his return to Seoul in the fall of 1932, joined Kubota in requesting the development rights, which were granted in April 1933 by Governor General Ugaki Kazushige.[157]

It has been argued that Noguchi became alienated from Mitsubishi because the zaibatsu resented losing the rights to the Changjin. But a detailed examination of the process by which Noguchi obtained the Changjin water rights indicates that Mitsubishi Bank withdrew from supporting Noguchi because the bankers were more conservative and worried about his implementing his plans for the Changjin River project. Noguchi insisted that he could earn profits, even during the Depression; that led to their quarrel. At the same time, Mitsubishi bankers, independent of their feelings about Noguchi's particular plans, also felt that they could not make such large loans in 1933.[158]

CHANGJIN RIVER RIGHTS AND PUBLIC POLICY

The development of the Changjin River was intimately tied to the industrialization policies of the Government General of Korea. The Government General strongly desired to have that river developed, and Noguchi, persuaded by Kubota, strongly wished to develop it. The confluence of interests of business and government in this case as in others expedited implementation of Noguchi's business plans.

The rapid development of North Korea began in June 1931 with the arrival of former War Minister Ugaki Kazushige as Governor General, following his departure from Japan after the March (1931) Incident, an aborted coup d'état in which Ugaki had been set up by right-wing young military officers to assume the reins of government. Since 1919, Korea had been exploited as a rice-producing region for Japan; after 1930, and particularly after Ugaki's arrival, the official slogan was "Industrialization."[159] Ugaki favored industrial development for both defensive and economic reasons.[160]

Ugaki's policies, including encouragement of promising individual entrepreneurs through the extension of credit, permission to construct, and granting of land and water rights, fostered a rapid development of certain industries during the 1930s. Growth was accelerated after 1935, concurrent with increased generation of electricity. Industrial development, as Tables 31 and 32 indicate, did not follow what could be called a more natural course of widespread expansion in numerous areas of typical consumer and business demand. Rather, Japanese planners used colonial policy to foster growth in areas they deemed necessary. Thus, the growth in size of the chemical industry, a "necessary" industry, was correspondingly greater than growth in other industries, like the machinery industry. It is true that inexpensive, readily available electricity facilitated the start-up of chemical manufacture, while the expense of machinery production impeded the start of the machinery industry, but this does not explain the difference in output. Rather, a large machinery industry was not "necessary" in the prewar Korean economy; Japanese machinery became good enough to replace Western-made machinery in high-technology industries in the late 1920s, and colonial policymakers in Korea were not about to undercut the emerging Japanese industry by encouraging Korean production for local use. As demonstrated by the colonial authorities' *encouraging* of rice production, this *discouraging* of machinery production shows clearly that the industrialization of Korea was geared to Japan's domestic needs, whether in the service of social order or economic development.

Encouragement of industries deemed necessary by the Government General was also the cause of a major demographic shift in the peninsula. The north became increasingly industrial, and the south gave over a larger percentage of its land to agriculture during the colonial period. Per capita output increased in the industrial north, while it decreased in the south. Noguchi's enterprise, the major chemical firm of the north, bore prime responsibility for the shift in population and heavy industry toward the north. Thus, business strategies of an individual firm affected the development pattern of an entire colony. Conversely, the growth of the north is a good index of the development of Noguchi's firm.

Growth accelerated after 1935, when the effects of Ugaki's industrialization policies began to take hold. The Governor General's stress

TABLE 31 Growth in Market Values of Output in Selected Industries in Korea, 1911–1940 (million Korean yen)

Year	*Textiles*	*Machines*	*Chemicals*	*Foods & Beverages*	*Metals*	*Printing*	*Gas & Electricity*	*Total*
1911	0.3	0.2	0.5	0.7	1.0	0.6	1.0	21.5
1914	0.8	0.2	0.8	3.4	2.1	1.0	1.7	36.4
1919	13.8	3.5	2.8	11.9	16.4	2.9	1.2	244.2
1924	21.1	3.2	6.7	29.3	16.5	5.1	8.3	294.0
1929	38.4	6.3	17.8	42.9	20.4	10.0	16.4	403.0
1930	33.8	5.0	25.0	36.2	15.3	8.2	6.4	307.5
1931	24.5	3.6	32.3	31.1	16.1	8.8	16.1	293.9
1932	30.8	3.6	35.8	39.5	21.5	9.7	11.1	344.8
1933	38.9	4.2	52.6	46.8	29.2	10.0	11.0	388.3
1934	49.9	6.6	69.0	59.4	41.3	11.3	12.8	488.6
1935	71.4	8.5	119.2	74.1	21.3	12.8	39.8	658.9
1936	90.6	9.6	163.9	88.9	28.4	13.1	40.0	765.0
1937	123.0	10.9	269.0	112.2	45.3	16.4	40.1	1,016.0
1938	156.6	21.5	319.9	143.5	86.9	17.0	–	1,209.8
1939	193.7	48.6	462.0	177.8	131.7	19.4	–	1,513.6
1940	231.3	71.4	653.0	203.4	144.4	19.2	–	1,736.9

Source: Table A, Sang-chul Suh, *Growth and Structural Changes in the Korean Economy 1910–1940,* pp. 162–163.
Note: Totals include products not separately listed in Table 31.

TABLE 32 Growth in Population and in Output in Korea, 1930–1931 and 1939–1940 (million Korean yen)

	1930–1931			*1939–1940*		
	South	*North*	*South/North*	*South*	*North*	*South/North*
Population (millions)	13.9	6.5	2.14	15.6	7.9	1.97
Commodity						
Textiles	11.1	2.9	3.79	64.1	12.9	4.95
Metals	1.8	0.8	2.33	6.0	49.6	.12
Machinery	7.1	1.5	4.67	44.7	17.2	2.60
Ceramics	3.4	8.6	.40	17.4	47.1	.37
Chemicals	13.9	19.9	.70	80.6	399.2	.20
Wood	2.6	0.8	3.07	10.4	8.2	1.27
Printing	9.0	0.7	12.65	18.5	2.9	6.33
Foods	30.7	15.1	2.03	126.2	70.6	1.79
Others	28.6	13.0	2.20	101.1	40.0	2.53
Total	108.2	63.3	1.71	469.0	647.7	0.72

Source: Suh, p. 141; p. 132.

on increasing the generation of electricity was particularly helpful to Noguchi. The benefits of the relationship between the two men, which dated from 1927 when then War Minister Ugaki first went to Korea, flowed both ways.[161] Ugaki recognized in Noguchi a bold man he could entrust with development of his favored projects, especially with construction of harbor facilities capable of being used for military purposes, and with little aversion to risk-taking.[162] The fact that the deceased elder brother of Ichikawa Seiji, Noguchi's oldest associate, had been a close personal friend of Ugaki helped bind the Governor General emotionally as well as rationally.[163] Noguchi needed not only official approval of development rights but also financial assistance. After the Mitsubishi Bank withdrew its support, Noguchi turned to official banks for long-term financing. Access to these funds was expedited by General Ugaki, who had clout as Governor General as well as having several bank officials among his friends.[164] In later years, Chōsen Chisso received important military contracts, aid in research and development, and other forms of government assistance as an outgrowth of the close personal ties Ugaki had had with Noguchi, even after the former's departure from his post.

When Noguchi's wife and children moved from Hiroshima to Tokyo in 1928, he began to spend increasing amounts of time in Seoul. His Chōsen Hydroelectric and Chōsen Chisso plant at Hungnam were completed during those years. Hungnam began producing inexpensive fertilizer, and production expanded. By December 1930, he had increased his Korean capital to 60 million yen, two-thirds of the entire worth of Nitchitsu in Japan and Korea.[165] He was clearly an appropriate partner for Ugaki's stated hopes for industrialization.

By most accounts, Noguchi was a close associate of Ugaki; perhaps their relationship was based not only on a brief meeting in 1927 in Korea but also on their shared ties with the influential financial community in Ōsaka.[166] In particular, Noguchi's old friend and longtime member of the Nitchitsu Board of Directors, Hori Keijirō (former President of Ōsaka Shōsen Kaisha), was also a close acquaintance of Ugaki. Noguchi's own aggressive style of industrial investment probably helped earn him Ugaki's support.

One of Ugaki's first acts on hearing of his appointment as Governor General was to become acquainted with the Japanese who had invested in Korea. Bringing together a group of senior executives to discuss planning, he turned to Kimura Kusuyata of Mitsubishi Gōshi Kaisha and asked what Mitsubishi was doing with its valuable rights to develop the Changjin River.[167] Kimura replied that Mitsubishi was making use of its rights and had begun to invest in the project. Ugaki was anxious to have the river developed, so he agreed to grant the company one more year to report whether Mitsubishi would develop the project in earnest.[168]

Ugaki was, as the legally absolute colonial administrator in Korea, acting within his authority. Decisions of Governors General could not be overturned, even in Tokyo. Under this authority, Ugaki had begun to issue regulations, and, on 18 February 1932, he called for controls on the electric industry in Korea based on studies by Political Affairs Officer Imaida Kiyonori.[169] Controls had been discussed since the end of the Taishō period, and the arrival of Ugaki, a well-known advocate of rationalization and controls, further encouraged the idea of government regulation of industry.

These regulations and their various amendments and additions during the 1930s permitted government control of all electric companies and regulation of transmission of electricity to customers. They mandated Government General approval of all blueprints for plants before companies could receive permission to construct. Furthermore, the regulations gave the Government General the right to approve or reject all upper-level management and technical employees of electric companies.[170] Finally, the country was to be divided into four regions with plans for transmission within those regions to be coordinated by the Government General.[171] National-policy companies (*kokusaku kaisha*) controlled by government officials could be established for transmission of power generated by private firms. There was no barrier, then, to the Government General's reclaiming Mitsubishi's water rights, but Ugaki was eager to have Mitsubishi continue developing the Changjin and decided to give the company the opportunity to profit from the investment cited by Kimura. Soon thereafter, Mitsubishi gave up its rights to the Changjin. Kubota, who had been discussing the Changjin rights with other officials, strongly believed that the Government General would grant Nitchitsu the rights if Mitsubishi should relinquish them.[172] When Mitsubishi managers could not reach a consensus about the Changjin project in 1932, they reported to Ugaki: "After investigative research and internal discussion, we do not expect to develop [the river] in the near future."[173]

Ugaki reported that, after Mitsubishi voluntarily withdrew, Noguchi was immediately at his door, requesting the right to develop the Changjin River. Ugaki replied equivocally, reminding Noguchi that the Changjin River project would be larger and more costly than the Pujon. Ugaki quizzed Noguchi about his financial ability to carry out the project should he lose Mitsubishi aid. Temporarily discouraged, Noguchi requested ten days to respond to the conditions Ugaki established. He took twenty days.[174]

Fortunately for Noguchi, Kubota was conducting negotiations on his own. Kubota was aware that the Governor General was trying to convince private capitalists to develop hydroelectricity and, in particular, the electricity of the Changjin River. The Pujon project was a large-scale, successful project, but it was dedicated to providing electricity for Hungnam. Ugaki needed electricity in Seoul. Seoul was 400 kilometers from the Changjin River, and Kubota reckoned Chōsen Chisso would be able to retain a significant percentage of the electicity for its own use. Kubota, without bringing in Noguchi, began exploratory talks with the President and head engineer of Seoul Electric Company (Keijō Denki Kaisha), and the latter advised him to begin studying the Changjin. These studies convinced him that Mitsubishi's projected cost of contruction was too high at 120,000,000 yen, and that his company could bring the plant on line for about 20,000,000 yen.[175]

It was at that juncture that Kubota urged Noguchi to go to the Changjin. Noguchi's enthusiastic return from the mountains occurred right after Mitsubishi handed over its rights. Mitsubishi managers reportedly commented, when returning the development rights, that Korea needed the project and that "whoever can do it should get the rights."[176] On return from his trip, Noguchi went to Ugaki's office. There is little evidence to indicate that Mitsubishi's abandonment of the project specifically inspired Noguchi's visit; Noguchi was so enthusiastic about his trip north that he probably would have requested the Changjin rights in any case. The timing of the petition may have been coincidental.

Kubota, meanwhile, was negotiating with Political Affairs Officer Imaida Kiyonori and with the Government General's Electricity Section head Imai Raijirō. Imai, who had become Kubota's friend in the course of working on regulating the electric industry,[177] conveyed to Kubota Imaida's deep concern that, if Nitchitsu assumed the Changjin rights, Noguchi would not have the capital for construction because Mitsubishi would abandon Nitchitsu. Kubota replied that Mitsubishi's estimates were too high, and that Noguchi, with a personal fortune of 3 to 4 million yen, did not need Mitsubishi for the project. He could get it started with his own funds, if need be.

In the event, Noguchi did not have to rely either on Mitsubishi loans or on his personal wealth in 1933. While he was required to pay Mitsubishi for the transfer of the water rights on the Changjin and to pay them back 25,000,000 yen in previous loans, apparently Mitsubishi did not demand all payments immediately. To be sure, a substantial amount was repaid right away, but, six months later, in October 1933, Nitchitsu still owed Mitsubishi Bank 7,816,000 yen. (Some of this amount was from a 1929 loan of 8,750,000 yen which was to have been paid off by June 1936.) Nitchitsu's balance sheet indicates that the company repaid 13,384,000 yen in loans between the first and second halves of 1933. This probably represented part of Nitchitsu's repayment to Mitsubishi.[178] The rest of the amount Nitchitsu owed Mitsubishi was not repaid until later, which meant that Nitchitsu continued to be in Mitsubishi's debt in the middle and later part of the decade; when economic conditions improved in Japan and the colonies, Mitsubishi Bank was one of three private banks (Sanwa and Aichi were the other two) noted in a company memo of June 1936 as making large loans to Nitchitsu. Thus, Mitsubishi's withdrawal from assisting Nitchitsu in 1933 was not as damaging as is often alleged, although the blow was, of course, quite severe. Paying back 13,000,000 yen was no simple task.

One way in which funds were obtained to repay the loan was by making use of the relationship between Nitchitsu and its subsidiary, Chōsen Chisso. Chōsen Chisso repaid 8,000,000 yen in loans to its parent company Nitchitsu in the second half of 1933, presumably to permit Nitchitsu to repay Mitsubishi. Chōsen Chisso could pay such an amount because it had large reserves from profits and especially from the 5,000,000 yen Changjin Hydroelectric (Chōshinkō Suiden) deposited with Chōsen Chisso in June 1933.[179] This flexibility in moving funds from one subsidiary to another was clearly helpful to Nitchitsu.

The success of the Changjin project was important to the Government General both politically and financially. The Governor General remarked that the Industrial Bank (Shokusan Ginkō) and the

Bank of Chōsen were established for the purpose of aiding industrial development in Korea, precisely the type of investment Noguchi was attempting.[180] Clearly, he felt, Noguchi should use all the facilities to which his position entitled him. With the authority of the Governor General behind him, Noguchi had little trouble obtaining loans and floating bonds. Most helpful in underwriting his bond issues were various bankers at the Industrial Bank, Yūki Toyotarō at the Japan Industrial Development Bank, (Nihon Kōgyō Ginkō) and Katō Keisaburō at the Bank of Chōsen.[181] In 1935, bonds underwritten by these banks were used for two major purposes—paying off the Mitsubishi loan and starting up production at Changjin.

It is not entirely clear how much each of the various methods of accumulating funds contributed to the repayment of Mitsubishi. Internal reserves, as noted above, clearly played a role.[182] But how Nitchitsu took advantage of its ability to raise funds through selling bonds is less obvious. Two Nitchitsu managers, Shiraishi and Ichikawa, agreed in their memoirs that the Japan Industrial Development Bank underwrote 60,000,000 yen in bonds in 1934, and that the funds thereby acquired were put to use repaying Mitsubishi.[183] Sugita Motomu, a member of the Accounting Division in 1933, recalled a somewhat different method of repayment. He said that Nitchitsu repaid Mitsubishi by using a basket of short-term loans from other banks. Bonds were issued to obtain the funds to pay back those banks for their short-term loans.[184] In both these scenarios, semi-governmental banks had to help with the issuance of bonds. Thus, connections with officials were significant.

In the following years, the government-managed Bank of Chōsen, Japan Industrial Development Bank, and Industrial Bank continued to underwrite bond issues and make regular commercial loans to Noguchi's enterprise. The loans were supposed to be of relatively short duration, but apparently not all were paid back quickly and so became long-term loans. Table 33 indicates which projects were the objects of official interest and therefore of Bank of Chōsen funding. Each of Noguchi's companies was a separate corporation within the Nitchitsu group (*kontsuerun*). Until 1937, most funds were funneled through the main fertilizer company, Chōsen Chisso. For the next few years, when Chōsen Chisso no longer required such large-scale loans, Chōsen Oil, a strategically important subsidiary, received larger short-term loans. Nitchitsu, the holding company for all the Korean operations, separately received Bank of Chōsen loans through the bank's Tokyo office. When Chōsen Chisso was merged with the parent company in 1941,[185] the amounts loaned to Nitchitsu surged, as did all loans from official and semi-official banks to the firms called new zaibatsu. Development of new technologies by various subsidiaries can be correlated with increased loans to those subsidiaries, which underscores the relationship between technology-intensive industrialization and colonial policy.

Noguchi's access to the Bank of Chōsen was facilitated by his introduction to Bank President Katō Keisaburō, a friend of Governor General Ugaki and a Japanese resident of Korea who, according to some sources, shared Ugaki's goal of developing colonial Korea.[186] There are several reasons why Noguchi and Katō found it mutually profitable to cooperate. Katō may have been impelled by ideological reasons. If Katō took it as his "special mission" (*tokushu shimei*) to lend to prospective developers, as some Japanese authors noted during the war, then he would have readily accepted Ugaki's warm endorsement of Noguchi Jun.[187] A commercial bank might have spent more time investigating the potential profitability of each bond issue or loan, ascertaining whether or not, for instance, the Korean market could absorb the large amount of ammonium sulfate Noguchi's enterprise could eventually produce.

For a government institution, however, political and strategic concerns affected the perceived viability of the loan. A wartime study indicated that strategic concerns were appropriate considerations for determining government assistance to industry, and that nitrogenous fertilizer, the basis for munitions production, was a strategic product. Tax policy certainly favored fertilizer production, so it stands to reason that lending policies should as well. The need to encourage nitrogenous fertilizer production helped the government justify its financial support for the construction of the hydroelectric plant on the Changjin River.[188] Katō Keisaburō was interested in government involvement in projects deemed politically necessary and offered the help of his semi-governmental bank.

TABLE 33 Bank of Chōsen Loans to Selected Subsidiaries of Nitchitsu, 1930–1943 (million yen)

Semiannum	*Chōsen Chisso*	*Nitchitsu*	*Hungnam Development*	*Yalu River Development*	*Chōsen Oil*	*Chōsen Fats and Oils*
1930.2	3.000	n.a.	–	–	–	–
1931.2	1.500	0	0.600	–	–	–
1932.1	3.800	0.800	0.600	–	–	–
1932.2	5.000	0	0	–	–	–
1933.1	5.000	0	0	–	–	–
1933.2	13.848	0	0	–	–	–
1934.2	0	n.a.	3.205	–	–	1.340
1935.1	n.a.	n.a.	3.000	–	–	1.096
1935.2	0	0	3.000	–	0.035	2.188
1936.1	0	0	3.000	–	1,367	n.a.
1936.2	10.050	1.000	2.550	0	5.802	4.015
1937.1	1.292	n.a.	2.550	0	8.675	2.553
1937.2	3.942	0	2.550	3.875	13.973	3.539
1938.1	4.400	5.000	2.550	1.600	10.731	n.a.
1938.2	15.000	2.000	2.550	3.550	3.641	n.a.
1940.2	n.a.	n.a.	n.a.	22.000	n.a.	n.a.
1941.1	n.a.	15.521	0	25.500	n.a.	n.a.
1941.2		15.031		18.715		
1942.1		19.169		n.a.		
1942.2		21.376		13.939		
1943.1		48.583		10.570		
1943.2		53.143		10.570		

Source: Chōsen Ginkō, *Shōkeisanshō*, 1930–1943, vols. XLIII–LXVIII. Set incomplete, unpaginated.

Note: Figures are amounts outstanding to the subsidiaries of Nitchitsu on 31 June (first semiannum) or 31 December (second semiannum). They do not represent new loans.

Other, less ideological reasons may also have persuaded Katō of the importance of cooperating with Noguchi's expansionary plans. Perhaps he viewed assistance to Noguchi Jun as an important investment in strengthening the independence of his bank, the Bank of Chōsen, vis-à-vis Tokyo.[189] The Bank of Chōsen faced serious financial problems during the late 1920s. In the wake of the liquidity crisis which forced the bankruptcy of the similarly government-managed Bank of Taiwan, both colonial banks were placed under careful observation by the Ministry of Finance in Tokyo. The presidents of both banks had to spend half their working time in Tokyo and half in their colonial headquarters. Katō Keisaburō wished to avoid the stifling effects of intense scrutiny by Tokyo, and felt that a major success in lending in Korea would underscore the effectiveness of the Seoul office as well as provide an indisputable justification for his remaining in Korea. Katō therefore used the Seoul office to lend large amounts of money to Noguchi, in whom he had great faith. Later loans to subsidiaries of Nitchitsu were delivered through different branches of the Bank of Chōsen; as but one of four Japanese international banks and one with numerous branches in other Japanese cities and colonial locations, loans from other local branches were quite usual. It was initially significant, however, for Katō to underscore the importance of the Seoul headquarters by establishing the first loan to Noguchi there.

A third reason for Katō's assistance was Noguchi's memorable personality. It was, perhaps, no less compelling a consideration than the desire to help one's nation or the maneuvering for independence in decision making. Bankers at the Bank of Chōsen dealt easily with Noguchi. He generally negotiated for loans in person, rather than assigning a representative, which was surely impressive to the bankers.

Noguchi seemed to be a good risk, and the bank president authorized loans to him. But Noguchi had to offer collateral, which he had not been required to do with long-term loans from the Aichi, Mitsubishi, or Yamaguchi Banks between 1927 and 1929.[190] The entrepreneur put up 37,000 of his own shares in Nitchitsu as well as a portion of the fertilizer output of Chōsen Chisso as collateral.[191] The shares were worth 1.85 million yen at par value or 2 to 3 million yen on the stock market in the early 1930s. This amount of collateral entitled Noguchi to borrow up to 40 million yen, which was one-third of the total 120 million yen lent out by the Bank of Chōsen in the late 1930s.[192] He did not hesitate. His first Bank of Chōsen loan after acquiring the Changjin rights was obtained in June 1933, and was for 5,000,000 yen.[193]

Other short-term loans to Noguchi came from the Industrial Bank (Shokusan Ginkō), although it was not as important as the Bank of Chōsen as a source of loans for his firm. The Industrial Bank, unlike the Bank of Chōsen, had no problems with liquidity during the 1920s and was therefore not subject to restrictions on its activities. But it was also an official bank, with responsibilities mandated by law. It was most active in lending money for real estate, farmers' cooperatives and, later, industrial ventures. One of those ventures was Noguchi's sardine-processing plant. Both the Bank of Chōsen and the Industrial Bank also served as agents for other Japanese banks. Japanese banks, like the Industrial Development Bank, could be assured that the Korea-based agents of the Bank of Chōsen and the Industrial Bank would transfer loans, pay necessary taxes and fees, and offer various customer services on their behalf to Noguchi and other borrowers.

In addition to borrowing money to expand in Korea, Noguchi discontinued payment of dividends in Chōsen Chisso in order to reinvest profits. The 5-percent dividend paid out when Chōsen Chisso's profits were 8.8 percent in the first half of 1933 was discontinued in the second half of 1933, although profits then climbed to over 30 percent as Hungnam's output of cheaply produced ammonium sulfate increased.[194] (See Figure 5.) By contrast, investors in the Nitchitsu Group or any of its parts in Japan continued to receive their dividend payments throughout the 1930s. Profit rates for Nitchitsu ranged between 11 and 14 percent during the period from 1933 to 1937, and dividends were commensurate, ranging from 8 to 10 percent.

Some historians believe Chōsen Chisso's discontinuation of dividend payments represented a strategy to save money in 1933 when funds were needed to pay off Mitsubishi. But Chōsen Chisso was

FIGURE 5 Profit Rates: Nitchitsu and Chōsen Chisso, 1930–1938

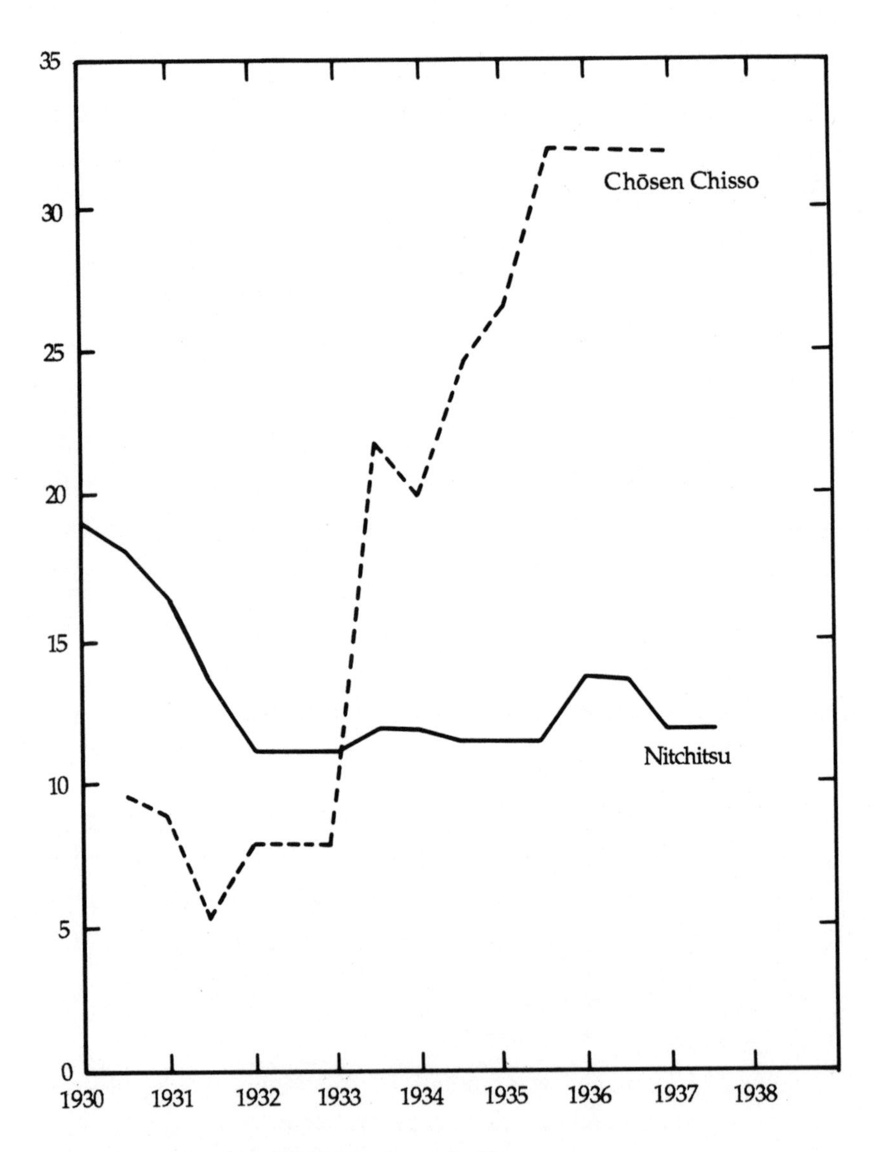

Source: Shimotani Masahiro, "Nitchitsu kontsuerun," p. 69.

almost wholly owned by Nitchitsu, and Changjin Hydroelectric was wholly owned by Nitchitsu.[195] So that explanation is unsatisfactory. It is more likely that dividends were discontinued for tax reasons. Rather than paying dividends, Chōsen Chisso, like all Nitchitsu subsidiaries except Asahi Bemberg, paid Nitchitsu a 5-percent commission for marketing its products (after 1934) and a 5-percent fee for procuring raw materials and resources (after 1937).[196] Each of these fees represented Nitchitsu controls over subsidiaries. For example, when Nitchitsu collected the marketing commission, the parent company could note what was selling and could therefore make suggestions to Chōsen Chisso on how to improve sales. But these fees had an additional purpose: tax benefits. The fees, which were equivalent to the dividends Nitchitsu would otherwise have received from Chōsen Chisso, could be deducted from income as part of the cost of production. Dividends were not deductible from income for tax purposes, so substituting commissions and procurement fees for dividends was good business strategy.

Creative financing techniques—new forms of borrowing, dependence on official and semi-official sources of funds, and flexible accounting procedures—permitted the surplus Noguchi needed for development of technology and diversification of products in Korea during the 1930s. These techniques were all developed as Nitchitsu moved into Korea, and were generally dependent on Nitchitsu's colonial venue.

DEVELOPMENT OF THE CHANGJIN RIVER PROJECT

The Changjin project, Noguchi's largest to date, had two parts, corresponding to the colonial government's policies on electricity. One was Changjin Hydroelectric, founded in May 1933, which generated electricity and was wholly owned by Nitchitsu and Nitchitsu-related individuals (and Chōsen Chisso, at its inception); Imaida's 1932 policy report had encouraged private investment in generation. The other part was Chōsen Electricity Transmission (Chōsen Sōden KK), founded in May 1934. Nitchitsu-related interests owned 155,000 of that company's 300,000 shares.[197] Chōsen Electricity Transmission, which was to transmit 160,000 kilowatts of electricity to Seoul and Pyongyang, was established at the insistence of Governor General Ugaki when granting Noguchi the rights to the Changjin River. Several other power-transmitting companies were involved in setting up Chōsen Electricity Transmission, including Western Chōsen Consolidated Electric (Nishisen Gōdō Denki), Kumgangsan Electric Railroad (Kingōsan Denki Tetsudō KK), and Hamnam Consolidated Electric (Seinan Gōdō Denki), in addition to the Oriental Development Company (Tōyō Takushoku).[198]

Under the 1932 regulations, transmission of electricity was to be under government control, and, although the management of Chōsen Electricity Transmission was composed of men from the private sector (Noguchi, Kubota, and Ōshima Eikichi were Changjin Hydroelectric representatives), it was a "proxy for government management" (*kokuei daikō*).[199] The interest of government in the transmission of electricity explains why Noguchi founded Changjin Hydroelectric and Chōsen Electricity Generation as separate companies. In response to a question by Ōhashi Shintarō, President of Seoul Electric, as to why two companies were established rather than one when that merely doubled the taxes the usually tax-evading Nitchitsu would have to pay, Kubota noted that controls were mandated for transmission, not generation.[200] The company clearly wished to remain as free as possible to follow its own strategies, unencumbered by government regulations, even if it meant higher taxes. Maintaining the company's autonomy came first, and permitting the state to benefit from involvement in running the company was secondary, even for a known patriot like Noguchi.

Kubota, Managing Director of Changjin Hydroelectric under Noguchi's presidency, worked hard to get his company started. Not only was he responsible for negotiating with electric companies like Seoul Electric for their purchase of electricity from Chōsen Electricity Transmission Company, but he also oversaw construction, built support services for the plant, and purchased land. Construction of the dam and other structures went smoothly, following the experience gained at the Pujon project. Land purchase was a bit more tricky. Mitsubishi had purchased land valued at 1,500,000 yen, which Kubota acquired easily; acquiring the rest of the necessary land was more difficult.[201] But contruction began soon, in any case. The first stage was begun by June 1933 and was completed by October 1935. In November, Governor General Ugaki threw the switch starting the plant.[202] The next three stages were completed at approximately one-year intervals in 1936, 1937, and 1938 (see Table 34).[203]

Changjin Hydroelectric was highly successful. When the twelve largest electric companies in Japan averaged—depending on method of calculation used—7.9 or 9 percent in profits in late 1936, Changjin earned 24.7 or 31.8 percent in April 1937. (The lower figure represents profits as a percentage of capitalization, the higher as a percentage of paid-up capital.)[204] This success led Kubota to request an increase in capitalization during one of Noguchi's visits to Changjin. Noguchi asked how much Kubota needed, and Kubota said 50,000,000 yen, which would have constituted a large increase over the initial 20,000,000 yen capitalization.[205] Even the initial capitalization had taken some time to pay up. Starting in 1936, shareholders had to begin paying for their shares.[206] But this was relatively easily done in the context of flexible accounting among subsidiaries. That is, Changjin held loans from Nitchitsu and simply converted the loans toward partial payment of the par value of the shares. That, in turn, was also easy, since Chōsen Chisso, which had held almost half of the shares in 1934, had transferred that ownership to Nitchitsu in 1935 as payment for earlier loans made in another context, independent of Changjin. Kubota needed more funds for construction and for acquisition of shares in related companies, and Noguchi agreed. But Noguchi's telegram authorizing increased capitalization, which should have read, "Increase capital to 50,000,000 yen" (*Gosenman en ni zōshi*), was mistakenly transmitted as "Increase capital by 50,000,000 yen" (*Gosenman en o zōshi*). At a stroke, Changjin Hydroelectric's capital was authorized to increase to 70,000,000 yen, an error Noguchi chose not to correct.[207]

As the reservoir on the Changjin River filled up, Noguchi redirected his attention to the Hochon River.[208] Noguchi had been inter-

TABLE 34 The Building of Changjin Hydorelectric, 1935–1938

Stage	*Year Completed*	*Vertical Drop (meters)*	*Generative Capacity (kw)*	*Cost (million yen)*
1	1935	407	144,000	35
2	1936	281	112,000	6
3	1937	116	42,000	6
4	1938	94	36,000	5.4
			334,000	55.4

Sources: Chōsen Denki, p. 258; Yoshioka Kiichi, pp. 236–238. Yoshioka states that the generative capacity of stage 4 was 34,200 kw.

ested in the Hochon since before Kubota had persuaded him to apply for the rights to the Changjin. Kubota had temporarily dissuaded him from starting construction, but, by May 1937, the Hochon project with its multiple dams and reservoirs was begun. The project was to be part of the Changjin Hydroelectric Company. The bulk of the construction was completed by 1940. Unlike the Changjin project, the Hochon project was required to transmit two thirds of its electricity to outside users; still Nitchitsu was left with more than enough. The costs of construction, especially of the transmission lines, were significantly higher than at the Changjin project. These costs would presumably be recouped through outside sales, although the amount generated failed to live up to expectations. As the Hochon project got underway, it was noted that the name of the company–Changjin Hydroelectric–was then inaccurate, and, in February 1938, the generating company comprised of the Changjin and the Hochon projects was renamed Chōsen Hydroelectric, Inc. (Chōsen Suiryoku Denki, KK.)

Governor General Ugaki was pleased with progress at Changjin. He even visited the plant while it was under construction and, praising it, he said he wished to have a vacation house built there.[209] The summer home was to play an important role in Nitchitsu's affairs. Ugaki and his colleagues in government and the military would later meet with Nitchitsu men and discuss policy in that house.

High profits during World War I gave Noguchi Jun the reserves he needed for upgrading the level of his firm's technology. But the decade after World War I brought numerous changes to the business environment in Japan, and these fundamentally altered the ways in which the new technology would be implemented. Noguchi encountered increasing competition from other domestic manufacturers as well as from importers of fertilizer. Shōwa Fertilizer and Dai Nihon Fertilizer, two companies to be discussed in Chapter 6, began production of synthetic ammonium sulfate using electrochemical technology. Not only were they competitors with Nitchitsu in the Japanese fertilizer market; they also faced the same problems in acquiring capital, resources, and technology as Noguchi. Shōwa Fertilizer responded to market conditions and capital problems by developing technology invented in Japan. Dai Nihon rapidly diversified in new areas of investment to alter its longstanding policy of reliance on a limited range of products.

The electrochemical technology used by Nitchitsu required electricity. Resource requirements affected the company's investment decisions. The need to lower costs in the face of highly competitive imports in conjunction with the tightening of supplies of resources impelled Noguchi's search for cheaper inputs. This led to his quest for electricity in Korea. But this search, in turn, affected his company's access to capital from Mitsubishi Bank. Acquiring new sources of capital led Noguchi to cooperate with authorities in colonial Korea. And contacts with the authorities permitted great growth and resultant diversification. The expansion of technology that took off after completion of the Changjin project unbound Prometheus in the next decade.

Burgeoning investment energy lay at the heart of Japanese economic imperialism in Korea after 1925.[210] Most scholars writing in English during the last three decades have played down specific economic motives or an explicit plot by government and business leaders for Japan's takeover of Korea as a colony in 1910.[211] Peter Duus has noted, for example: "While business leaders supported political expansion, they did not regard the acquisition of new territory as necessary for the expansion of private business interests abroad."[212] That is, the net effect of the involvement of Japanese investors, whether farmers or businessmen, may have been economic imperialism, but few believed political imperialism necessary for investment at the turn of the century. The motives for the acquisition of Korea as a colony were fundamentally political.[213]

Most scholars agree that, after colonization, Japan actively encouraged increased agricultural output in Korea to benefit the mother country.[214] I would add that they encouraged increased industrial output as well during the last two decades of colonial rule. But, while government policies may have encouraged development, they could not require it; the authorities needed private investors. The dynamics of colonial economic penetration cannot be understood without an analysis of what motivated specific investors whose priority was the service of company strategy. To be sure, public policy could be formulated to tighten the bond between government interests and company strategy. But it was the companies that invested in Korea, together with colonial officials who made policy, that actually carried out economic imperialism.

Most authors dealing with the later colonial period treat the results of economic imperialism—the reorientation of the economy, imbalances in wealth, growth of some sectors at the expense of others—and give little attention to the incentives for economic imperialism carried out by businesses. At least one scholar, Duus, does analyze the motives of business leaders in Japan's economic penetration of Korea in the pre-colonial period, and finds, as does Wray in his discussion of the business activities of the shipping company N.Y.K. in China, that Japanese businessmen in both areas were generally most interested in furthering company interests. Wray takes issue with Duus's characterization of business leaders as "falling into line" behind the Japanese government's military policy toward the colony, preferring to emphasize the primacy of business interests.[215] In the later colonial period, business leaders had no choice but to operate within the colonial context, and, as we have seen, Nitchitsu interacted effectively with its environment, both exploiting and shaping it. Indeed, Nitchitsu would have been unable to exploit the opportunities for developing Korea as effectively in the absence of Japan's colonial control. By the same token, Japanese political authorities would have been unable to develop Korea without private investors. It is no exaggeration to equate economic imperialism with the fulfillment of business goals.

FIVE

Nitchitsu and the Japanese Empire

Noguchi Jun's harnessing of electricity at the Changjin River permitted an acceleration of technological development in the Japanese empire. Access to plentiful and cheap resources was crucial to technological breakthroughs, and, in the case of Nitchitsu, inexpensive electricity was necessary for the company to survive competition as well as to experiment with a variety of energy-dependent processes and products. The development of these new products relied on creating new technology—the science, its applications, and the machinery to apply it. But this new technology was, in turn, dependent on factors outside the company and its research staff. These factors included having a market for the company's products, having access to capital, and operating in a business climate conducive to research, production, and sales. If these conditions for investment in technology were met, then the new technology would be perfected, with a host of other processes developed in its wake. These processes could be more technologically advanced derivatives based on the primary technology or they could be "back inventions" of processes to expedite the production of the primary industry. For example, one of Nitchitsu's "forward" derivatives was the process for coal liquifaction, procedurally related to ammonia synthesis; one "back invention" was the creation of a modern machinery industry to supply Nitchitsu's needs. Without access to plentiful electricity, chemical companies would have been hard pressed to survive, let alone serve as a trigger industry spurring a wide variety of products and processes.

This wide variety of products and processes was the heart of the progressive diversification. Diversification and expansion occurred rapidly and successfully during the five years after Changjin Hydroelectric was founded. Growth was so extensive that the company structure changed. Particularly illuminating is the relationship between the parent company and its subsidiaries. Even as diversification of production excited the imaginations of scientists at Nitchitsu, financial decisions were becoming as important as advances in production in furthering company success. But financial transactions could not have played such an important role in company strategy without advances in technology. That is, technological breakthroughs produced sophisticated diversification. New processes were developed in a rapidly growing number of subsidiaries, which not only provided Nitchitsu with a vast variety of products to support its manufacturing network, but also offered the parent company the opportunity to maximize profits by flexible accounting procedures within the conglomerate. As a result, financial decisions and transactions played an increasingly large role in company planning.

Product development was the major method of furthering company interests in the period before 1937; after 1937, financial transactions were more significant because they were freer of outside controls than production. To be sure, an even wider variety of products appeared after 1937, but most were urged on Nitchitsu by the military for strategic reasons rather than being developed by Nitchitsu as an obviously profitable outgrowth of already marketed products. Of course, the products of strategic diversification had a market, the military, just as civilian products had a civilian market. That was an important incentive to production. And access to resources and capital was often expedited by the military's desire to develop a certain technology—a reversal of the pattern before 1937 in which access to capital and resources expedited development of technology. As interested individuals outside the company, such as the military or colonial leaders in northeast Asia, began to influence technological investment decisions, production came under increasingly strong outside control. Financial transactions within Nitchitsu became the area of business in which management's ability to plan strategy was least circumscribed, and this accelerated Nitchitsu's transition to *konzern* status. At the same time, the quantity as well as the diversity of output could no longer be handled to Nitchitsu's satisfaction by trading companies working on contract. Nitchitsu increased its role in distribution and marketing throughout the 1930s and 1940s by adding subsidiaries in those fields. The financial relationship between Nitchitsu and these subsidiaries was such that the *konzern* integrated production, distribution, management and finance, creating what Alfred Chandler has called the "modern industrial corporation."[1]

LAYING THE GROUNDWORK FOR DIVERSIFICATION, 1932–1937

Cooperation between the government and the chemical companies strengthened during the late 1930s, since demand for at least one important product of the industry—munitions—was rising in Japan. Cooperation could be particularly effective, from the Japanese government's point of view, because chemical companies produced commodities either necessary for mobilization or else easily made into necessary products. Officially planned product conversion could be further facilitated if a company's top managers were as willing as Noguchi to follow government orders. Noguchi was, in fact, so willing to cooperate with the government that he occasionally encountered resistance from company managers who were more concerned about finances.[2] Cooperation between the company and the colonial authorities during the mid-1930s established the foundation for more extensive cooperation later. It should be kept in mind, however, that, until 1937, and to a much smaller extent after that date, cooperation occurred when and if it served company strategy.

By 1937, Noguchi's conglomerate had moved beyond production of fertilizers alone to become a multi-faceted chemical company worth 500 million yen and producing fats and oils, refined metals, explosives, carbide, coal-tar derivatives, soybean derivatives, and bleach and soda (see Table 35).[3] Some types of diversification were more profitable than others. Indeed, a 50-year employee recalled that the most successful types were the simplest modifications of processes on hand. More complicated technology that diverged significantly from products and processes on hand, he noted, was more likely to fail.[4]

Successful new areas of production that built on technology already developed permitted exchange of data and information among the subsidiaries of the firm. These exchanges included the twelve major conferences of scientists and engineers from 1931 to 1943, noted above, designed to foster creative exchanges of ideas.[5] In holding such formal scientific conferences, Nitchitsu resembled two other large and successful chemical conglomerates, I.G. Farben in Germany and Imperial Chemical Industries (I.C.I.) in Great Britain.[6]

Nitchitsu scientists were interested not only in pure science but also in applications of their interrelated technologies. Until 1937, most types of products were connected to other types through manufacturing processes. Figure 6 indicates the relationships within Chōsen Chisso in 1935 and Table 36 indicates the proportion of company investment in each category of production in comparison with the investments of other new zaibatsu.

GLYCERINE AND EXPLOSIVES IN KOREA

The first type of diversification of product was the manufacture of explosives and the intermediate product necessary for explosives manufacture, glycerine. Glycerine could be made from fish oil. The waters around Korea were teeming with sardines.[7] For that reason alone, an opportunistic investor like Noguchi might be expected to

TABLE 35 Product Diversification at Chōsen Chisso, 1930–1941

	Fats, Oils Fertilizer	*Fats, Oils Explosives*	*Carbide*	*Soda, Food*	*Metals*	*Products from Yong'an*
1930	Ammonium sulfate					
1931	Ammonium phosphate					
1932	Ammonium sulfate	Hardened oil Glycerine Fatty acids				Tar
1933				Caustic soda Soda	Gold Silver Lead	Methanol
1934	Compound fertilizer Superphosphate					Formalin
1935			Carbon			Phenol
1936		Nitric acid Dynamite Ammonium nitrate explosive	Carbide Calcium cyanamide	Soda ash Soy MSG		
1937					Iron	
1938						
1939					Magnesium	
1940				Margarine	Aluminum	
1941			Acetone Butane Acetate			Gasoline

Source: Watanabe Tokuji, *Gendai Nihon sangyō hattatsushi,* p. 389.

Figure 6 Productive Relations in Chōsen Chisso, 1935

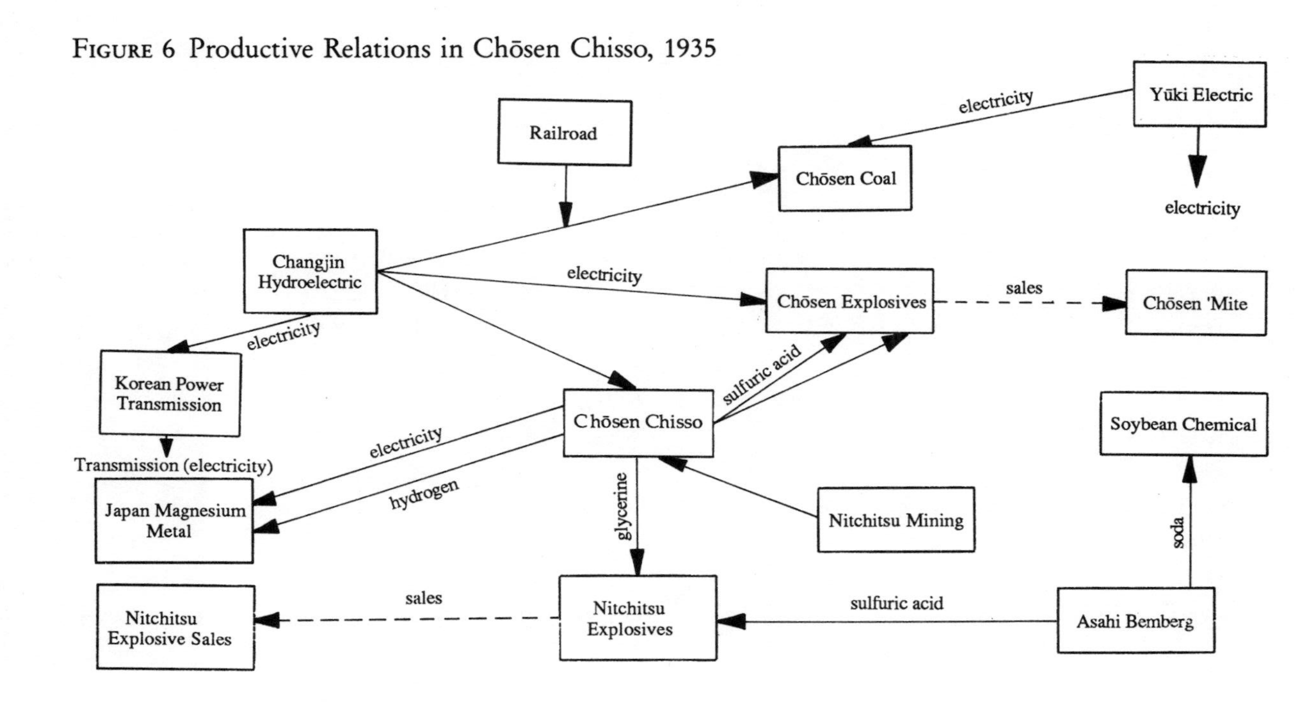

Source: Shimotami Masahiro, "Nitchitsu kontsuerun," p. 66.

TABLE 36 Proportion of Investment in Various Industrial Areas—New Zaibatsu, 1937
(million yen; figures in parentheses represent % of firm's investment)

Product	*Nissan*		*Nitchitsu*		*Mori*		*Nissō*		*Riken*		*Total*	
Finance	1.2	(0.3)	–	(–)	–	(–)	–	(–)	–	(–)	1.2	(0.1)
Heavy Industry	268.9	(56.8)	15.2	(7.7)	48.8	(34.4)	20.2	(23.9)	16.0	(52.4)	369.1	(39.8)
Mining	170.8	(36.1)	8.2	(4.2)	15.8	(11.2)	18.2	(21.6)	–	(–)	213.2	(23.0)
Metals	–	–	4.2	(2.1)	32.9	(23.2)	1.5	(1.8)	10.4	(34.3)	49.1	(5.3)
Machines	84.8	(17.9)	2.7	(1.4)	–	–	0.4	(0.5)	5.5	(18.1)	93.5	(10.1)
Ships	13.2	(2.8)	–	(–)	–	–	–	(–)	–	(–)	13.2	(1.4)
Chemicals	99.7	(21.1)	108.0	(54.6)	29.2	(20.6)	60.6	(71.5)	2.5	(8.3)	300.0	(32.4)
Electric & Gas	6.3	(1.3)	63.9	(32.4)	61.4	(43.3)	0.7	(0.8)	–	(–)	132.5	(14.3)
Spinning	–	(–)	–	(–)	–	–	–	(–)	0.5	(1.6)	0.5	(0.1)
Food & Beverage	69.6	(14.7)	–	(–)	2.5	(1.7)	1.5	(1.8)	4.4	14.4)	78.0	(8.4)
Other	27.7	(5.8)	10.5	(5.3)	–	–	1.7	(2.0)	7.1	(23.3)	47.1	(4.9)
Total	473.4	(100.0)	197.7	(100.0)	141.9	(100.0)	84.7	(100.0)	30.6	(100.0)	928.6	(100.0)

Source: Watanabe Tokuji, *Gendai Nihon sangyō hattatsushi*, p. 372.

try to find a way of cheaply exploiting the resource. Sardines could also be used in conjunction with products he was already making: Ground sardines could be mixed into compound fertilizer and the fish oil could form the basis for glycerine for use in the explosives Noguchi was producing at Nobeoka. The benefits of exploiting these sardines were obvious.

Explosives were not the only use for glycerine. Nitchitsu scientists researched various products related to glycerine, including soaps and margarine. Though manufacture of these consumer products was not the reason Noguchi originally set up his glycerine laboratory, it proved highly profitable.[8] Soap was, of course, primarily a consumer product, but even soap had military applications. During the last year of World War II, when supplies of petroleum products were low, the Japanese Navy began to use a soap product called "olefin" in place of airplane-lubricating oil.[9] Thus, the production of fats and oils at Hungnam had a variety of uses.

Glycerine appeared so potentially profitable that Noguchi began work on his fish-oil operation in October 1931, and the following June began daily production of two tons of glycerine.[10] His initial efforts were not without their problems, though. For example, the glycerine made at Hungnam caused (unintended) explosions when first used in explosives manufacture at Nobeoka. English glycerine had to be substituted while scientists repeatedly carried out analyses of the glycerine made at Hungnam. Eventually, they discovered the impurities that had caused the explosions and removed them, permitting Hungnam glycerine to be used at Nobeoka.[11]

After a few years of shipping most of his Korean glycerine to Nobeoka, where it was made into explosives and then shipped throughout Japan and the empire (including Korea), Noguchi and his managers decided that local production of explosives in Hungnam was more practical. Although Japanese policy in Korea forbade production of explosives for civilian use, Governor General Ugaki waived the restriction in 1934, both because he respected Noguchi and his ablity to accomplish large-scale industrial projects and because he encouraged construction, mining, and industrial development.[12] Noguchi founded Chōsen Chisso Explosives, Inc. (Chōsen

Chisso Kayaku KK) in April 1935 and began construction immediately in a relatively isolated area about 4 kilometers from Hungnam. Ten high-ranking engineers were brought over from Nobeoka, including Miyamoto Shōji, Managing Director, and Kariya Susumu, Vice-President for the Explosives Section.[13] The plant was built quickly and began operating in October 1936. In November, the first stage of the electric generating plant at Changjin was finished, supplying the plentiful and inexpensive energy necessary for nitric acid manufacture at Hungnam.[14]

Manufacture of explosives was a good example of intra-company cooperation in Noguchi's empire. Trained scientists and engineers sent from Nobeoka used electricity from the Changjin; ammonia, oxygen, glycerine, and nitric acid from Hungnam; and glycol from the nearby Pon'gung plant. The gaseous materials were transported by pipeline.[15] The operations were aided by interests outside the company as well. Mitsubishi Trading (Mitsubishi Shōji) supplied a 10,000-ton tank.[16] In addition, military scientists were actively involved in research in explosives, especially in the development of new types of detonators. Hungnam produced insufficient lead and mercury to keep up with the demand of the detonator department for these metals. New refining techniques had to be developed for the new types of compounds used as detonators. The Government General assisted Noguchi in obtaining raw materials for making detonators.[17] In addition to studying detonators in their own laboratories, Nitchitsu purchased several other firms in Japan involved in explosives and their components in 1936.[18] By the end of the decade, Nitchitsu and its Korean subsidiary, Chōsen Chisso Explosives, were large producers of explosives, at first mainly for industrial use. The Korean plant made numerous types of explosives including carlite, black powder, and ammonium nitrate, but concentrated on dynamite. Production of the dynamite was particularly large—40 tons per day.[19]

Noguchi had to overcome some notable problems in establishing his plant in Korea. In particular, the cold winters of North Korea created inappropriate weather conditions for producing explosives. One of his top managers, Kariya, insisted that only Noguchi's diligence permitted him to surmount the problems of construction.[20]

On the other hand, marketing the explosives presented no problem. As early as June 1927, Noguchi had founded a company with the improbable name of Chōsen 'Mite, Inc. (Chōsen Maito KK, an abbreviation for dynamite). Chōsen 'Mite began marketing explosives in Korea years before they were manufactured in Korea and years before Nitchitsu Explosives Sales, Inc. (Nitchitsu Kayaku Hanbai KK), its counterpart in Japan, was founded. In 1934, the Ōsaka-based Nitchitsu Explosives Sales absorbed the much smaller Korea-based Chōsen 'Mite.[21] The merged sales firm would sell explosives made only by Nitchitsu or, after 1936, Chōsen Chisso Explosives. Nitchitsu had marketed its own fertilizer during the company's earliest years, but soon turned that job over to other distributors. The establishment of these two explosives sales companies, among the first subsidiaries in the expanded, diversified Nitchitsu, indicates that Nitchitsu was beginning to expand vertically by creating functionally different subsidiaries.

Much of the output was sold to entrepreneurs or government interests intent on creating an industrial base in Korea. Some was exported to Manchuria or North China as well as lesser amounts to French Indochina, Thailand, and Iran.[22] Some was being sold to the military for munitions use. Although the exact amount sold for munitions during the manufacturing firm's first few years is uncertain (explosives used to help create an economic infrastructure were as important strategically as those used in gunpowder or bullets in any case), Chōsen Chisso Explosives was designated by the military as contributing to Japan's national defense—and therefore susceptible to government controls—as early as 1938.[23] By the early 1940s, Nitchitsu, with its Nippon Chisso Explosives and Chōsen Chisso Explosives, was Japan's largest civilian producer of arms and munitions.[24] From a 2-ton daily output of glycerine, Noguchi's fats and oils plant at Hungnam grew to a large-scale explosives operation producing 5 times as much glycerine and almost 18,000 tons of explosives by 1941, using technology derived from ammonia technology.[25] Moreover, soap products using glycerine earned large profits at Chōsen Chisso Explosives.

In sum, diversification to produce glycerine was highly successful.[26] In 1942, Chōsen Chisso Explosives doubled its capitalization from 10,000,000 yen to 20,000,000 yen. All 200,000 new shares were held by Nitchitsu. On the other hand, Nitchitsu Explosives, initially 90 percent owned by Nitchitsu, was increasingly owned by Chōsen Chisso during the 1930s. A new stock issuance of 110,000 shares in April 1940 was entirely held by Chōsen Chisso. By October 1940, Chōsen Chisso held 88 percent of Nitchitsu Explosives shares.[27] This is a good example of Nitchitsu's transfer of stock ownership among subsidiaries to maximize company profits.

SYNTHETIC FUELS AND THE NAVY

Noguchi's other important area for diversification in Korea in the early 1930s was in coal-based products. Chōsen Chisso managers began planning a coal-distillation plant at Yong'an, about 4 kilometers from Hungnam in July 1931, and low-temperature carbonization, which yielded coal tar, was first carried out there in June 1932.[28] Noguchi's initial purpose in creating this carbonization facility was to produce oil from coal so as to relieve Japan's dependence on foreign petroleum.[29] Soon, methanol, formalin for use in plastics, and other related products with strategic potential were coming off the line as well.

Investing in the extraordinarily sophisticated technology for synthetic-fuel production appeared impractical to many observers in the early 1930s. Japan could import from the United States all the petroleum it needed. Thus, Noguchi's decision to upgrade his level of technology from carbonization to liquifaction of coal was significant.[30]

Technologically, the leap to liquifaction of coal was not as great as it may seem. Indeed, advances in ammonia technology had helped lay the groundwork for synthetic-fuel production. This was true in other countries as well, where such advances have been credited with setting the stage for more sophisticated fuel production. In Germany, for instance,

> (the) breakthrough in synthetic ammonia led to three important developments in the search for synthetic petrochemicals: it perfected the production methods to produce large quantities of hydrogen from coal at low cost, it taught the Germans the technique of industrial operations at high pressure, and, perhaps most important, it made them familiar with catalysis.[31]

Researchers at Nitchitsu and other laboratories had also been experimenting with coal liquifaction, but few had the ability to begin large-scale production. The Japanese Navy, seeking fuels independent of foreign sources, was among those who experimented; they finally succeeded in synthesizing fuel in 1932.[32] The Navy then, rather than producing fuel on its own, sought private investors to develop the technology.

Nitchitsu's move into synthetic-fuel production began by fits and starts. When Nitchitsu got access to German technology for coal liquifaction in the late 1920s, the process was not economical, so the company made finding an economical production method a priority.[33] During the previous two years, Tashiro Saburō had been working on coal liquifaction in the Fuel Laboratory of the Ministry of Commerce and Industry. By the time he was hired by Chōsen Chisso in May 1931, he had four years' experience. With Kudō Kōki and an assistant, Tashiro initiated serious research in coal liquifaction at Hungnam.[34]

As research progressed, production of synthetic fuel became less improbable. Synthetic fuel, converted from coal or tar (in a process called hydrogenation) by adding large amounts of hydrogen at high atmospheric pressure, at low temperature, and in the presence of a catalyst, seemed a natural product for Chōsen Chisso, since the Navy expressed a strong interest in its manufacture. But no one anticipated profits, and, until 1943, they were quite right.[35] In fact, some employees recall synthetic-fuel operations as a "great waste" (*idai naru rōhi*).[36] The only justification for investment in hydrogenation was Noguchi's patriotism; as he put it, his concern was, "What next will be the enterprise most useful to the nation?"[37]

From June 1932 until March 1935, when Chōsen Coal Industries

(Chōsen Sekitan Kōgyō KK) was founded, Chōsen Chisso worked on synthetic fuel and related products. Land, plant, and machinery to set up a separate facility were donated by Chōsen Chisso in December 1937. Like other new companies, Chōsen Chisso was plagued with problems. Machinery frequently failed. High-quality hydrogen was difficult to obtain; Kudō Kōki, who had directed construction, retired; and Navy personnel assigned to the factory impeded effective work. On the positive side, the Bank of Chōsen underwrote the synthetic-fuel operations, and machinery problems led to successful research in high-pressure machinery, thereby significantly advancing Japan's technology in that area.[38] In December 1940, the company changed its name to Chōsen Artificial Oil (Chōsen Jinzō Sekiyu), reflecting its increasing emphasis on synthetic fuel as the new facility geared up for production.[39]

The synthetic-fuel plant combined the work of Navy and Nitchitsu researchers. The method initially employed Nitchitsu's technology for high-pressure manufacturing and the Navy's technology for catalysis.[40] The involvement of the Navy should have benefited the private company; in fact, it caused many problems. First, the military men brought into the plants at Chōsen Artificial Oil and Jilin Artificial Oil (Kichirin Jinzō Sekiyu, founded in 1939 in Manchuria) received no training for their positions. Some were retired military men who simply drew pay and got in the way of the researchers.[41] Second, a secretive Naval establishment stifled dissemination of ideas. A particularly absurd case occurred when Army and Kempeitai (military police) officials wished to view the plant. They were denied access pending Navy approval. Third, the Navy's concern with security led them to distrust Professor Ōshima Yoshikiyo, Noguchi's long-time advisor and head of the Chōsen Artificial Oil plant. The Navy had Ōshima supervised, thus compromising his ability to perform.[42]

The worst problem, however, was the rivalry between Navy and civilian managers within the firm over which products to make and which processes to use. This was particularly evident after 1940, when researchers at Chōsen Artificial Oil discovered a catalyst for the manufacture of synthetic fuel that worked better than the one they had been using at the Navy's suggestion. The method used at Chōsen Artificial Oil was considered to be an original Japanese method, although all hydrogenation methods were based on the pioneering work of the German chemist Friedrich Bergius, recipient of the Nobel Prize in 1931.[43]

The Navy temporarily backed away from daily supervision of the coal-liquifaction process for synthetic fuel when Chōsen Artificial Oil adopted its new catalyst. In 1941, the Navy authorities decided they were more interested in methanol production than in synthetic fuel and informed Chōsen Artificial Oil that they would have two years, until October 1943, to shift from oil to methanol manufacture.[44] The plant at Yong'an had been producing methanol successfully since 1933, so no one anticipated problems with the upcoming technological shift.

But an unfortunate event precipitated new Navy interference. In the summer of 1942, the gas-producing furnaces at Yong'an broke down. During the course of repairs, it was noted that the gas could be used to make methanol according to the Navy's technology, which differed from Chōsen Artificial Oil's production method. The Navy decided to take advantage of this opportunity to make methanol. Thus, a year before the originally planned shift to methanol, the Navy gave an order (*meirei*) to Yong'an to begin its production. Yong'an scientists wished to use their own successful methods; the Navy insisted on installing its own technology. The production equipment kept breaking down, despite the assistance of Navy researchers in setting it up. Finally, the Navy relented and permitted Chōsen Artificial Oil to continue producing synthetic oil while Navy technicians repaired the faulty methanol equipment. Then new problems arose. A particular heavy-duty heat coil was required for synthetic-oil manufacture, but the Navy authorized only an ordinary coil, which Chōsen researchers insisted would burn out within a week. The researchers were right. After much wrangling, synthetic-fuel production was started up again in July 1942, and for approximately one month Chōsen Artificial Oil made 100 tons of synthetic oil per day.[45] Then the Navy finished its repairs of the damaged machinery and ordered Chōsen to cease producing oil and again take up production of methanol by the Navy method, previous equipment accidents notwithstanding.[46]

The Navy presence within the plant became increasingly evident; controls tightened, Naval officers took executive positions, and civilians were warned to avoid opposing Navy researchers.[47] By the spring of 1944, the Navy had effectively eroded civilian control of production by legally changing the status of the Yong'an plant to a National-Policy Company.[48] The Navy was also extensively involved in fuel and other munitions production in Noguchi's plants in the Hungnam area. Military involvement imposed a great deal of secrecy on Nitchitsu operations. Several plants were known only by code names, and several forbade access even to Nitchitsu and Chōsen Chisso employees without special clearance. Most notable in this regard was Nitchitsu Fuel Industries (Nitchitsu Nenryō Kōgyō), founded in July 1941 to produce isooctane from synthetic butanol for use as airplane fuel for the Navy. Wholly owned by Nitchitsu but run in cooperation with the Navy, the "NA" plant (so-called because "N" stood for nitrogen, Noguchi, and Navy and "A" stood for airplane fuel) was extremely secretive. Set up to satisfy Navy needs, this plant, like the synthetic-oil plants, showed no profit until 1944.[49] And it was not until the day after the war ended in 1945, in anticipation of the surrender of these munitions plants, that military men relinquished control to civilians.[50]

In the end, Navy involvement in research and development of synthetic oil, methanol, and butanol was a burden to Nitchitsu. Only two factors ameliorated this otherwise unfortunate situation. First, Nitchitsu obtained loans more easily when it could show Navy backing for a particular product. The official Industrial Development Bank never hesitated to lend funds to Noguchi for projects bearing the seal of the Navy Ministry. One Nitchitsu accountant called this easy access to funds "ordered financing" (*meirei yūshi*).[51] Second, Chōsen Chisso's investment share in Chōsen Artificial Oil had decreased substantially by 1943, which may have diluted the impact of Navy involvement in research and development. In January 1941, Chōsen Chisso owned over 99 percent of all shares in Chōsen Artificial Oil. By September 1943, Imperial Fuel Development, Inc.

(Teikoku Nenryō Kōgyō KK) owned 50 percent of all shares. After 1939, Imperial Fuel began to replace Chōsen Chisso as a major lender to Chōsen Artificial Oil, and, by August 1941, Imperial Fuel had extended more loans than Chōsen Chisso.[52]

Chōsen's sister company in Manchuria, Kilim Artificial Oil, faced similar problems. Manchukuo Governor General Hoshino Naoki (later head of the Cabinet Planning Board), Assistant Director of the Industrial Section Kishi Nobusuke (later Vice-Minister of Munitions and a postwar Prime Minister), General Ueda Kenkichi of the Kwantung Army, and Army Chief of Staff Major General Itagaki Seishirō were among admirers of Noguchi's synthetic-fuel venture.[53] They had urged construction of a synthetic-oil plant in Manchuria. Most top managers at Nitchitsu opposed the move, but Ichikawa Seiji agreed with the Manchukuo government request, so in 1939 Nitchitsu built a synthetic-fuel plant at Jilin.[54] This plant was managed by Nitchitsu, although they held just 30 percent of the initial stock offering of 2,000,000 shares. (Imperial Fuel held 20 percent and the Minister of the Manchukuo Economic Section [Manshūkoku Keizaibu] held 50 percent.) By 1942, the military-dominated Manchukuo government and Imperial Fuel each held 35 percent, while Nitchitsu still held 30 percent, making Nitchitsu the minority owner. The following year, Nitchitsu relinquished management responsibilities to the South Manchurian Railway (Mantetsu), as problems with construction, technology, and expenses for related companies beset the firm and as Japan's southward advances brought oil supplies into Japanese reach in any case.[55] Jilin Artificial Oil was still under construction when the war ended.[56]

In sum, Nitchitsu's experience with production of synthetic oil was problematic for several reasons. Unlike glycerine and explosives, there was no immediate civilian market, and the munitions market for synthetic oil developed slowly. Moreover, because of the strategic nature of the product, the military and the military-dominated government of Manchukuo were involved in both investment and management of Nitchitsu's companies. This drawback probably outweighed the benefits achieved in improved access to capital. But, while starting production involved problems, Nitchitsu did invest in synthetic oil, so the factors for investment must be examined. Investment was promoted by the facilitated access to capital, strong encouragement from outside the company (that is, government pressure), and the relative ease of diversification due to the technological similarities in the production methods of ammonia and synthetic oil. The primary physical problems Nitchitsu faced initially were inadequate machinery and resources. The necessary coal was mined by another subsidiary of Nitchitsu, but its quality was low.[57] Only later did the plant acquire a higher grade of coal from Manchuria. The Japanese-made machinery was initially unreliable because neither of the machine-tool makers, Hitachi and Kōbe Steel (Kōbe Seikōjo), had much experience with high-pressure machinery.[58] Nevertheless, the world's third synthetic-fuel plant started producing in February 1938 (the others were at Leuna in Germany and Billingham in England). It eventually employed 3,000 factory workers and 6,000 coal miners.[59]

A remarkable contrast to Nitchitsu's experience with development of synthetic fuel was its investment in magnesium, another product with strategic uses. Noguchi was initially uninterested in magnesium refining. But Governor General Ugaki's policy of fostering the refining of metals in Korea, including copper, zinc, and iron ores, was aggressively sold to Noguchi by Ugaki's Chief of the Mining Bureau, Kōtaki Motoi.[60] Noguchi, with his large supply of electricity, was the perfect target for government planners.[61] But Noguchi saw neither purpose nor profit in refining metals. Refining magnesium was not as closely related to electrochemical production as some other processes, and Noguchi struggled to find any justification for the new demand. He requested a government subsidy to carry out the Mining Chief's demands. Although Kōtaki rejected this request, Noguchi eventually agreed to establish a refinery. Perhaps because Nitchitsu had been so hesitant, the government and military refrained from further pressure on the company over this issue.

Despite the strategic applications of magnesium, Nitchitsu was largely unencumbered by military involvement in daily operations. Furthermore, until the height of World War II, foreigners were extensively involved in the Nitchitsu-dominated magnesium company.

Founded in June 1934, Japan Magnesium Metals, Inc. (Nippon Maguneshiumu Kinzoku KK) was principally owned by Nitchitsu. But one-third of its shares were owned by the American Magnesium Metals Corporation, part of the Kaiser group, and small holdings of 100 shares were in the hands of individual foreigners.[62] Technology was developed by the Austrian Fritz Hansgierg who had ties to the company.[63] Management at Japan Magnesium was international; Noguchi was President, a certain Alfred E. Gordicke was Vice-President, and the directors included Shiraishi, Ōishi, and Kaneda as well as men named Konrad Erdmann and Bernard Moore.[64] Although the plant was capable of producing two-fifths of the world's supply of magnesium and was by 1938 the world's largest magnesium plant, it was not financially sound and faced repeated losses. Added losses were incurred when Nitchitsu, pleased with the initial successes of the magnesium production, attempted electrolytic steel refining in 1936.[65] In contrast to the synthetic-oil production, the military maintained their distance from the magnesium plant and must have tolerated the remarkably large involvement of foreigners, including enemy foreigners, into the World War II period. And Nitchitsu, the nationalism of its management notwithstanding, maintained its foreign ties because of the company's own needs for capital and technology, which apparently surpassed management's patriotism, at least until 1943.

EXPANSION OF CONSUMER PRODUCTS ON THE EVE OF WORLD WAR II

In addition to diversifying into munitions, Noguchi and his colleagues in Chōsen Chisso decided to expand fertilizer capacity in the mid-1930s. Hungnam had always produced fertilizers other than ammonium sulfate, but, in the mid-1930s, non-nitrogenous fertilizers like superphosphate of lime and compound fertilizers began to be produced in larger amounts. Around the same time, Noguchi noticed that a large part of Manchuria's soybean crop was being exported to Germany for use in chemical manufacture. Not content to permit a valuable crop to be exploited by scientists of another nation, Noguchi founded Soybean Chemical Industries, Inc. (Daizu Kagaku Kōgyō KK) at Pon'gung in April 1935. Access to soybeans was easy because of Japanese control of the state of Manchukuo.[66] The company made not only the ever-popular, and therefore profitable, food additives derived from soybeans but also worked out a method for synthesizing acetone, which was related to consumer products like carbide, fuel, and plastics, and also was particularly important in the production of isooctane for military use.[67] The method was developed under the research directorship of Tashiro Saburō from Chōsen Coal Industries, the coal liquifaction subsidiary.

The additional plant space at Pon'gung, about 4 kilometers from Hungnam, also permitted Noguchi to return to his first type of chemical production, carbides, calcium cyanamide, and related products, as well as ammonia by the Casale method.[68] According to his friends, it was natural that Noguchi would try to return to carbide products. His first achievements were in that area and, furthermore, calcium cyanamide was a good, if expensively produced, fertilizer. He sent a man who had been with him since the time of his earliest work thirty years before, Shimada Shikamitsu, to help set up the cyanamide works.[69] The Pon'gung plant was merged with Chōsen Chisso in June 1936 and then sold to Nitchitsu in May 1937.[70]

By 1936, Nitchitsu's main Korean subsidiary was capitalized at 70 million yen and was producing more than 400,000 tons of ammonium sulfate, 40,000 tons of superphosphate, 46,000 tons of calcium cyanamide, and 360,000 tons of other fertilizers per year.[71] After the merger, no other Japanese chemical firm could approach Chōsen Chisso's output, and only two other firms in the world surpassed it. By 1937, even the newly expanded Chōsen Chisso plants were producing at their capacity. The outbreak of the China War in that year meant that domestic ammonia producers would soon be required to sell some of their product for manufacture of explosives rather than fertilizer. The Korean company supplied the unfilled demand for fertilizers in Japan. This benefited Chōsen Chisso greatly. Indeed, despite the extensive diversification already completed by mid-decade, fertilizers continued to supply most of the profits (and income) of Chōsen Chisso. In 1931, fertilizers accounted for 97 percent of Chōsen Chisso's income; by 1937 they had dropped only to 70 percent, and by 1941, to 60 percent. Selling prices for fertilizers were consistently above production costs, although production costs began to rise continuously after 1936, when many of Chōsen Chisso's tax reductions ended.[72]

Mid-decade was not, however, a period of unfettered growth for Chōsen Chisso and other fertilizer producers. As noted above, producers had made repeated attempts to control sales through cartels and other organizations during the late 1920s and early 1930s. By 1936, government controls tightened. In November 1936, the Important Fertilizer Industry Control Law (*Juyo hiryōgyō tōseihō*) was passed. This law required producers to form a Control Association, a producers' cartel organized as a corporation, empowered to set production quotas and prices. Twelve companies formed this cartel, and Mori Nobuteru of Shōwa Fertilizer acted as administrative head.[73] The Chemical Fertilizer Control Association had 56 member companies in 1942, and was joined by 6 other Control Associations in the chemical industry. (These were the Nitric Acid Control Association, with 20 members; Sulfuric Acid, with 57 members; Soda, with 25; Carbide, with 17; Organic Synthetics, with 16; and Coal Distillation, with 40 members.)[74] Import-export controls on ammonium sulfate were imposed in March 1937, and, in October, the Ministry of Commerce and Industry required that manufacturers report continuing production of nitric acid through the Temporary Measures Concerning Import and Export Quality Law (*Yūshutsu-nyū hintō ni kansuru rinji sōchi ni kansuru hōritsu*). In sum, Chōsen Chisso continued to produce a large amount of fertilizer profitably, but production and distribution decisions were affected significantly by exogenous factors like government controls.

THE YALU RIVER PROJECT

The Yalu River, which separated Japan's colony, Korea, from its dependency, Manchukuo, was still undammed in the mid-1930s. But it had become the object of great fascination for bureaucrats, military men, and businessmen. In September 1936, shortly before his scheduled return to Japan to try to form a Cabinet, Governor General Ugaki journeyed to his vacation home at the Changjin. Ugaki's guest was General Koiso Kuniaki, head of the Chōsen Army and former Chief of Staff of the Kwantung Army in Manchukuo. In the shadow of Noguchi's monumental development at the Changjin, the two generals discussed electrification of the Yalu with Kubota Yutaka, who was in charge of the Changjin project while Noguchi was in Ōsaka. Kubota apparently interested them in developing the Yalu, but the two had a few reservations. They worried that, since the project was "international"—albeit between two Japanese holdings—problems would arise. Furthermore, Manchukuo was run by military men hostile to private capital. How would private investors fare in relating to Manchukuo officials? Koiso knew the men in Manchukuo well because of his former posting there, so he believed he could work with them, but he certainly had some concerns. In addition, it was not clear which authorities, the Japanese in Korea or in Manchuria, controlled the water-development rights to the river they shared.

At the same time, two prominent Japanese civilians in Manchukuo, Abe Kōryō and Shiina Etsusaburō, visited the Changjin project. Impressed, they decided Noguchi should develop the Yalu. Abe was a civil engineer with no academic credentials who owed his rise to prominence to his having been a highly favored advisor to Taiwan Governor General Gotō Shinpei. In 1936, he was head of the Manchurian Industrial Research Association (Manshū Sangyō Chōsakai) and known as a *rōnin*. Shiina Etsusaburō was a noted authority on economic planning and head of the Manchurian Mining and Industry Section. After the trip by the two men to the Changjin, Abe traveled to Japan where he approached Noguchi while the latter was vacationing in Beppu. Abe implored Noguchi to develop the Yalu.

When he said, "You're the only one who can develop [the Yalu]," Noguchi told him to see Kubota.[75] Twenty years earlier, Kubota Yutaka had written his senior thesis on the harnessing of hydroelectric power on the Yalu. Kubota was thrilled to begin planning electrification of the Yalu, but worried that Noguchi appeared initially less enthusiastic. His report to Noguchi was, therefore, conservative. He noted that the Yalu was capable of generating 50,000 kilowatts, although he was privately aware that a much larger project was possible. Noguchi claimed astonishment that Kubota would bother him with a report for so minor a project; Kubota explained the reason for his conservative estimate and revised his projections upward to a more realistic 300,000 kilowatts.[76]

Kubota next had to persuade Manchukuo military officials of the importance of developing the Yalu. Unlike the civilians Abe and Shiina, Kwantung Army officers were ideologically opposed to private capital, as Kubota's discussions with Lt. Col. Akinaga Tsukimi of the Kwantung Army's Economic Supply Staff indicate. (Later, as a lieutenant general, Akinaga headed the Cabinet Planning Board.) Akinaga made an inspection tour of the Changjin facility soon after Noguchi told Kubota he could continue with his plans for developing the Yalu. Akinaga was to study the Changjin as a model for the Kwantung Army's planned project on the Songhwa River. Kubota gave Akinaga a personal tour. Akinaga liked what he saw and asked Kubota about the source of investment capital for the project. He did not hide his belief that all capital was somehow squeezed from the government and that Kubota must have been very fortunate to receive so much from the Government General. Kubota countered with a simple Adam Smith-style analysis of capitalism, saying that investors invested because they trusted entrepreneurs, who also sold bonds and received bank loans. Akinaga was not swayed by Kubota's arguments in favor of capitalism, but he liked him and believed Kubota and Noguchi to be unusually able men.[77]

Akinaga's support helped Kubota. When the Kwantung Army needed expert analysis of their planned Songhwa project, they called in Kubota. He was not informed in advance of the agenda the Manchukuo authorities wished to address at that meeting, but he guessed correctly and gave a skilled performance, discussing the merits of various dam sites and the ability of the Manchukuo authorities to manage such a project. The Chief and the Assistant Chief of Staff of the Kwantung Army, Generals Itagaki Seishirō and Imamura Hitoshi, as well as government officials and colleagues of Abe, attended Kubota's presentation, after which General Itagaki held a banquet in Kubota's honor. Kubota took the opportunity to request a private meeting with Itagaki to discuss the Yalu project; this had not been addressed at the earlier meeting. The two projects were fundamentally different—Songhwa was a Manchukuo government project, and Kubota's role at that time was merely to do a feasibility study, while the Yalu project would involve investment of private capital, which the Manchukuo authorities generally opposed. Kubota could not predict Itagaki's response.

Itagaki listened attentively and promised to study Kubota's proposal. Because the Yalu flowed between Manchuria and Korea, Itagaki's approval was necessary. By that same evening, Itagaki sent a positive response via Imamura. The official approval from the civilian administrators in Manchukuo, Hoshino and Shiina, was then a foregone conclusion. On the Korean side, Ugaki had already left his post as Governor General, but his successor, Minami Jirō, had not yet taken up residence in Korea. Thus, when Kubota returned to Ōsaka to report his progress to Noguchi, he continued on to Tokyo so that Ugaki might introduce him to Minami.[78]

Through adroit negotiation with government officials, Kubota, who like other salaried managers was taking on greater responsibility for the firm's external relations as Noguchi aged, was able to secure government support for Nitchitsu's investment in capital. This was particularly important in the case of the Yalu River. The project was "international," so approval had to be received from both sets of officials. It was a massive project, and the high costs meant that extensive loans would be required; government approval could help secure those loans. The government's attitude was significant for another reason. Approval by the Manchukuo authorities marked a transition for Nitchitsu. Before 1937, exogenous factors—most important, government approval and aid—had helped influence Nitchitsu's investment decisions. But only in the case of synthetic oil had government encouragement swayed Nitchitsu to invest in production when the conditions for investment dictated by company strategy had not favored it. After 1937, however, most of Nitchitsu's investments were undertaken at the request of military authorities. Many of the products in which Nitchitsu invested did not have a close technological relationship with products already manufactured at Nitchitsu, and none was profitable. The Yalu project may be seen as transitional in that its Korean side, while contributing to Korean industrial development and benefiting from government loans, was essentially initiated to advance company goals, whereas its Manchukuo side was initiated to advance public goals and was controlled and financed by the state.[79] Furthermore, Nitchitsu interests were permitted to use one-third of the electricity transmitted to Korea, that being 50 percent of the electricity generated. (The other 50 percent went to Manchuria.) This meant that Nitchitsu's direct benefits from investment in generation of electricity were decreasing: The company retained 100 percent of the electricity from the Pujon; 50 percent from the Changjin; 33.3 percent from the Hochon; and 16.6 percent from the Yalu.[80]

Dealing with the Manchukuo government was not initially easy. Disputes over the location of the headquarters of the Yalu River generating facility unsettled a January 1937 meeting attended by representatives of the Manchukuo government, the Kwantung Army, the Korean Government General, and Noguchi and Kubota. The location of the main office would determine whether the company was controlled by the state or by private investors. Noguchi, comparing the squabble to two parents fighting about their child's surname and thereby preventing his registration in elementary school, proposed establishing two companies. These were to be Chōsen Yalu River Hydroelectric and Manchurian Yalu River Hydroelectric (Chōsen and Manshū Ōryokō Suiryoku Hatsuden).

The two companies would be twins in almost every way. Both were founded on 18 August 1937; both had the same top managerial staff made up of men from Nitchitsu, the Manchukuo government, and the Oriental Development Company (another investor); and both were capitalized at 50,000,000 yen.[81] The sources of financing were different, but both hydroelectric companies were considered by Nitchitsu management to be subsidiaries of Chōsen Electricity Transmission and Changjin Hydroelectric.[82] The meaning of "subsidiary" was clearly different in each case, as the Manchukuo government's authority was far stronger in Manchurian Hydroelectric. For example, Chōsen Chisso managers were required to work on the Kwantung Army's project on the Songhwa River if they wished to receive electricity from the Yalu.[83]

Investment was different on each side of the river. On the Manchurian side, the Manchurian Development Bank (Manshū Kōgyō Ginkō) funded, through loans, 70 percent of the capitalization of Manchurian Hydroelectric. Changjin Hydroelectric funded 20 percent and Chōsen Electricity Transmission, 10 percent, in the last two cases through purchase of shares. On the Korean side, Changjin Hydroelectric's purchase of shares contributed 20 percent of the capital, with Chōsen Electricity Transmission funding 10 percent and the Oriental Development Company funding 20 percent. This dismayed Chōsen Chisso's management, who realized that Oriental Development would be entitled to a management position as a direct investor. Furthermore, Oriental Development had a special relationship with the Government General of Korea, one that was even closer than Chōsen Chisso's; Chōsen Chisso managers were concerned that such close ties might further limit managerial options in decision making.[84] Later, Changjin bought Oriental Development's shares, and the latter company lost its auditor's seat on the Yalu companies' boards of directors. The other 50 percent of the initial capital came from the Bank of Chōsen. (Later the Industrial Development Bank of Japan also funded the Yalu project.) Through Ugaki's connection, Noguchi and Kubota had earlier been introduced to the new President of the Bank of Chōsen, Matsuhara. He gave Changjin Hydroelectric a 25,000,000-yen loan, unsecured, to be used for the Korean Yalu project. In return, the Bank received shares of the Korean Yalu company.

Investment was not the only thing handled differently on opposite sides of the river. According to Kubota's reminiscences, Changjin Hydroelectric "represented" the Korean side, while the state of

Manchukuo "represented" the Manchurian side.[85] That is, the company was viewed as private in Korea and as a state enterprise in Manchukuo. Construction began in October 1937, soon after the companies were founded and financed.[86] To construct the world's second largest dam (after the Grand Coulee) advanced technology was necessary. Railroad lines had to be built to bring materials to the dam site, located about 100 miles from the mouth of the Yalu River at Pyonganpukto.[87] The terrain was wild and hilly, and complicated tunneling was required. Kubota contracted with three construction companies to help build the railroads, which would have to transport a variety of materials and machinery, including the world's largest crane.[88]

As the China War of 1937 had already begun, a sense of urgency impelled the builders, who extended the railway deep into Manchuria. But, although work moved quickly, the railroad could not be used effectively. Foreign-exchange controls in Japan and the empire prevented Noguchi from buying the Czech engines and rolling stock he needed. He circumvented government controls with the help of the government authorities in Manchukuo. The Assistant Chief of Manchukuo's Industrial Division, Iwao Nobusuke, sold off a quantity of Manchurian soybeans, which Noguchi marketed through Mitsubishi Trading to earn foreign exchange.[89]

The large Yalu project required an extraordinary amount of cement; 800,000 tons were needed at the main dam site alone. Onoda Cement Company was contracted to build a cement plant with an annual capacity of 180,000 tons.[90] In addition to the high costs of cement, costs of railroad construction began to exceed expectations.[91] Machinery costs also mounted. The sophisticated machinery needed to generate electricity at the Yalu project had never been built in Japan for civilian use, so Noguchi had ordered machinery from Siemens. The outbreak of World War II in Europe prevented its delivery. Kubota turned to the Vice-Minister of the Navy for assistance in acquiring the Navy's methods for forging steel for special machinery; special steel was then forged for him by Nippon Steel (Nippon Seikō). The machinery was made by Tōshiba.[92]

A particularly thorny problem involved the moving of 100,000 residents whose homes would be destroyed by the new dam. This included Korean farmers and salt workers and Japanese officials, teachers, and other white-collar workers. Farmers were promised land higher in the hills, but, of course, this caused extreme hardship as families were forced off ancestral lands. Salt workers were given new jobs. The Japanese settlers were posted to new positions. All were financially compensated. The moving of people not only raised humanitarian issues; it also added to the company's financial concerns.[93]

The company's financial problems were addressed by the Industrial Development Bank, where Kubota had two key contacts: Vice-President Kawakami Hiroichi and Securities Department Head Kurusu Hisao. These men put together a syndicate of large banks like Mitsui, Mitsubishi, Yasuda, Sumitomo, Sanwa, and Daiichi, with the Industrial Development and Chōsen Banks as managers of the consortium, to lend money to the Yalu project. The syndicate members were initially somewhat wary of lending. It was uncommon in Japan to issue bonds for construction costs; bonds were usually used to cover operating costs, with the finished project as security. At the suggestion of Abe Kōryō, Kubota invited the elderly top executives of the syndicate banks to a screening of a movie he had been making on the construction of the Yalu project, an unprecedented use of film "advertising." The film was persuasive, and the syndicate responded with a loan of 100,000,000 yen.[94]

By the summer of 1941, construction was completed. Noguchi Jun was becoming increasingly incapacitated, so Kubota stood in his place at the opening ceremony, attended by Japanese dignitaries from Korea and Manchukuo, for the first stage of the Yalu project on 26 August 1941. The huge dam was the first of three stages, all completed by June 1942. The three together generated 600,000 kilowatts, or 38 percent of the hydroelectric power generated in Korea.[95] Since Noguchi Jun was gradually removing himself from daily operations, the Yalu project brought the 47-year-old Kubota to the attention of the Japanese military authorities in Northeast Asia. Soon he would be asked to lead Noguchi's company expansion throughout East and

Southeast Asia in support of Japan's war effort. The Yalu project was indeed a turning point in company autonomy.

NITCHITSU AND STRATEGIC EXPANSION, 1937–1945

Japan's escalation of the China conflict after 1937 affected management decisions at Nitchitsu and its subsidiaries. First, military purchases caused Nitchitsu to produce munitions in increasing amounts. Second, greater production of munitions caused Nitchitsu to reduce or suspend production of certain consumer products, either because management found it lucrative to do so or because civilian or military authorities required a shift to munitions. The best example was the closing down of rayon production at Asahi Bemberg after the last half of 1938. No priority for materials or financing was given to consumer production, so Nitchitsu planned to merge Asahi Bemberg with the nearby Nitchitsu Explosives, a company that required the same raw materials as the rayon plant and was rapidly growing. The merger occurred in April 1943 when the three foreigners among Asahi Bemberg's managers left Japan.[96]

A third effect of Japan's military escalation was the geographic expansion of Nitchitsu; the company literally followed the Army and Navy as they extended Japanese control in East and Southeast Asia. This dispersal of company administration permitted local military authorities to exert greater influence over local company affairs. Fourth, Nitchitsu had earlier begun to manufacture other strategic products in addition to chemicals, the company's sole products during its first three decades, but after 1937 a large variety of technologically unrelated products were added to Nitchitsu's repertoire. Previously, diversification was undertaken to benefit the company, and each case of modification of existing processes or products was judged in terms of its profitability; after 1937, however, the military helped make the decisions to invest and based its demands on strategic needs. Thus, the widening conflict in East Asia enabled the military to influence investment decisions.

The decision to invest in production, then, was principally affected by what we have called exogenous factors. Other usual considerations, such as access to capital and resources, of course continued to be important factors in decision making. But, because capital and resources were more readily and cheaply obtained with government help, discussion about their availability was somewhat muted. Profitabilty as an investment motive receded when government orders and assistance increased.

Not only were management decisions affected, but also the structure of management itself. For example, one result of Japan's widening war—geographic expansion of the company—produced a diffusion of decision making. Several mechanisms were built into the company's structure which minimized the effects of this diffusion, in particular interlocking management (see Table 37), ownership by various subsidiaries of other subsidiaries' shares, and Nitchitsu's own ownership of subsidiaries' shares. Nitchitsu's ownership of shares of subsidiaries, which increased as its subsidiaries proliferated, had an additional effect. That is, the percentage of securities among Nitchitsu's holdings increased because the value of Nitchitsu's fixed assets grew more slowly.

A last factor, completely unrelated to exogenous events, also affected management during the 1937–1945 period. Noguchi Jun, the founder and leader whose personality dominated management, suffered a cerebral hemorrhage in February 1940 and died in January 1944. Noguchi had integrated his Korean and Japanese operations merely by spending one week each month in Korea.[97] That procedure had ended. Nitchitsu's management was therefore quite different during the war years from the previous three decades.

Nitchitsu expanded throughout East Asia during the war. Company growth appeared greater in the empire than in Japan proper because of the shutdown of rayon production, which had accounted for a significant percentage of both investment and product in the Japanese part of the conglomerate. But the whole enterprise grew. In March 1937, Nitchitsu was made up of 27 companies whose total capital was 290,000,000 yen. By April 1941, these figures had increased to 64 companies and 899,000,000 yen. By December 1945, mergers

TABLE 37 Management at Nitchitsu Subsidiaries, 20 April 1940

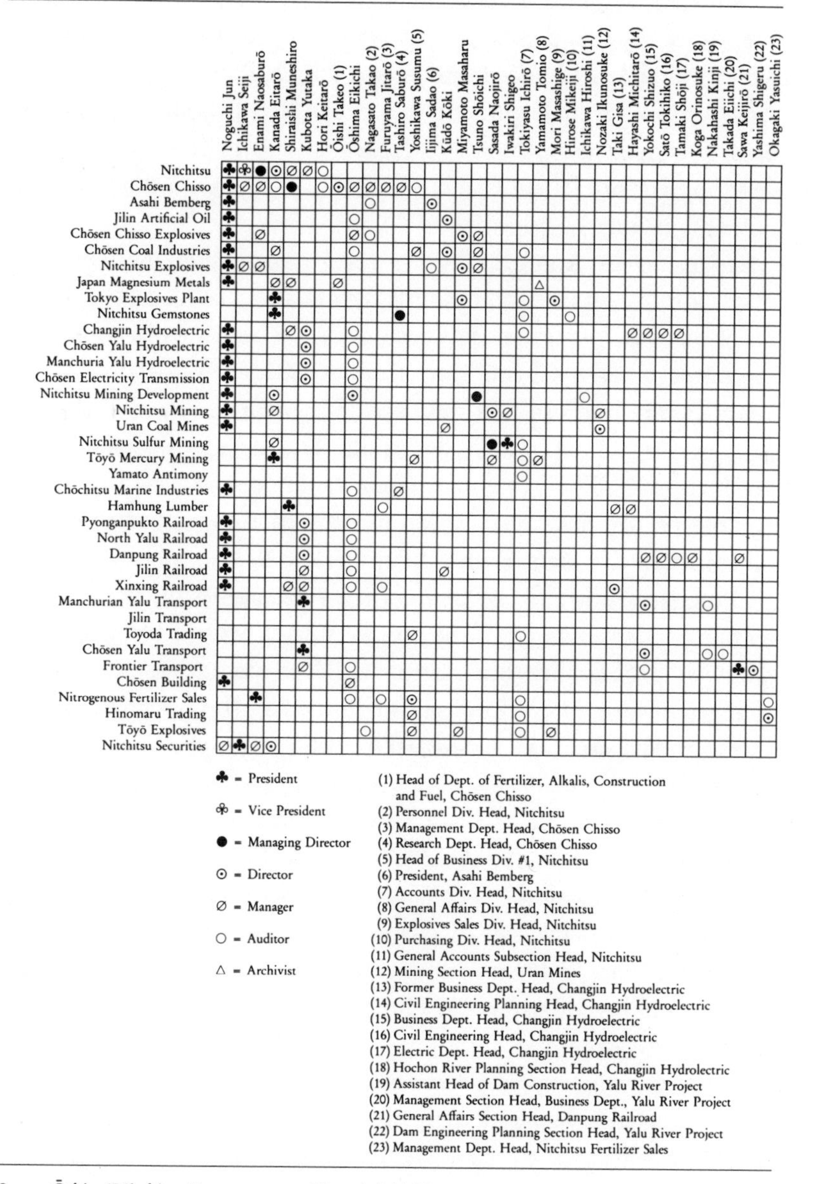

	Noguchi Jun	Ichikawa Seiji	Enami Naosaburō	Kanada Eitarō	Shiraishi Muneshiro	Kubota Yutaka	Hori Keitarō	Ōishi Takeo (1)	Ōshima Eikichi	Nagasato Takao (2)	Furuyama Jitarō (3)	Tashiro Saburō (4)	Yoshikawa Susumu (5)	Iijima Sadao (6)	Kūdō Kōki	Miyamoto Masaharu	Tsuno Shōichi	Sasada Naojirō	Iwakiri Shigeo	Tokiyasu Ichirō (7)	Yamamoto Tomio (8)	Mori Masashige (9)	Hirose Mikeiji (10)	Ichikawa Hiroshi (11)	Nozaki Ikunosuke (12)	Taki Gisa (13)	Hayashi Michitarō (14)	Yokochi Shizuo (15)	Satō Tokihiko (16)	Tamaki Shōji (17)	Koga Orinosuke (18)	Nakahashi Kinji (19)	Takada Eiichi (20)	Sawa Keijirō (21)	Yashima Shigeru (22)	Okagaki Yasuichi (23)
Nitchitsu	♣	♧	●	⊙	∅	∅	○																													
Chōsen Chisso	♣	∅	∅	○	●		○	⊙	∅	∅	∅	∅	○																							
Asahi Bemberg	♣									○				⊙																							
Jilin Artificial Oil	♣								○							⊙																						
Chōsen Chisso Explosives	♣		∅						∅	○						⊙	∅																				
Chōsen Coal Industries	♣			∅					○				∅		⊙		∅			○																	
Nitchitsu Explosives	♣	∅	∅											○		⊙	∅																				
Japan Magnesium Metals	♣			∅	∅			∅													△																
Tokyo Explosives Plant				♣													⊙					○		⊙															
Nitchitsu Gemstones				♣								●								○			○														
Changjin Hydroelectric	♣				∅	⊙			○											○									∅	∅	∅	∅						
Chōsen Yalu Hydroelectric	♣					⊙			○																												
Manchuria Yalu Hydroelectric	♣					⊙			○																												
Chōsen Electricity Transmission	♣					⊙			○																												
Nitchitsu Mining Development	♣			⊙					⊙								●								○													
Nitchitsu Mining	♣			∅															⊙	∅						∅												
Uran Coal Mines	♣														∅										⊙												
Nitchitsu Sulfur Mining				∅														●	♣	○																	
Tōyō Mercury Mining				♣									∅					∅		○	∅																
Yamato Antimony																				○																	
Chōchitsu Marine Industries	♣								○			∅																									
Hamhung Lumber					♣						○																∅	∅									
Pyonganpukto Railroad	♣					⊙			○																												
North Yalu Railroad	♣					⊙			○																												
Danpung Railroad	♣					⊙			○																		∅	∅	○	∅			∅		
Jilin Railroad	♣					∅			○						∅																				
Xinxing Railroad	♣				∅	∅			○		○																⊙									
Manchurian Yalu Transport						♣																						⊙				○				
Jilin Transport																																				
Toyoda Trading														∅						○																	
Chōsen Yalu Transport						♣																						⊙				○	○			
Frontier Transport						∅			○																			○						♣	⊙	
Chōsen Building	♣								∅																										
Nitrogenous Fertilizer Sales			♣						○		○		⊙								○																○
Hinomaru Trading														∅						○																⊙
Tōyō Explosives											○			∅			∅				○		∅													
Nitchitsu Securities	∅	♣	∅	⊙																																

♣ = President

♧ = Vice President

● = Managing Director

⊙ = Director

∅ = Manager

○ = Auditor

△ = Archivist

(1) Head of Dept. of Fertilizer, Alkalis, Construction and Fuel, Chōsen Chisso
(2) Personnel Div. Head, Nitchitsu
(3) Management Dept. Head, Chōsen Chisso
(4) Research Dept. Head, Chōsen Chisso
(5) Head of Business Div. #1, Nitchitsu
(6) President, Asahi Bemberg
(7) Accounts Div. Head, Nitchitsu
(8) General Affairs Div. Head, Nitchitsu
(9) Explosives Sales Div. Head, Nitchitsu
(10) Purchasing Div. Head, Nitchitsu
(11) General Accounts Subsection Head, Nitchitsu
(12) Mining Section Head, Uran Mines
(13) Former Business Dept. Head, Changjin Hydroelectric
(14) Civil Engineering Planning Head, Changjin Hydroelectric
(15) Business Dept. Head, Changjin Hydroelectric
(16) Civil Engineering Head, Changjin Hydroelectric
(17) Electric Dept. Head, Changjin Hydroelectric
(18) Hochon River Planning Section Head, Changjin Hydrolectric
(19) Assistant Head of Dam Construction, Yalu River Project
(20) Management Section Head, Business Dept., Yalu River Project
(21) General Affairs Section Head, Danpung Railroad
(22) Dam Engineering Planning Section Head, Yalu River Project
(23) Management Dept. Head, Nitchitsu Fertilizer Sales

Source: Ōshio, "Nitchitsu Kontsuerun no kigyō haichi," between pp. 80–81.

had brought the number of firms down to 44, but capitalization had grown to 1.6 billion yen.[98] Hungnam had increased its work force and its output, and new facilities and subsidiaries were added in Korea. Nitchitsu advanced into North China, Taiwan, and Southeast Asia. Ventures were begun in Sumatra and briefly considered in Vietnam. Of great interest both to Nitchitsu and to the Navy was the development of Hainan Island. We shall consider each of these developments in turn.

Expansion in Korea took two forms: increasing investment in plants already on line and building new facilities to fill new orders or carry out new processes. Between July 1939 and July 1941, invested capital at Chōsen Chisso grew from 62,500,000 yen to 77,500,000 yen.[99] Ammonium sulfate output at Hungnam increased by 145,000 tons in 1941 alone—which was about 11 percent of total Japanese production.[100] The Hungnam plant assumed a central role in the company's organizational structure; this role continued even after Chōsen Chisso's merger with Nitchitsu on 26 December 1941.

Hungnam continued to be important despite its disappearance as an independent company. The merger with Nitchitsu had nothing to do with profitability but rather with the company's response to changes in tax policy. To raise wartime taxes, the Finance Ministry investigated past tax records. The Ministry discovered Chōsen Chisso's practice of paying commissions to Nitchitsu for marketing and acquisition of materials—which lowered Chōsen Chisso's taxable profits—in lieu of paying dividends, and declared that Nitchitsu had attempted tax evasion. The merger was permitted by the Finance Ministry in 1941 in place of payment of back taxes.[101]

One limiting problem at Hungnam was labor. By 1940, Koreans had outnumbered Japanese at Hungnam. Military conscription rates for Japanese employees were high and continued to rise throughout the war, reaching almost 40 percent by 1945.[102] In the last days of the war, Hungnam managed to retain a work force of approximately 45,000 workers. But this included at least 2,059 students; 2,029 Korean Industrial Patriotic Association members, many of whose rural compatriots died while being marched to Hungnam; 1,027 convicts; 350 English and Australian prisoners of war; and 5,000 unskilled laborers and 1,000 soldiers used in tunnel construction after July 1945.[103]

Other Korean facilities expanded as well (see Table 38).[104] Chōsen Chisso Explosives doubled its capitalization from 10,000,000 yen to 20,000,000 yen between 1941 and 1943, and its profits tripled during the same period. Chōsen Artificial Oil (Chōsen Coal) showed no profits until the period between October 1943 and March 1944, but its capital increased 9-fold in the first months of 1943. Changjin Hydroelectric, consistently profitable throughout the war, showed remarkable growth in capital from 5,000,000 yen in 1935 to 70,000,000 yen in 1938, 150,000,000 yen in 1940, and 201,000,000 yen in 1942.[105] Changjin's immense size permitted it to function much as Chōsen Chisso did in relation to Nitchitsu; that is, it held shares of other subsidiaries and was able to help transfer funds within the Nitchitsu group to maximize profitability. Nitchitsu's loans to Changjin were rolled over into shares on several occasions, permitting easy increases in capitalization. Each of these subsidiaries contributed to strategic requirements, and military orders encouraged their expansion.[106] Explosives, synthetic fuel, and the varied products of the Hungnam plant had obvious military uses, but what of an electrical generating company? In addition to the production requirements at Nitchitsu-related plants in Korea, Changjin Hydroelectric (including both Yalu companies) filled energy needs for the Manchukuo government and military as well as private enterprises in Korea producing magnesium, aluminum, steel, zinc, and lead.[107]

Several new companies emerged in Korea in addition to the expanding "older" firms. Nitchitsu Gemstones (Nitchitsu Hōseki) was created in July 1936 on the advice of Fritz Hansgierg, the Austrian consultant to Japan Magnesium Metals. The gems produced there were a major component in military-use ball bearings.[108] Chōsen Marine Products (Chōsen Suisan Kōgyō), wholly owned by Chōsen Chisso, was established in July 1937 (capitalization: 1,000,000 yen) to produce sardine oil, the raw material for explosives. By 1941, the sardine catch decreased dramatically despite the need for oil, but the company's losses were relatively small.[109]

Another area in which Noguchi had long been interested was

TABLE 38 Nitchitsu Structure, January 1942

President
Vice President
Managing Director
Directors

- I. General Affairs Division
  - A. Archives Section
    - 1. Legislation Subsection
    - 2. Stocks Subsection
  - B. General Affairs Section
    - 1. General Affairs Subsection
    - 2. Patent Subsection
    - 3. Communications Subsection
- II. Personnel Division
  - A. Employees Section
  - B. Labor Section
- III. Accounting Division
  - A. Accounting Section
    - 1. First Accounting Subsection
    - 2. Second Accounting Subsection
    - 3. Third Accounting Subsection
    - 4. Financial Affairs Subsection
    - 5. Securities Subsection
- IV. Business Division #1
  - A. Fertilizer Section
  - B. Pharmaceutical Section
- V. Business Division #2
  - A. Fats, Oils, and Foods Section
  - B. Metals and Materials Section
- VI. Business Division #3
  - A. Explosives Section
  - B. Tokyo Office Section
- VII. Construction Materials Division
  - A. Business Section
  - B. Machinery Section
  - C. Materials Section
- VIII. Raw Materials Division
  - A. Raw Materials Section
    - 1. Supply Subsection

Table 38 *(continued)*

- IX. Control Division
  - A. Control Section
  - B. Data Section
- X. Southern Division
  - A. Southern Headquarters
    - 1. Business Department
    - 2. Planning Department
  - B. South Seas Business Office
    - 1. General Affairs Department
      - a. General Affairs
      - b. Labor
      - c. Resources
      - d. Finances
      - e. Hospital
      - f, g, h. Three branch offices
    - 2. Construction Department
    - 3. Harbor Department
    - 4. Mining Department
    - 5. Engineering Department
- XI. Minamata Plant
  - A. Business Section
    - 1. General Affairs Subsection
    - 2. Research Subsection
    - 3. Personnel Subsection
    - 4. Treasurer Subsection
    - 5. Accounting Subsection
    - 6. Purchasing Subsection
    - 7. Warehouse Subsection
    - 8. Shipping Subsection
    - 9. Security Maintenance Subsection
    - 10. Insurance Subsection
  - B. Inorganic Products Section
    - 1. Carbide Subsection
    - 2. Nitrogen Subsection
    - 3. Electrolysis Subsection
    - 4. Synthesis Subsection
    - 5. Sulfuric Acid Subsection
    - 6. Ammonium Sulfate Subsection
    - 7. Nitric Acid Subsection

TABLE 38 *(continued)*

- C. Organic Products Section
    1. Acetic Acid Subsection
    2. Anhydrous Acetic Acid Subsection
    3. Acetone Subsection
    4. Cellulose Acetate Subsection
    5. Plastics Subsection
    6. Nipolite Subsection
    7. Minarane Subsection
    8. Analysis Subsection
- D. Machinery Section
    1. Planning Subsection
    2. Construction Subsection
- E. Electricity Section
    1. Planning Subsection
    2. Transmission Subsection
    - 3–9. Electricity generating facilities listed, each a separate subsection.
    - 10. Electricity Transmission Lines Subsection

XII. Seoul Branch Office
- A. General Affairs Department
- B. Business Department

XIII. Tokyo Business Office
- A. Business Section

XIV. "Chissolite" Research Laboratory

XV. Hungnam Plant
- A. Office of the Head of the Plant
- B. General Affairs Department
    1. General Affairs Section
    2. Accounting Section
    3. Supplies Section
    4. Security Guards Section
- C. Welfare Department
    1. Labor Section
    2. Training Section
    3. Company Housing Section
    4. Supply Office
    5. Hospital
- D. First Production Department
    1. Gas Section
        - a. Electrolysis Subsection

TABLE 38 *(continued)*

b. Hydrogen Subsection
c. Nitrogen Subsection
d. Gas Subsection
e. Synthesis Subsection
f. Catalysts Subsection
g. Coke Subsection
2. Sulfuric Acid Section (5 subsections)
3. Ammonium Sulfate Section
a,b. First and Second Subsections
4. Ammonium Phosphate Section
a, b. First and Second Phosphatic Acid Subsections
c. Ammonium Phosphate Subsection
d. Superphosphate of Lime Subsection
e. Second Packaging Subsection
5. Fats and Oils Section
a. Hardened Oil Subsection
b. Glycerine Subsection
c. Fatty Acids Subsection
d. Soap Subsection
e. Third Packaging Subsection
E. Second Production Department
1. Aluminum Section (7 subsections)
2. Zinc Section (2 subsections)
3. Carbon Section (6 subsections)
4. Steel Section (2 subsections)
F. Third Production Department
1. Soda Section
a. Soda Subsection
b. Chlorine Subsection
c. Ammonium Choride Subsection
d. Seasonings Subsection
2. Ammonia Section
a. Pon'gung Electrolysis Subsection
b. Pon'gung Hydrogen Subsection
c. Pon'gung Synthesis Subsection
3. Carbide Section
(9 subsections, including calcium cyanamide, acetylene, "Chissorundum")
4. Organic Synthesis Section
a. Methanol Subsection

TABLE 38 *(continued)*

- b. Butanol Subsection
- c. Acetone Subsection
- d. Glycol Subsection
- 5. Steam Section (5 subsections)
- G. Research Department (3 sections)
- H. Construction Department
  - 1. Planning Section
  - 2. Engineering Section (8 subsections)
  - 3. Pon'gung Planning Section
  - 4. Pon'gung Engineering Section (5 subsections)
  - 5. Architecture Section (3 subsections)
  - 6. Civil Engineering Section (3 subsections)
- I. Electricity Department
  - 1. Business Section
    - a. Electric Lighting Subsection
    - b. Telephone Subsection
  - 2. Electricity Section (6 subsections concerned with distribution of electricity)
  - 3. Pon'gung Electricity Section (5 subsections)
- J. Generation of Electricity
  - 1. Generation Section (7 subsections)

XVI. Yong'an Plant
- A. Business Section
  - 1. General Affairs Subsection
  - 2. Warehouse Subsection
  - 3. Purchasing Subsection
  - 4. Accounting Subsection
- B. Hospital
- C. Supply Office
- D. Production Section
  - 1. Hydrogenation Subsection
  - 2. Paraffin Subsection
  - 3. Methanol Subsection
  - 4. Pharmaceutical Subsection
  - 5. Resins Subsection
  - 6. Testing Subsection
- E. Engineering Section
  - 1. Planning Subsection
  - 2. Engineering Subsection
  - 3. Electricity Subsection
  - 4. Civil Engineering Subsection

TABLE 38 *(continued)*

XVII. Kiyomizu Plant
- A. Business Section (4 subsections, including security guards)
- B. Engineering Section (7 subsections, including coal, electricity, acetylene)
- C. Hatoiwa Mine

XVIII. Mining Division
- A. Mining Section
  1. Business Subsection
  2. Mine Affairs Subsection
- B–E. Four mines listed, each with subdivisions for business, labor, engineering, etc.

Source: Ōshio Takeshi, "Nitchitsu Kontsuerun no kigyō kōzō," pp. 114–115.

mining. Chōsen Mining Development (Chōsen Kōgyō Kaihatsu, founded in 1937 and capitalized at 10,000,000 yen) produced gold, silver, copper, and most of Japan's nickel. Nitchitsu bought half the shares of Chojin Sulfur Mining (Kusatsu Iō Kōgyō) in 1936 and assumed a dominant role in the company's management. The sulfur was used as a raw material in sulfuric acid production.[110]

As Nitchitsu's Korean operations grew, additional office space was required. Nitchitsu expanded vertically by establishing, in May 1936, Chōsen Building (Chōsen Birudingu), a company that rented office space. The 2,000,000-yen company was 85 percent owned by Chōsen Chisso and 12 percent owned by Nitchitsu.[111]

Among the remaining companies created by Nitchitsu in Korea one deserves particular notice. Nitchitsu Fuel Industries (Nitchitsu Nenryō Kōgyō) was created at the request of the Navy in 1941 to produce butanol for use in isooctane aviation fuel. The company was capitalized at 30,000,000 yen and came on stream by May 1942. But it remained highly secret, like the synthetic-oil plants also ordered by the Navy. The Navy, according to an internal memo of June 1941, was to play a cooperative role in management, but all investment capital came from Nitchitsu. No profits were earned until late in 1944.[112] In sum, Korean expansion was extensive although not necessarily lucrative.

North China, near to Korea and occupied by the Japanese military early in the Sino-Japanese War of 1937, was the obvious next site for expansion. As early as 1938, Kubota had considered damming the Yellow River to generate electricity for several industrial projects in North Korea. Bearing an introduction from General Koiso Kuniaki of the Chōsen Army, Kubota approached the Japanese military authorities in Beijing, who showed him the Yellow River and coal deposits at Datong. Kubota's interest was sparked.[113]

Four years later, in September 1942, North China Nitrogenous Fertilizers (Ka-hoku Chisso Hiryō) was established, with its main office in Beijing and its major production facility in Taiyuan. Nitchitsu's investors had been studying resources in the Shandong Peninsula, intending to make ammonium chloride, but military authorities changed Nitchitsu's plans. The North China Army undertook the task of planning the region's economy and, in February 1942, decided an ammonium sulfate plant was necessary.[114] They pressed Nitchitsu to produce ammonium sulfate, so, in April 1942, a research team set out to investigate Shanxi province. The team settled on Taiyuan, with its ample supplies of water, gypsum, and coal, as a good site. Nitchitsu decided to develop a private joint venture there with the North China Development Corporation (Kita-Shina Kaihatsu KK) and explained its plans to the military authorities in May. Then the team returned to Beijing, where they reported to the Asian Development Bureau (Kō-A In), a Japanese government bureau, and the Agriculture and Forestry Division (Nōrinbu), an arm of the Japanese Agriculture and Forestry Ministry.[115] The joint venture was capitalized at 40,000,000 yen (10,000,000 yen paid up); Nitchitsu and North China held 95 percent of all shares; the remaining 5 percent were bought by individual investors.[116] Raw materials were allocated by the Asian Development Bureau and had to be bought by Nitchitsu in Tokyo and Ōsaka, a complicated procedure. The plant was briefly in operation before being bombed in the last months of the war.

As the war expanded southward, Japan's munitions requirements increased. But it was not until February 1943 that Nitchitsu established Taiwan Nitrogen Industries (Taiwan Chisso Kōgyō) as a wholly owned subsidiary. Taiwan had long been a Japanese colony, and Noguchi had decided to investigate Taiwan in January 1939, but no action was taken until 1942. Plans were then made to produce explosives for use in Taiwan, the Philippines, and Malaya, and construction began after the company's incorporation in 1943. As in Japan, construction was regulated by the Japanese Home Ministry, but at least materials could be obtained with greater ease than under the rigid allocation system at home. Supplies were got from various branches of the Nitchitsu empire, including the Asahi Bemberg plant and Hungnam.[117] Nevertheless, production was delayed until September 1944 due to shortages of key raw materials. By then, the plant had come under military control. The plant's top managers were drafted and served their tours of duty as managers, arriving at work in uniform. Thereafter, military orders had to be followed.[118] This was probably a more effective way for the military to influence management decisions than inserting a few military men from outside the company into management, as was done earlier in synthetic-oil production. It was more efficient to give skilled managers commissions than to try to turn untrained military men into managers.

Nitchitsu investments continued to move southward in the wake of Japan's military advances. In 1938, after the Japanese occupation of Canton, Kubota traveled there with Abe Kōryō in a Navy airplane. From Canton, Kubota hoped to go to Hainan Island in the Gulf of Tonkin, where he expected to find a rich source of iron ore. A French expedition to the island had reported finding iron ore there a decade earlier. Kubota calculated that plentiful electricity would be needed for mining and refining ore, and began to consider developing electricity on Hainan. But his plans were necessarily tentative, since the Japanese military had not yet widely acknowledged its intention to seize Hainan. But Hainan was strategically located, and its conquest was a necessary component of Navy planning for Japan's advances in Southeast Asia. Thus, Kubota anticipated that the military planned to take Hainan and was ready in 1940 to take Noguchi to Hainan. Investigating the island from the ground and from the air, they decided to send in a team led by Nakahashi Kinji, son of Nakahashi Tokugorō, Nitchitsu's first President, to scout for good hydroelectric generating locations. The team was accompanied by a large Navy escort, since the Japanese never completely overcame the resistance on the island.[119] While investigating, the team found a superb source of iron ore, estimated to total 400,000,000 tons of 65-percent pure iron.[120]

With the completion of the investigation, two other companies, Japan Steel (Nippon Seitetsu) and Japan Steel Pipe (Nippon Kōkan), hoped to join the production efforts. Noguchi, wishing to keep the project in Nitchitsu's hands, sought to persuade the two companies as well as his own managers to permit Nitchitsu to develop Hainan on its own.[121] In the midst of these negotiations, in February 1940, he collapsed with a cerebral hemorrhage.

He had convinced his managers, however, and they followed through on the Hainan project. While Nitchitsu men were planning development of Hainan, the Navy also showed an interest. Initially wishing to construct a small thermoelectric plant to generate electricity for lighting, the Navy altered its plans to encourage developing large-scale hydroelectricity. It was while construction was under way that the great deposits of high-grade iron ore were discovered, leading the Navy to urge Nitchitsu to implement plans to develop the iron resources. In October 1942, Nitchitsu Hainan Development (Nitchitsu Kainan Kōgyō) was founded, initially capitalized at 50,000,000 yen.[122] Although the expected output of the mines was only 500,000 to 1,000,000 tons, the Navy requested 3,000,000 tons of ore. It was recognized, even before the founding of the company, that this would require construction of a harbor for oceangoing ships and a railroad to transport the ore overland to the port. Military interference made construction more expensive than it should have been. Kubota had selected a harbor site based on economic calculations and even appealed to civilian officials to persuade the military to accept his proposal. The Navy argued that his choice of site could not be adequately defended, and insisted on building a harbor in another location. The Navy's harbor site was farther from the mine than Kubota's and therefore required a more expensive railroad, but, since the Navy was footing the bill, Kubota did not push his own petition aggressively. Nevertheless, Kubota's original harbor site was eventually accepted, and construction was undertaken there in the summer of 1942.[123]

Funding for this unusually expensive construction was not difficult to obtain. The Japan Industrial Development Bank offered Kubota 200,000,000 yen without collateral because, Bank President Kawakami said, questions about borrowers' ability to repay became moot when the fate of the nation hung in the balance.[124] The Hainan project was undertaken because the Navy requested it; Kawakami was reassured of the project's viability because of the Navy's involvement. In January 1945, Kubota received a check for 80,000,000 yen from the Navy Ministry. The war was going badly for Japan, but Kubota signed the check over to the Industrial Development Bank in an attempt to pay off part of his company's loan before defeat.[125] In the end, the Hainan project was a qualified success. Although unable to produce the 3,000,000 tons of ore the Navy requested, the Hainan mines did attain an output of 1,000,000 tons and had a stockpile of 500,000 tons.[126] But much shipping had been destroyed, and Kubota's visits to Home Minister Fujiwara yielded no additional ships to transport the ore.[127] Ultimately, the expansion of Nitchitsu's productive capacity would fall victim to limited transportation facilities.

Both the Army and the Navy urged Nitchitsu to continue expansion in Southeast Asia. Kubota investigated developing hydroelectricity in Vietnam through his contact with a Japanese-French businesswoman there. No action was taken to implement the plans during the war, but, after the war, Japanese engineers followed up on the earlier research.

The situation was somewhat similar in Sumatra and Java. In late 1941, the Army Ministry informed Kubota of the Army's imminent invasion of Sumatra and requested that Nitchitsu prepare teams of researchers to accompany the military. This they did, and, before the end of the war, Nitchitsu had begun to produce aluminum, soda, carbide, and electricity in Southeast Asia. But most efforts were incomplete before the war ended.[128] Many Japanese technicians and engineers left their posts permanently as newly independent governments took over half-finished projects in their countries. Others stayed on quietly. In Indonesia, for example, Tamaki Shōji reported that he later took part in a joint Japan-Indonesia project which in 1982 completed an electricity-generating facility and an aluminum plant begun four decades earlier. Many Japanese had similar experiences in other colonial areas. To be sure, many Japanese at Hungnam, for instance, were arrested and spent years in Siberia at the end of the war. But others were like Tamaki, who remained in Seoul as a consultant after the war and was even called in by the Americans during the Korean War to restore interrupted electric service.[129]

During the last months of the war, manufacturing of all products was stopped except those with immediate application to munitions. Work on most unfinished plants ceased, so most activity was concentrated in Korea where the plants—for steel and explosives—had long been completed. Production was taken underground, so new plants and tunnels had to be built. Construction work was therefore classified as strategic in Korea and was unified under Kubota's direction.

With the end of the war, Nitchitsu's extensive investments throughout Asia were lost to the company. In 1940, Nitchitsu was Japan's 6th largest corporation—Chōsen Chisso was the 22nd and Asahi Bemberg the 51st—but by 1955 it had sunk to 99th position, and by 1972 it had dropped from the list of the top 100 corporations.[130]

NITCHITSU AS A "KONZERN"

Nitchitsu's expansion in the 1920s and 1930s, led by its dynamic founder and his equally enterprising successors, was extraordinarily extensive. The company's geographic expansion was matched by its capital growth, which paralleled its diversification of production and diffusion of basic technology. In total capital, the Nitchitsu conglomerate was almost in the same league as the zaibatsu (one-fifth the size of Mitsui, one-fourth the size of Mitsubishi, and half the size of Sumitomo) and larger than all but one of the so-called "new zaibatsu" (*shinkō* zaibatsu)—Nissan, Mori, Nihon Soda (Nissō), and Riken. It has generally been accepted that Nitchitsu became a "new zaibatsu" as it grew to large proportions in the 1930s and as exogenous conditions accelerated the trend toward the preponderance of securities over fixed assets among Nitchitsu's total holdings. Recently, several scholars have analyzed the application of the term "new zaibatsu" to companies like Nitchitsu and Nissan. Some have found it wanting, although some scholars have used it interchangeably with "new *konzern*" (*shinkō kontsuerun*), another term for which no universally agreed-upon definition exists.[131]

The term "new zaibatsu" was coined by Japanese journalists in the 1930s to describe a phenomenon just then becoming noteworthy. These journalists noticed the rapid growth in the previous decade of a half dozen firms, several of which specialized in chemicals. Their size called to mind the established zaibatsu. The journalists agreed that a few differences existed between the two types of businesses, though not so many as to disqualify the epithet "new zaibatsu." These differences were in ownership of shares (one family held most shares in an old zaibatsu, shares were usually offered publicly in the new); in origin and educational background of the companies' founders (leaders in the old zaibatsu began in finance or sales, leaders in the new had advanced training in high technology); in areas of concentration (finance, retail sales, and mining for old zaibatsu, chemicals, heavy industry, and electricity for new); in entrepreneurial spirit (the old zaibatsu were more conservative, the new more risk-taking and interested in colonial expansion); and in government contracts (the old were often associated with the political parties, the new with militarists and the "reform bureaucrats" of the 1930s).[132]

In recent years, several scholars have disagreed with their journalist predecessors, arguing that these differences make the term "new zaibatsu" unclear. For example, zaibatsu are commonly associated with family ownership and control of subsidiaries, a practice not common among the new *konzerns*, thus, they find the term "new zaibatsu" misleading. These scholars prefer the German term *Konzern.*[133] Others are disturbed by the above definition's emphasis on the immediate post-World War I period as an era of extraordinary growth for the new zaibatsu.[134] One writer calls for a new term to replace "new zaibatsu" and "new *konzern*,"[135] while another uses both freely.[136] And one defines Konzern without using the term "zaibatsu." It is this use which is most applicable to Nitchitsu as analyzed in my research. Shimotani Masahiro says that Nitchitsu became a *konzern* when it developed three characteristics: capital concentration through the company's possession of numerous shares of other companies; a company structure determined by this concentration of capital; and involvement in a variety of areas of production.[137]

Konzern as used in this book builds on Shimotani's definition. That is, I suggest that Nitchitsu was transformed from a chemical company to a *konzern* in the mid-1930s because:

1. Technological developments permitted diversification to a wide variety of related products.
2. Access to capital and resources accompanied by the stimulus of conditions outside the firm permitted even wider diversification to unrelated (and not-always-profitable) products.
3. Managers were most focused on strategy to further the company's success and not on personal gain—this was true of Nitchitsu from the beginning, despite Noguchi's personal possession of a significant share of total capital.
4. The company had brought under its control all stages of manufacture, transportation, and marketing of products, making it a modern vertically integrated firm.
5. The head company's assets were predominantly shares of its subsidiaries rather than fixed capital.
6. Subsidiaries were controlled by the head company.

Most of these conditions could be applied equally to so-called old and new zaibatsu. The following chapter will discuss several new chemical companies as well as the entry of the old zaibatsu into chemical production. The dynamics of development of all these companies have striking similarities, which suggests the applicability of the term *konzern* to all firms in a similar state of development.

While it is true that the original old zaibatsu leadership did not spring from the technology departments of Japan's universities and did not constitute a "technology-clique" style of management (*gijutsu shūdan-teki keiei*) as did the leaders of the new zaibatsu,[138] these differences are not germane to our definition of the *konzern*. That is to say, the technological background of Nitchitsu's leadership was certainly significant in determining the types of investment they advocated and their boldness in decision making, but it was not crucial to Nitchitsu's transformation to a *konzern.* Technological considerations had helped determine investment at all stages of company development, not just during the 1930s. In fact, the technological interests of Nitchitsu's managers kept the focus of company strategy more on continuing production—instead of increasing the emphasis on finance—than might otherwise have occurred. Nitchitsu's persistent strength was in manufacturing; it was the profits of manufacturing in Korea that permitted the company to invest in subsidiaries, that is, to develop as a *konzern.*

Loans to these subsidiaries and ownership of their shares began to occupy a dominant position among Nitchitsu's assets in the late 1920s, as is evident in Table 30. Through these two types of financial transactions—loans and shareholding—the Accounting Department of the parent company could exert fairly strong control over the subsidiaries. During the early 1940s, the predominance of loans and securities among Nitchitsu's holdings declined. This did not mean that Nitchitsu was paring down its investments in the areas represented by its subsidiaries. Rather, the two largest subsidiaries were brought into Nitchitsu itself through mergers.[139] This decreased the paper Nitchitsu held and correspondingly increased its fixed capital. (Chōsen Chisso merged with Nitchitsu in 1941, and Asahi Bemberg and Nitchitsu Explosives merged in 1943 to form Nitchitsu Chemical [Nitchitsu Kagaku].) Needless to say, control over the production of these firms did not diminish by these mergers; although it may have appeared that Nitchitsu's control of capital declined, it did not. Nitchitsu was still a *konzern.*

Because of Nitchitsu's close control of its subsidiaries—either by 100 percent ownership or by domination of the subsidiaries' boards of directors in cases where ownership was less—subsidiaries could be viewed as similar to branch plants, differing only in that they were set up to take advantage of tax benefits, of their managers' will to be independently successful, and of decentralization of day-to-day operations. Therefore, the reduction in the number and total value of Nitchitsu's subsidiaries did not affect Nitchitsu's ability to act as a *konzern* involved in a wide variety of productive areas. Control was what was significant, and, the mergers of the 1940s notwithstanding, control was usually exercised through the subsidiaries.

The subsidiaries were, indeed, closely integrated with the parent company. Generally speaking, each function undertaken by the subsidiaries was overseen by the appropriate department or division in the main office. Personnel matters for all subsidiaries were handled by Nitchitsu's Personnel Division. Regular employees (*shain* and *junshain*) were hired by Nitchitsu and transferred among subsidiaries.[140] Daily management activities, including the filing of offical reports, were handled by subsidiaries themselves, but important managerial decisions were made by the General Affairs Division at Nitchitsu's head office.[141] Similarly, Nitchitsu's management controlled subsidiaries by supplying commodities to and marketing the finished products of all subsidiaries except Asahi Bemberg.[142] As we have seen already, the 5-percent fees Nitchitsu assessed for those services were deducted from the subsidiaries' incomes, thereby reducing taxes. But, equally significant, the handling of these two functions permitted Nitchitsu a large degree of control. Nitchitsu's marketing of the subsidiaries' products informed the parent company of sales trends—which made its advice on planning production useful to the subsidiaries—and its supplying of resources kept the parent company informed of the subsidiaries' choices of areas of investment.[143]

It is clear that control was a prime goal of these marketing and supplying policies because Nitchitsu excused subsidiaries from paying fees when they showed superior performance. It is unlikely that companies that performed well would have been released from paying fees had tax savings or transfer of some of the profits to the parent company through the 5-percent commission been the company's principal justification for these policies. Nitchitsu was more willing to relinquish its financial benefits than its control of subsidiaries.

The most effective means of controlling subsidiaries was the Accounting Division's supervision of all subsidiaries' finances.[144] This took several forms. First, the Accounting Department could examine loan requests. This permitted Nitchitsu to study subsidiaries' planning for use of lent funds, which involved it in management of the subsidiaries.[145] Second, financial control of subsidiaries was aided by the parent company's ownership of their shares, which was facilitated by the

March 1937 creation of Nitchitsu Securities (Nitchitsu Shōken). Although theoretically a separate corporation, Nitchitsu Securities was managed by Nitchitsu's Accounting Division.[146] Its function was to transfer shares of subsidiaries among the subsidiaries and the parent company in order to consolidate Nitchitsu's management. This paralleled the parent company's flexible use of subsidiaries' shares to maximize profits. Nitchitsu Securities' consolidation attempts were possible because Nitchitsu shares were not publicly owned, like those of other new zaibatsu, but were, for many years, principally owned by Noguchi Jun (see Table 39).

It must be noted that ownership by Noguchi Jun was not equivalent to family ownership in the old zaibatsu. Noguchi did not place his relations in ranking positions in the firm or its subsidiaries. His own considerable personal wealth was used to establish two institutions—one for research, one for education—when he became incapacitated in 1940. The Noguchi Research Laboratories (Noguchi Kenkyūjo) were set up with 25,000,000 yen in funds from Noguchi, and the Chōsen Scholarship Foundation (Chōsen Shogakkai) received a 5 million-yen start-up grant.[147] Noguchi's illness only removed Noguchi from daily management; it did not enrich his family or bring them into management above the longstanding and experienced professional managerial staff.

The empire Noguchi had built collapsed at the end of World War II because its vast overseas extensions were lost. But, at its height, Nitchitsu and its subsidiaries were consistently pioneers of new technologies. To be sure, government agencies helped determine the types of technologies developed, but management continued to debate the viability of investment in terms of access to capital and resources. The structure of the company evolved during the 1930s toward the form known as the *konzern* (see Table 40), but this did not diminish the managers' desire to make decisions relating to manufacturing and control of production. Rather, it implied that managers devised new strategies—a focus on financial methods—to achieve long-held goals—profitability and autonomy. This is an economic motivation evident among people of different classes, cultures, genders, and eras, and, in the case of Nitchitsu, managers changed the structure of the modern high-technology firm to maintain the founder's goals.

TABLE 39 Largest Shareholders in Nitchitsu, 1927–1937 selected years

1927		
Noguchi Jun	139,000	(15.4%)
Kondō Shigeya	42,000	(4.7)
Aichi Bank	33,140	(3.7)
Iwasaki Hisaya	30,000	(3.3)
Morobe Colonial Development	29,000	(3.3)
Noguchi Jun	28,076	(3.1)
Ichikawa Seiji	26,769	(3.0)
Chōsen Hydroelectric	20,000	(2.2)
Tokyo Marine Fire Insurance	20,000	(2.2)
Kagami Yoshiyuki	17,754	(2.0)

1931		
Noguchi Jun	226,000	(12.6)
Hungnam Development	94,000	(5.2)
Kondō Shigeya	73,800	(4.1)
Aichi Bank	73,400	(4.1)
Iwasaki Hisaya	59,000	(3.3)
Ichikawa Seiji	46,000	(2.6)
Tokyo Marine Fire Insurance	40,000	(2.2)
Morobe Colonial Development	29,730	(1.7)
Morobe Seiroku	29,730	(1.7)
Kagami Yoshiyuki	28,108	(1.6)

1935		
Noguchi Jun	182,500	(10.1)
Hungnam Development	137,500	(7.6)
Iwasaki Hisaya	59,000	(3.2)
Kondō Shigeya	36,900	(2.0)
Ichikawa Seiji	30,000	(1.7)
Nakahashi Buichi	30,000	(1.7)
Tatsuma Yoshio	24,000	(1.3)
Nita Sadao	23,500	(1.3)
Hiromi Nisaburō	21,480	(1.2)
Tokyo Marine Fire Insurance	20,000	(1.1)

1937		
Hungnam Development	406,000	(10.1)
Noguchi Jun	287,000	(7.2)
Nitchitsu Securities	244,801	(6.1)
Ichikawa Seiji	68,000	(1.7)
Kondō Shigeya	67,900	(1.7)
Imperial Military Mutual Insurance	60,860	(1.5)
Iwasaki Hisaya	60,000	(1.5)
Tatsuma Yoshio	48,000	(1.2)
Hiromi Nisaburō	42,780	(1.1)
Nakahashi Buichi	42,000	(1.0)

1938		
Hungnam Development	459,500	(11.5)
Nitchitsu Securities	244,719	(6.1)
Noguchi Jun	233,500	(5.8)
Ichikawa Seiji	68,000	(1.7)
Kondō Shigeya	60,900	(1.5)
Iwasaki Hisaya	60,000	(1.5)
Imperial Military Mutual Insurance	57,660	(1.4)
Yasuda Savings Bank	55,380	(1.3)
Tatsuma Yoshio	48,000	(1.2)
Hiromi Nisaburō	42,780	(1.1)

Sources: For 1927, 1931, and 1938, Ōshio Takeshi, "Nitchitsu Kontsuerun no seiritsu to kigyō kin'yū," p. 122; for 1935, Shimotani Masahiro, "Nitchitsu Kontsuerun," p. 64; and for 1937, Watanabe Tokuji, *Gendai Nihon sangyō hattatsushi,* p. 64.

Note: 1.8 million shares were held by 7,324 shareholders in 1935; 4 million shares were held by 12,737 shareholders in 1937.

TABLE 40 Sources of Income, Nitchitsu, 1930–1944
(unit: 1,000 yen; %)

	1930		*1931*		*1932*		*1933*		*1934*	
	1	*2*	*1*	*2*	*1*	*2*	*1*	*2*	*1*	*2*
Product Sales	7,967 (88)	5,570 (62)	5,631 (51)	5,467 (56)	4,053 (44)	6,103 (64)	6,269 (63)	6,536 (65)	4,587 (48)	4,204 (47)
Electric Rates	82 (1)	77 (1)	84 (1)	138 (1)	134 (1)	82 (1)	85 (1)	280 (3)	490 (5)	477 (5)
Dividends			1,762 (16)	1,975 (20)	2,169 (24)	1,885 (20)	1,949 (20)	1,701 (17)	1,538 (16)	1,461 (16)
Interest on Loans	1,002 (11)	3,103 (35)	1,742 (16)	1,896 (20)	1,502 (16)	1,427 (15)	1590 (16)	1,364 (14)	1,089 (11)	1,422 (16)
Commissions, fees (Other)	36 (0)	236 (3)	1,809 (16)	223 (2)	1,299 (14)	98 (1)	64 (1)	129 (1)	1,776 (12)	1,395 (16)
Total	9,089 (100)	8,988 (100)	11,030 (100)	9,701 (100)	9,159 (100)	9,597 (100)	9,959 (100)	10,103 (100)	9,483 (100)	8,961 (100)

	1935		*1946*		*1937*		*1938*		*1939*	
	1	*2*	*1*	*2*	*1*	*2*	*1*	*2*	*1*	*2*
Product Sales	4,592 (49)	4,949 (53)	4,444 (43)	4,827 (45)	5,495 (42)	9,628 (51)	11,940 (57)	11,118 (51)	15,147 (58)	16,040 (58)
Electric Rates	479 (5)	361 (4)	273 (3)	121 (1)	638 (5)	487 (3)	388 (2)	361 (2)	213 (1)	315 (1)
Dividends	1,520 (16)	1,583 (17)	1,697 (17)	2,082 (19)	2,507 (19)	3,503 (19)	3,748 (18)	4,144 (10)	4,129 (16)	4,286 (15)
Interest on Loans	1,204 (13)	1,086 (12)	1,025 (10)	1,037 (10)	1,152 (9)	1,740 (9)	1,673 (8)	1,936 (9)	2,332 (9)	2,820 (10)
Commissions, fees (Other)	1,583 (17)	1,335 (14)	2,809 (27)	2,705 (25)	3,235 (25)	3,502 (19)	3,120 (15)	4,142 (19)	4,460 (17)	4,399 (16)
Total	9,379 (100)	9,316 (100)	10,250 (100)	10,773 (100)	13,029 (100)	18,797 (100)	20,870 (100)	21,704 (100)	26,283 (100)	27,862 (100)

TABLE 40 *(continued)*

	1940		*1941*		*1942*		*1943*			*1944*
	1	*2*	*1*	*2*	*1*	*2*	*1*	*2*	*3*	*1*
Product Sales	18,985 (60)	20,022 (56)	19,374 (50)	21,459 (50)	87,399 (82)	81,452 (80)	71,296 (72)	73,354 (72)	62,254 (81)	102,099 (84)
Electric Rates	382 (1)	330 (1)	187 (0)	165 (0)	738 (1)	610 (1)	677 (1)	608 (1)	502 (1)	631 (1)
Dividends	5,083 (16)	6,461 (18)	7,620 (20)	7,872 (18)	6,609 (6)	7,950 (8)	14,896 (15)	9,477 (9)	753 (1)	10,929 (9)
Interest on Loans	2,493 (8)	3,205 (9)	3,760 (10)	4,869 (11)	6,980 (7)	6,495 (6)	6,921 (7)	5,096 (5)	3,598 (5)	5,174 (4)
Commissions, fees (Other)	4,628 (15)	6,048 (17)	8,128 (21)	8,274 (19)	5,260 (5)	4,945 (5)	5.077 (5)	13,091 (13)	10,122 (13)	3,265 (3)
Total	31,573 (100)	36,067 (100)	39,071 (100)	42,642 (100)	106,989 (100)	101,454 (100)	98,869 (100)	101,628 (100)	77,231 (100)	122,100 (100)

Source: Ōshio Takeshi, "Nitchitsu Kontsuerun no kin'yū kōzō," pp. 230–231.

Note: From 1930 till 1943, Period 1 covers the months from December of the previous year through May of the year in question, and period 2 covers the months from June through November. In 1943 period 3 covers the months December 1943–March 1944. In 1944, period 1 covers the months from April until September. Totals may be somewhat different from the sums because of rounding.

SIX

The Second and Third Waves: Widening the Circle of Investors in High Technology

When the two Japanese pioneers in the electrochemical industry, Noguchi Jun and Fujiyama Tsuneichi, parted ways in 1912, their investments took different forms. Their two models of investment were available as examples to other potential investors in chemical synthesis. Noguchi's greater success in manipulating the requirements for investment drew the attention of several firms during the 1920s, including those already involved in chemical manufacture at a lower level of technology or in production of unrelated goods.

As Noguchi had done, managers in these firms had to consider the requirements for investment: capital, resources, market, skilled managers, and an advantageous environment. Some firms fulfilled these requirements and found investment possible during the 1920s. Once they had invested in electrochemical technology, most discovered that diversification of their enterprises followed certain paths determined by the technology. Other firms were more hesitant. Some of their managers were interested in developing the technology of ammonia synthesis during World War I but were slow to invest because they had no commitment to taking risks in scientific investment, were unwilling to channel sufficient capital into chemical production, or lacked easy access to resources. Such inadequacies impeded the early entry of zaibatsu firms into ammonia manufacture, for example.

The zaibatsu firms were initially interested in ammonia synthesis during World War I; managers at both Mitsui and Sumitomo examined conditions for investment and determined that the market for nitrogenous fertilizers was good. Makita Tamaki, Director of General Affairs of Mitsui Mining, investigated German methods; Sumitomo's Suzuki Masaya was being introduced to electrochemical methods at about the same time, by Takamine Jōkichi, the pioneer in superphosphate and other chemicals.[1]

Takamine Jōkichi had maintained his ties with America's General Chemical Corporation, which modified the Haber method and began to produce synthetic ammonia in the United States in 1913. He interested the leadership of Sumitomo in experimenting with ammonia manufacture by the General Chemical method, and the Sumitomo managers persuaded their counterparts at Sankyō Company of the value of investment in such an expensive venture. With Takamine acting as intermediary, Sumitomo and Sankyō planned to acquire the General Chemical method for a joint investment of 1.8 million yen and 1.2 million yen, respectively.[2] Concerned about making such a large investment in a complex process, Sumitomo's management asked Mitsui and Mitsubishi to join them in developing the process. Mitsui's Makita Tamaki was most interested. With the conclusion of World War I, however, these potential investors shifted their sights from the American method just as technicians from the four firms were preparing to inspect General Chemical's Laurel Hill plant in November 1918. They decided to study the Haber method instead, because it was the original process of ammonia synthesis. The French military authorities occupying Badische's Oppau plant opened the plant up to an inspection team of technicians from four Japanese firms in May 1919, and the team wasted no time getting to Europe.[3]

One month earlier, in April 1919, Toyama Tamotsu of the Yokohama Chemical Research Laboratory had begun negotiating with Badische for the Haber method. Representatives of seven other firms and laboratories interested in ammonia synthesis—Sumitomo, Mitsui, Mitsubishi, Sankyō, Nihon Chemical Industries, Dai Nihon Artificial Fertilizer, the Watanabe Chemical Research Laboratory, and the Yokohama Chemical Research Laboratory—soon joined him to form the Eastern Nitrogen Association (Tōyō Chisso Kumiai).[4] The Association continued the negotiations and worked out a contract with Badische which stipulated the cost of a license permitting manufacture of a negotiated amount of ammonium sulfate, fees for annual renewal of that license, and royalties for each ton of fertilizer produced. The Association estimated that these fees would come to 30 million yen in initial payments plus 2.5 million yen in annual costs for the right to produce the stipulated 100,000 tons annually.[5] Moreover, construction costs were estimated to reach 30 million yen. Assuming a 15-year period of operation for the proposed plant, consistent production at its 100,000-ton capacity, and constant depreciation, fixed costs (licensing fees plus construction costs) would be at least 85 yen per ton in addition to costs for energy, labor, staff salaries, and materials. German prices were about 280 yen per ton and Japanese prices about 350 yen per ton in 1919, but they were falling, and eventually dropped to 145 yen and 220 yen, respectively, by 1921. Fixed costs of 85 yen would have made the cost of fertilizer about 55 yen per ton more than the average; that was clearly prohibitively expensive. The contract could not be carried out under those terms.[6]

The Haber license soon became available by another route. With the end of the war, the Japanese government offered the Badische patent to private investors. In July 1921, for a nominal fee of 9,500 yen for registration of license transfer, the Tokyo municipal government "sold" the license to the Eastern Nitrogen Association.[7] Although the license appeared cheaper in 1921 than it had been in 1919, it continued to require large payments to the German company. It is not clear why the royalties were no lower under the 1921 terms than under the 1919. The Japanese consortium held a license seized by their own government, but each member of the group would have to pay about 8.5 million yen in royalties for its share of the license, estimated to cost 68 million yen over a 15-year period.[8] None could afford the investment, which was high compared to Noguchi's 1-million-yen Casale license. Therefore, the members of the Eastern Nitrogen Association could not immediately begin to produce ammonium sulfate with the license they nominally held.

Several additional factors have been cited for the Association members' reluctance to risk investment.[9] First, Badische President Karl Bosch, en route to Japan to help install machinery for production, was forced to return suddenly to Germany from the United States on hearing news of an accidental explosion at the plant at Oppau. And, second, an overnight doubling of domestic output of ammonium sulfate, which could be accomplished by construction of a 100,000-ton-capacity plant, would elevate supply temporarily far above demand, thereby depressing prices. (Consumption from all other sources, domestic and foreign, was less than 200,000 tons in 1921.) The second factor is quite plausible, but the first, Karl Bosch's return to Germany, seems a bit far-fetched. It is more likely that increased dumping of ammonium sulfate by English and American manufacturers anxious to maintain their postwar growth in a Japanese market insufficiently protected by tariffs made investment unattractive to the members of the Eastern Nitrogen Association. Whatever the reason, the members demurred so long in implementing the license that the chance to use the patent profitably for production had passed by 1923 when German imports resumed.

As we have seen in Chapter 4, the Eastern Nitrogen Association transformed itself into a nitrogen-importing association in 1923. Most members, therefore, had little incentive to continue studying the question of investment in ammonia synthesis and in fact made competition more difficult for producers like Nitchitsu. But not all members of the Association were guilty of impeding new production. Suzuki Shōten, for instance, was a leader in importing technology and began construction of manufacturing plants in the early 1920s. Sumitomo and Mitsui began manufacture by the end of that decade. But the need to invest in manufacturing in order to profit from chemicals was muted by the companies' ability to earn profits from importing. An inexpensive license and ready availability of resources were conditions not yet attained by the members of the Eastern Nitrogen Association. None was ready, therefore, to invest in the risky technology in the early 1920s. Favorable conditions permitting the zaibatsu to invest in advanced technology were created by 1930, and, in conformity with expectations, the zaibatsu did invest.

THE SECOND WAVE: THE NEW ZAIBATSU IN ELECTROCHEMICALS

By the 1930s, the major new zaibatsu were heavily involved in manufacturing chemicals, and most specialized in electrochemicals. Nitchitsu and its subsidiaries continued to dominate electrochemical production, but other manufacturers played an increasingly important role. Two companies are particularly interesting: Dai Nihon Fertilizer, Inc., which diversified production and began making electrochemicals in the decade before merging with Nissan, and Shōwa Fertilizer, Inc. (Shōwa Hiryō KK, later Shōwa Denko), which pioneered use of the advanced electrochemical technology developed in Japan.

DAI NIHON FERTILIZER

Dai Nihon Fertilizer, Japan's earliest producer of chemical fertilizers, had long specialized in superphosphates, as discussed in Chapter 1. By gradual acquisition of rival firms and expansion of already existing plant, Dai Nihon dominated Japan's superphosphate market. Each Dai Nihon facility resembled the others; similar processes were carried out in each branch of the firm. Unlike the pattern characteristic of Nitchitsu after the 1910s, Dai Nihon's geographically scattered parts did not complement each other productively or technologically but were relatively uniform.

By the mid-1920s, the almost half-century-old company was ready to modify this conservative pattern. Venturing at first only into other fertilizers and not yet into a wide range of chemicals, Dai

Nihon was beginning to test the effectiveness of diversification. Developments occurred in two major areas, one more innovative than the other: in caustic fertilizers and in synthetic ammonium sulfate.

"Caustic" fertilizer was a compound fertilizer gaining acceptance in Europe where the German I G Farben (successor to BASF) was producing it.[10] Planning and construction of Dai Nihon's caustic fertilizer plant were completed by 1928. The company mounted an extensive advertising campaign, and the initial sales of 33,000 tons in 1928 had risen to 222,000 tons by 1935.[11]

Dai Nihon's other, and scientifically more important, diversification of product was in synthesis of ammonium sulfate by the Fauser method. Although Dai Nihon began synthesis five years after Nitchitsu, its management had been attracted to the possibility of producing ammonium sulfate at least as early as 1917. In that year, the firm invested heavily in Hokuriku Electric, Inc. (Hokuriku Denryoku KK) with the expectation of turning its electricity to use in synthesizing ammonium sulfate by the cyanamide method. The plan failed, but, two years later, Dai Nihon joined the Eastern Nitrogen Association in the hope of acquiring the Haber-Bosch method. During the 1920s, the company was forced to buy 30,000 tons of ammonium sulfate per year to use as a component in its fertilizers. Ammonium sulfate was clearly a product the firm would want to supply for itself.

Researcher Tanaka Jūichi, who joined Dai Nihon after its merger with Kantō Soda, had been studying methods of synthesis. He found a kindred spirit in Dai Nihon's head of research, Furukawa Seiji, and together the two took off on a tour of inspection of available European patents in December 1924. In January 1925, company headquarters wired them in Europe to inform them of restrictions that various methods might have.[12] Some methods were legally off-limits: Nitchitsu had the Casale method; the Eastern Nitrogen Association had the Haber-Bosch; Sumitomo controlled the General Chemical method; and Suzuki Shōten used the Claude. This made the Fauser method most attractive to Dai Nihon. Furthermore, the Fauser method was available for a much lower price than the Haber-Bosch; Fauser royalties were only about one tenth the Haber-Bosch royalties.[13] Tanaka Jūichi and Furukawa Seiji heard of the relatively unknown Fauser method from a Parisian consultant of Dai Nihon. Fauser was working for the Italian Montecatini company, and the Japanese had to negotiate through that company's president for the rights to the license. In the weeks after the two Japanese scientists had heard from their French contact, Mitsui agents also discovered the process and competed with the two researchers for its acquisition. But the Italian company president, like his countryman Casale in the latter's negotiations with Noguchi years before, was apparently more impressed by the enthusiastic young researchers. He decided to sell Dai Nihon the patent.[14]

Fauser's method was, like Casale's, a process for ammonia manufacture involving electrochemical reactions to produce hydrogen. When Tanaka returned in September 1925 to become head of the new nitrogen division at Dai Nihon, his first responsibility was to find a plant site with a good source of electricity. He found that Toyama fit his specifications and signed a 20-year contract with Japan Electric Power Co. in Toyama to purchase 10,000 kilowatts of electricity at 1 sen per kilowatt. Other considerations favoring the location were its railroad accessibility, availability of land for construction, and abundance of pure water.[15] After several setbacks, the plant at Toyama was ready for operation by April 1928. Capacity was 25,000 tons per year, sufficient to supply Dai Nihon's own requirements for compound fertilizer. Because the contracted amount of electricity from Japan Electric Power would permit up to 50,000 tons of output, several managers proposed raising production. But Dai Nihon was unable to bring its fertilizer costs below 249 yen per ton, the selling price at that time. Greater output might produce greater losses rather than economies of scale, several managers thought. Dai Nihon's competitors spent an average of 150 yen per ton on construction of plant and other set-up costs; in Dai Nihon's case, these costs were 240 yen per ton.[16] The decision to expand horizontally was vetoed.[17]

But that did not prevent Dai Nihon's diversification in other areas. In 1927, the company acquired the Kagami plant for 2 million yen from Nitchitsu when that giant ammonium sulfate maker discontinued the cyanamide method used at Kagami.[18] Around the same time, Dai Nihon took over the glycerine production of the failed

Suzuki Shōten.[19] Both the carbon production at Kagami, which formed the basis for plastics, and the glycerine production, which could be combined with the recently added ammonia to manufacture explosives, had important later applications. But each was acquired by purchase from another company which had originally invested in research and development. Dai Nihon bore the primary expense of developing only the technology for ammonia production. Indeed, with the development of the new ammonia technology Dai Nihon increased its capitalization substantially, from 12.6 million yen to 35 million yen.[20]

Dai Nihon began ammonium sulfate production just as world prices were quickly sinking. The company also faced potential problems in acquiring the electricity necessary for electrochemical production when restrictions began to appear in the 1930s. By 1930, company researchers began studying conversion to the gas method. Dai Nihon's study of appropriate processes is a classic case of carefully analyzed and implemented managerial decision making.

The gas method was generally more common in Germany and Austria, countries with abundant coal, and the electrochemical method more common in Sweden, Norway, and Switzerland. Electrochemical methods had much to recommend them; they produced pure hydrogen and were safe. The gas method was inherently dangerous, since the gas itself contained highly explosive traces of nitric acid. Foreseeing the kinds of problems with cost and supply that decided the case in favor of the gas method in Germany, however, Dai Nihon's Managing Director, Ishikawa Ichirō, authorized research in the gas process in 1930.[21] The previous year, Dai Nihon's management sought to build a second plant at Ube, where Ishikawa thought the gas method could be applied. The Ube City authorities as well as the Ube Development Company (Ube Kōsan) wished to develop a coal-based chemical industry themselves, and so Dai Nihon withdrew. When Ube failed even with the help of the South Manchurian Railway Company, however, Dai Nihon agreed to cooperate with the Ube interests and establish a Fauser method facility.[22] Under the direction of Dai Nihon's Ōyama Takekichi, the Ube Development Company was ready in February 1933 and began production in July 1934.[23]

Thus, when the electricity controls began to be implemented in the 1930s, Dai Nihon had experience with the gas method and was able to make a carefully planned shift. In 1935, Dai Nihon examined gas methods used in Austria, Germany, England, and Italy before discovering that a new I.G. Farben method was ideal for the type of coal used in Japan. By 1938, its research completed, Dai Nihon began production by the gas method.[24] Until the end of World War II, Dai Nihon produced almost as much gas-method ammonia as electrochemical. The company was clearly keeping all options open (see Table 41).

In 1936, the President of Nissan, Aikawa Gisuke, and Dai Nihon's President, Tanaka Eihachirō, met to begin merger talks. Merger would benefit both sides; the wealthy and politically favored Nissan was seeking a chemical company that could become its chemical division, and Dai Nihon was suffering a dearth of capital during the mid-1930s as wartime controls tightened.[25] The following year, the merger was completed, and Dai Nihon was temporarily renamed Japan Chemical Industries (Nihon Kagaku Kōgyō) before becoming Nissan Chemical Industries (Nissan Kagaku Kōgyō). Dai Nihon's glycerine plants were combined to form another subsidiary of Nissan, Japan Fats and Oils (Nihon Yushi). Together the two subsidiaries were capitalized at 79 million yen (62 million for Nissan Chemical and 17 million for Japan Fats and Oils). Only three of Nissan's subsidiaries were larger—Japan Mining, Hitachi Manufacturing, and Japan Marine.[26]

SHŌWA FERTILIZER

With an output in 1932 of ammonium sulfate second (in the Japanese empire) only to that of Chōsen Chisso, Shōwa Fertilizer, part of the Mori conglomerate, made significant advances in the chemical industry in Japan.[27] Chemicals occupied only one-fifth of Mori's investments, yet the Mori chemical division, Shōwa Fertilizer Co. (Shōwa Hiryō), was historically important. The first of Japan's ammonia producers to rely almost exclusively on Japanese inputs in technology and plant, Shōwa Fertilizer came about through the collaboration of two men, Suzuki Saburōsuke and Mori Nobuteru.

TABLE 41 Ammonia Production at Dai Nihon (Nissan Chemical), 1928–1945 (tons)

Year	*Electrochemical Method*	*Gas Method*	*Total Ammonia*	*Ammonium Sulfate*
1928	3,302	–	3,302	15,889
1929	7,529	–	7,529	24,509
1930	11,286	–	11,286	44,130
1931	13,044	–	13,044	48,198
1932	13,090	–	13,090	47,676
1933	13,265	–	13,265	45,206
1934	14,234	–	14,234	48,915
1935	16,251	–	16,251	55,156
1936	17,161	–	17,161	57,950
1937	17,211	–	17,211	54,284
1938	17,010	1,654	18,664	58,659
1939	14,567	15,876	30,443	97,786
1940	17,562	15,752	33,314	100,778
1941	18,613	15,161	33,774	90,284
1942	18,314	16,532	34,846	98,939
1943	17,792	13,271	31,063	85,434
1944	17,088	8,918	26,006	77,714
1945	17,318	4,387	21,705	70,364

Source: Nissan Kagaku, p. 321.

After the former's death in 1931, the firms affiliated with Shōwa Fertilizer came to be known as the Mori Group.

Suzuki Saburōsuke (1867–1931) and Mori Nobuteru (1884–1941) began their separate careers as processors and marketers of iodine and other products.[28] Suzuki's most noteworthy claim to fame was his development in 1909 of Ajinomoto monosodium glutamate, the formerly ubiquitous Japanese flavor enhancer.[29] But it was through their shared interest in iodine and generation of electricity that the two men solidified their working relationship. Tōshin Electric (Tōshin Denki KK) was founded in 1917 by the elder. Mori's iodine company had expanded too rapidly during World War I, and, when faced with possible bankruptcy in 1919, he merged his firm with Suzuki's electric company.[30]

The economic slump of the early 1920s left Mori and Suzuki with surplus generating capacity when their customers were forced to cut their use of power. As with others before them, this prompted a search for either new industrial customers or new uses for electricity. Some of the surplus was absorbed by Mori's new company, Japan Iodine, Inc. (Nihon Yōdo KK, the predecessor of Mori's aluminum works, Nihon Denkō), founded in 1926. Some was used by a new fertilizer company, Shōwa Fertilizer, founded by the two men in October 1928. The large new company (capitalized at 10 million yen) began production of calcium cyanamide in Niigata prefecture within months, just as Japan entered a serious recession. Mori was concerned that marketing the firm's product through the usual channels would be unwise during the recession, and decided to circumvent fertilizer merchants and sell directly to farmers through agricultural cooperatives.[31] This was a wise move for two reasons: A new company could not expect to receive the same consideration in marketing as established customers from the fertilizer merchants; and Mori was able to use farmers' cooperatives effectively when they became the politically expedient marketing method a few years later. Mori was thus able to minimize the effects of a sluggish market in 1929.

Mori Nobuteru was not satisfied with the calcium cyanamide he and Suzuki Saburōsuke manufactured. Though relatively uneducated compared to the early developers of ammonium sulfate, he was anxious to upgrade his technology to Nitchitsu's level. His purpose in advancing technology was influenced by a sense of patriotism as well as the more economic desire to invest in higher technology when encouraged by market conditions and ready availability of resources. Mori noted that Japan, weak in natural resources, should concentrate on products like ammonium sulfate which used resources found domestically. These resources included hydroelectricity, which he proposed to "make into a resource" (*suiryoku no genryōka*).[32] Thus, Mori was ready to cooperate with other firms when they suggested using Tōshin Electric's surplus power for fertilizer manufacture. Yamamoto Jōtarō, President of the South Manchurian Railway (Mantetsu), tried to form a consortium with Tōshin Electric and Daidō Electric to produce ammonium sulfate by the expensive German Uhde process.[33] Mantetsu was to put up 50 percent of the money, while Tōshin Electric and Daidō Electric were to supply 25 percent each. The two electric companies sent scientists to Europe to examine the process. Tōshin scientists were dissatisfied with the Uhde method, claiming it was too costly.[34] They dropped out of the consortium with Mantetsu and Daidō.

While Mantetsu and Daidō continued to study the Uhde method for possible implementation in Manchuria, Mori invited the head of the Electrochemical Division of the Tokyo Industrial Experimental Laboratory (TIEL), Kitawaki Ichitarō, to ascertain the feasibility of using the TIEL method commercially. Kitawaki was enthusiastic about Mori's overture. Scientists at TIEL had hoped that one of the large established zaibatsu would use the TIEL method for commercial production, but only Mitsui seems to have expressed even moderate confidence in the Japanese technology. And Mitsui's interest came too late, after Shōwa Fertilizer claimed the license. Mori's bid was welcome, therefore, even though his capital resources failed to match those of the zaibatsu.[35]

Mori found the price attractive. TIEL charged no license fee or royalties per ton of ammonia. Rather, the contract between TIEL and Mori stipulated that Shōwa Fertilizer pay for the privilege of using the process by a somewhat complicated profit-sharing formula. TIEL was to receive 10 percent of all profits, if profits were sufficiently high to permit 10 percent dividends.[36] If the process failed, the cost to Shōwa Fertilizer would be low; if it succeeded, the company would have to pay a reasonable fee to the government agency.[37] The government was so eager to promote the TIEL process that it absorbed the costs of its 10-year period of research and development. Construction costs were comparatively low—110 yen per ton capacity—but even that amount was too high for the company.[38] The cyanamide plant at Kanose cost 2,500,000 yen, about 20 percent of the firm's total authorized capital, leaving little paid-up capital for construction of a plant to use the TIEL method. By 1929, Mori had borrowed extensively and called in some money owed on shares, giving him access to sufficient capital to implement the TIEL process. He had extensive surveys for plant sites conducted through Nagano, Gumma, and Tokyo, and decided that Kawasaki, with its nearby electric source at Tsurumi, was ideal. Construction began in September 1929, and the plant was ready in February 1931.[39] By 1932, Mori had persuaded his shareholders to pay up almost to par value, and, in September of that year, a group of life-insurance companies lent Shōwa Fertilizer 6 million yen.[40]

The Kawasaki plant was carefully planned. TIEL sent a research team from Tokyo to see that specifications were followed. The team was headed by Yokoyama Būichi, who left TIEL and entered Shōwa Fertilizer to become director of the Kawasaki plant. Other members of the research team were Shibata Kentarō (later of Tōyō Kōatsu), Sōji Nobusane (later of Nihon Kasei), and Nakamura Kenjirō (later of Shōwa Fertilizer).[41] Not only did these scientists help Shōwa Fertilizer immeasurably, but their move to private industry from public service was an important step toward diffusion of electrochemical technology.

Also boosting Japanese industry was Shōwa Fertilizer's decision to use principally Japanese-made machinery and equipment. With the exception of one type of imported machine, all other equipment was purchased from Japan's leading machine-tool makers at the time, Kōbe Steel, Ishikawajima Shipbuilding, Hitachi Manufacturing, and Japan Steel.[42] The decision to use Japanese equipment was motivated by the same interests as the decision to use the TIEL process: desire to "buy Japanese" and, no less important, cost advantages.[43]

Shōwa Fertilizer managers were concerned about cost advantages, because they were indebted to banks and insurance companies, especially the Yasuda Bank and the Kōgyō Bank. Certain types of products put out by the Mori firms were particularly attractive to various institutions and individuals who could invest in the firm or could help it financially. For instance, refined aluminum interested Yasuda Bank Manager Kondō Seirō, and explosives attracted the naval authorities.[44] The Mori group's investment in aluminum was, with the exception of its electrical generating plants, the largest part of the conglomerate's investment. By 1937, the aluminum works were capitalized at 50 million yen, while the fertilizer works were worth only 30 million yen.[45] The Yasuda Bank alone lent over 10 million yen to Mori for aluminum; the Kōgyō Bank underwrote Mori's mining and explosives operations; only the fertilizer operations were in tight financial straits.[46] By being able to obtain funding for the other types of production, Mori was able to keep his fertilizer operations working. Furthermore, during the 1930s, Shōwa Fertilizer stayed afloat by diversifying in numerous profitable ways: The fertilizer firm moved into metal refining and carbide; it owned a large part of the Mori subsidiary Japan Oil Industries (Nihon Yuka Kōgyō); and it even built in Changjin, Korea, site of Noguchi Jun's electric plant.[47] Finally, Mori Nobuteru's business operations may have been assisted by his political ties. Mori loved politics and ran for and won Diet seats during the 1920s.[48]

For several reasons, then, the financially weak Shōwa Fertilizer survived into the 1930s to become the basis of a major chemical and aluminum industry. Its output of ammonium sulfate, while significant, was far below Nitchitsu's. But Mori was noteworthy in aggressively seeking out indigenous methods of production. He was also somewhat rare in that he shared with Noguchi a commitment to the electrochemical method of ammonia production even after others began to shift to the gas method in the 1930s. Like Noguchi, Mori had his own electricity, although this fact was obscured by his method of internal bookkeeping. That is, Shōwa Fertilizer actually contracted to purchase electricity from Mori's own Tōshin Electric rather than generate its own.[49]

Mori was considered to be one of the "Four Heavenly Kings" (*shitennō*) of the new zaibatsu. (The others were Noguchi Jun, Aikawa Gisuke, and Nakano Yūrei.) But he differed in some important respects from Noguchi's archetype. Though both started with electricity firms, had large debts to commercial banks and the Kōgyō Bank, had close ties with political and military men, moved into the development of Korea, and continued electrochemical production of ammonia even after the government began regulating electricity, Mori was not a scientist. Furthermore, his managerial style differed somewhat from the other three "Heavenly Kings." Members of his own and Suzuki Saburōsuke's families—whom Mori controlled tightly—ran the Mori conglomerate.[50] Despite these differences, Mori founded an important new firm, significant not only in Japan's business history but also in its technological contributions in chemical applications.

As we have seen in the cases of Dai Nihon and Shōwa Fertilizer, Noguchi's earlier success in implanting new technology was a model for subsequent manufacturers. Both new producers focused on manufacturing and had a more limited interest in such activities as marketing. Both were public joint-stock companies which also made active use of outside capital in the form of bonds and loans. Both used hydroelectricity as a source in electrochemical manufacture. The two companies differed in their origins: Dai Nihon was a chemical company with a long history of success, but its technology had remained fairly unsophisticated until its diversification in the 1920s; Mori's company was a new venture into electrochemistry.

When Dai Nihon and Shōwa Fertilizer decided to upgrade their technology by investing in electrochemicals, neither suffered major impediments in access to resources or capital. In this regard, the two "new zaibatsu" were more successful than the "old zaibatsu" which had also attempted to acquire electrochemical technology. Licensing costs were prohibitively high for the old zaibatsu, because they conservatively wished to use a well-proven expensive German method. Resources were also a problem for Mitsui, Mitsubishi, or Sumitomo. As other potential investors in electrochemicals, these firms would have to seek access to hydroelectricity from outside their firms. But this problem was eased by the late 1920s, when new electrochemical methods, using coal instead of electricity as a major resource, were developed. With ready availability of coal within their own enterprises, the opportunity for profitable investment improved.

THE THIRD WAVE: THE OLD ZAIBATSU IN ELECTROCHEMICALS

The entry of the zaibatsu into the field of electrochemicals constituted a third wave of investment. Within a decade of their first production of ammonia, the companies related to the original zaibatsu surpassed many of the earlier ammonia producers. Even Nitchitsu, the first successful ammonia manufacturer, felt the competition. Nitchitsu's Korean subsidiary, Chōsen Chisso Hiryō, maintained a virtually unassailable lead over other entrants in the field, but the Minamata plant, noted as "Nitchitsu" in Table 42, dropped to 8th place among Japanese makers of ammonium sulfate.[51]

Furthermore, with their entry into production of ammonium sulfate, the zaibatsu-connected producers undertook, as Nitchitsu had earlier, production of technologically related chemicals like methanol, formalin, nitric acid, and synthetic fuel, and expanded into unrelated areas like synthetic rubber. Their rapid growth and diversification in the chemical industry during the 1930s was possible because of their long-held interest in producing fertilizers. Diversification by the zaibatsu-related companies was virtually identical to that of Nitchitsu, with one important difference—the specific technology used. The technology influenced how each company diversified, not whether it would diversify. In the case of the chemical companies related to Mitsui and Sumitomo, expansion to a *konzern* occurred naturally as related products were spun off from ammonia synthesis. On the other hand, Mitsubishi's chemical company differed from other *konzerns*, whether they were new *konzerns* like Nitchitsu or older ones like Mitsui, in that its founders planned from the start to create an already well diversified company producing a variety of chemicals selected by its managers.

It is commonly asserted that the somewhat slower development of the new chemical technology by the zaibatsu contrasts with the more aggressive investment of Noguchi Jun and Mori Nobuteru. It is true that the third-wave companies developed later. But the conditions under which Mitsui, Mitsubishi, and Sumitomo leaders decided to launch their chemical enterprises resembled those favor-

TABLE 42 Ten Largest Producers of Ammonium Sulfate in Japan, 1937

Company	*Method*	*Capacity (tons/year)*
Chōsen Chisso	Electrochemical	460,000
	Gas	40,000
Shōwa Fertilizer	Electrochemical	150,000
	Gas	180,000
Oriental High Pressure	Coke oven gas	60,000
(Mitsui)	Hydrogen gas	217,000
Sumitomo Chemical	Hydrogen gas	210,000
Denka (Mitsui)	Cyanamide	102,000
Ube Chisso	Coal	200,000
Nissan Chemical	Electrochemical	89,000
Nitchitsu	Electrochemical	85,000
Asahi Bemberg (Nitchitsu)	Electrochemical	60,000
Nippon Steel	By-product	45,000

Source: Takashima Suekichi, *Kindai kagaku sangyō shiryō*, pp. 243–244.

Note: Capacity often exceeded output, and a ranking by output would therefore place several manufacturers in different positions in this table.

ing Noguchi's earlier move into electrochemicals. One such similarity was in the type of management. Technically literate managers held high-ranking positions within companies bearing the Mitsui, Mitsubishi, or Sumitomo name. They were interested, above all, in manufacturing and in promoting technology; although they were concerned about profits, these professional managers stressed growth and product development. They differed from those company directors who had been rewarded with management positions because they were large shareholders.[52] Their roles in the firm were primarily as scientists and developers. Makita Tamaki of Mitsui Mining, Suzuki Masaya of Sumitomo, and Ikeda Kamesaburō, Nishikawa Torakichi, and Yamada Sōjirō of Mitsubishi Mining are some examples of these professional managers.

Makita Tamaki, a graduate of Tokyo University's Faculty of Engineering (1895), was Fujiyama Tsuneichi's patron in 1912 when the latter established Denki Kagaku Kōgyō (Denka) with Mitsui assistance.

Makita's move into other chemical products, particularly coal tar and dyes, began when he convinced Mitsui officials to produce coal tar during the first decades of the twentieth century.[53] Sumitomo's Suzuki Masaya, a graduate of Tokyo University's Law Faculty, founded Sumitomo Fertilizer, a phosphate company, in 1913. Mitsubishi advocates of chemical manufacture included Mitsubishi Mining's Ikeda Kamesaburō, who conducted research in Europe into ammonium sulfate and aluminum in 1926; Nishikawa Torakichi of Mitsubishi affiliate Asahi Glass, who went to Mitsubishi Gōshi as a consultant and who helped plan coal liquifaction at Mitsubishi Mining; and Yamada Sōjirō, a close advisor of Mitsubishi Gōshi's President Iwasaki Koyata. Iwasaki Koyata was also instrumental in supporting chemical manufacture.[54]

The most important consideration in analyzing the timing of zaibatsu investment in chemicals was who was doing the investing. The case of Mitsubishi diverged somewhat from the pattern indicated by Mitsui and Sumitomo, but it shared many similarities. In all three cases, changes in company structure around the time of World War I made it technically inaccurate to say chemical investment stemmed from the "zaibatsu." During this period, manufacturing branches of each company's main office gained their legal independence as corporations, though they remained closely related to the partnerships (*gōmei* or *gōshi kaisha*) that were their parent companies. In each zaibatsu, it was from the newly independent mining companies that the impetus for chemical development came. And the reasons for investment, especially in Mitsui and Sumitomo, were similar to Nitchitsu's. As new technologies were developed, new products were spun off, to be manufactured in closely related subsidiaries. The process continued until the mining companies could be called "producing *konzerns*." Mitsubishi Mining, Asahi Glass, and Mitsubishi Gōshi managers made a conscious decision to start chemical manufacture with a greater degree of diversification, making Mitsubishi's pattern of expansion different, but mining engineers were also instrumental in Mitsubishi's decision.

In some cases, financial assistance from other parts of the zaibatsu was necessary for the mining companies to begin chemical manufacture, but the decision to invest and the planning for that investment were largely undertaken by mining-company managers who saw chemicals as technologically related to production already in progress. Indeed, the mining companies and not the zaibatsus' central offices have been called the "wombs" (*botai*) from which chemical production emerged.[55] Official zaibatsu designations of chemical companies as "direct subsidiaries" of the parent companies notwithstanding, the chemical companies were not diversification attempts.[56]

A second factor in determining timing of investment was the development of technology appropriate to the zaibatsu-related mining companies' resources. Noguchi's early and continued involvement had depended on his access to hydroelectricity. The Mitsui, Mitsubishi, and Sumitomo Mining Companies had resources—coal and coke and sulfuric acid—but, until the end of the 1920s, lacked the technology to use them for electrochemical manufacture. When they acquired the new technology, their access to resources was already assured, a favorable condition for investment and production.

Thus, just as in the cases of Nitchitsu and the second wave of investors, the third wave invested when managers interested in technology development acquired technology and a large degree of managerial autonomy. And, as in the earlier investors' cases, chemical manufacture itself tended to lead to the discovery of new types of technology which, when commercially developed, engendered subsidiaries and the formation of *konzerns*.

In addition to the intramural factors cited above, investment was also affected by conditions outside the firm. A third factor favoring investment—development of a consumer market—encouraged Nitchitsu in the 1910s and the zaibatsu-related companies, particularly Sumitomo, in the 1920s and 1930s. Noguchi recognized a potentially lucrative market when farmers first became interested in synthetic fertilizers, while the third-wave companies seized the opportunity to gain a large share of the market when fertilizer imports declined in 1932 as a result of changes in foreign exchange rates.

Fourth, exogenous political conditions, such as government policy and political climate, also affected these companies' decisions to invest in electrochemical production. An entrepreneur like Noguchi, who depended on close ties to government leaders for hydroelectric rights, based investment decisions on political as well as technological considerations. The third-wave companies used a technology less dependent on electricity and may have been, therefore, better insulated from certain exogenous factors than Noguchi or Mori because they did not have to obtain hydroelectricity-generating rights. But government controls on distribution and sale of fertilizers after 1937 as well as government contracts for chemicals used in munitions, restrictions on certain types of civilian production, and wartime subsidies of fertilizer manufacture affected all fertilizer and chemical companies alike. By the time the economy was subjected to wartime mobilization, however, the new entrants in the field of electrochemicals had established themselves as leading manufacturers. The manner in which the Mitsui, Sumitomo, and Mitsubishi Mining Companies invested in chemicals during the 1920s and 1930s shows important similarities with first-and second-wave companies with respect to pre-investment management concerns and market conditions as well as differences in response to exogenous political factors.

MITSUI

Mitsui leaders had been interested in chemicals for several decades before beginning manufacture of electrochemicals and indeed before the independent establishment of Mitsui Mining, Inc. (Mitsui Kōzan KK). Individuals like Masuda Takashi, a high-ranking manager at Mitsui Bussan, had been instrumental in Mitsui's negotiations with the Italian possessors of the calcium cyanamide patent obtained by Noguchi Jun and Fujiwara Tsuneichi. Though Masuda was bested by Noguchi and Fujiwara in the 1908 negotiations, he remained interested in electrochemicals. Six years later, Fujiyama's break with Noguchi offered Mitsui Bank investors the opportunity to support Fujiyama's new fertilizer company, Denka. During World War I, Mitsui Mining directors took up the Japanese government's request to develop a domestic dye industry to offset wartime losses of imports in dyes and related chemicals. (This was Miike Dye Industries [Miike Senryō Kōgyōjo] established in 1918, later renamed Mitsui Chemical Industries [Mitsui Kagaku Kōgyō] in 1941.) Behind many of these moves was Makita Tamaki, Director of General Affairs at Mitsui Mining.

Mitsui Mining, located in the Miike and Ōmuta area, engendered Mitsui's chemical industry. Part of the reason for the involvement of the mining company in the manufacture of chemicals and, in particular, of dyes, was the presence of Makita. But more immediately important was the proximity of many raw materials essential for chemical-dye production. Coal, coke, electricity, and sulfuric acid were readily available. One material used in dyes, nitric acid, had to be brought in from the outside, but Makita soon found a way to produce that product at Ōmuta. The most important material in nitric acid is ammonia; therefore, when the Haber-Bosch license for production of ammonia became available for use by Japanese private manufacturers during World War I, Makita applied for the license. Seventeen firms and private individuals had the same idea; as we have seen, 8 of them joined together in the Eastern Nitrogen Association (Tōyō Chisso Kumiai) to collaborate in working the license. But, when the holder of the patent (BASF) demanded extraordinarily high fees and royalties equaling 68 million yen over a 15-year period (8.5 million yen per Association member), Makita Tamaki refused to go along with such a costly venture.[57] Criticized as "uncourageous" at the time, his refusal to work the license for the exorbitant fee was probably wise. He knew that 8.5 million yen would strain the company's finances.[58] In the end, the attempted collaboration failed, but not before the Eastern Nitrogen Association chartered itself as a nitrogen-importing company.[59] Mitsui did not abandon dye production.

The failure of the attempted collaboration led to several years of inactivity in the development of ammonia at Mitsui Mining. While not aggressively seeking a patent to purchase, Mitsui Mining tentatively approached America's General Chemical Corporation to study that firm's production method. The American method was found somewhat inferior, having technical problems that never surfaced in the Haber method. The next move was made by another division of Mitsui. Mitsui Bussan leaders, concerned that between 11 and 26 percent of the ammonium sulfate they handled during the years 1922 to 1929 was from foreign sources, attempted, in 1926, to bring engineers from Mitsui Mining and Denka together with English scientists to obtain a reliable production method.[60] Nothing resulted from that effort.

The same year, Makita Tamaki, pursuing every possible method of securing a license without too many risks, appealed to Kodera Fusajirō, Director of the Tokyo Industrial Research Laboratory, for help in technology acquisition.[61] But again he was unsuccessful; in 1929, Shōwa Fertilizer, having beaten Mitsui Mining to the punch, used the TIEL license to begin production of synthetic ammonium sulfate. Three options remained open to Makita if he wished to succeed in synthesizing ammonia rather than relying on imports. He could develop a new type of technology at Mitsui Mining's own research facilities. He could continue to pursue acquiring technology in Europe or the United States. Or he could obtain a license from a Japanese company abandoning production of ammonia. The former two options were either too costly or not likely to develop in the near future. The third, surprisingly, became possible in the winter of 1928 when Suzuki Shōten's bankruptcy forced the divestiture of its two ammonia plants under construction.

Suzuki Shōten had invested a great deal of time and energy in establishing its two plants—Claude Method Nitrogen Industries (Claude-shiki Chisso Kōgyō) and Daiichi Industries (Daiichi Kōgyō)—but failed to produce any ammonia by the time of its bankruptcy. The Claude method used at the facilities was fraught with problems. It required sophisticated equipment capable of withstanding 1,000 atmospheres of pressure, 5 times as much pressure as produced by Haber-method synthesis.[62] Even its French developers had serious problems with it. Trouble began at Daiichi in June 1926, and in August 1927 the machinery exploded. Still in the testing stage in France when Suzuki Shōten acquired the license, the Claude method remained weak, even after Mitsui took over operation of the plants at Hikoshima and made substantial improvements.[63] But, despite the method's flaws, Mitsui Mining's Makita Tamaki saw it as a potentially profitable investment.

Suzuki Shōten was unfortunate in its development of new electrochemical technology, and its financial prospects were even less favorable. Rapid expansion during the 1920s supported by weakly secured loans from the Bank of Taiwan precipitated its growing debt and overextension. The Bank had accepted Suzuki Shōten's entire stock issue for Daiichi Industries and Claude Method as collateral for further loans. The bankruptcy left the Bank of Taiwan with control of both plants.

Some of the managers at each of the two plants began talks with different groups of potential purchasers: Daiichi managers met with Sumitomo and Dai Nihon Fertilizer; Claude Method managers, with Mitsui Mining. In the meantime, the Bank of Taiwan had placed Isobe Fusanobu, a director at Daiichi, in charge of disposing of the plants. Unlike some other Daiichi managers, Isobe opted for a bailout by Mitsui Mining. He then approached two middle managers at Mitsui, the head of Mitsui Mining's Meguro Laboratories and the head of the Industrial Affairs Section at Denka, who brought Isobe's suggestion to Makita Tamaki.[64] While Makita was considering Isobe's offer, a well-placed auditor at the Bank of Taiwan was courting Mitsui Gōmei. This auditor worked not only for the Bank but also for Suzuki Shōten, and was well aware of the condition and availability of the bankrupt company's fertilizer holdings.[65] By January 1928, the fertilizer plants of the failed company were placed in trust under joint management of Mitsui Mining and Denka.

Mitsui Mining supplied much of the capital for research and plant modification during the following year, while Denka supplied engineers from Ōmuta experienced in electrochemicals. Throughout the 1920s, Denka had continued to produce ammonium sulfate by the cyanamide method, by then an outmoded technology. Managers at Denka as well as the company's patrons at Mitsui had begun to consider converting to a more efficient method of ammonia synthesis.[66] Denka's assistance in managing the Claude-method plants at Hikoshima during 1928 provided the opportunity for Denka engineers to observe and learn new technology.

From January 1928 to January 1929, managers and top engineers from Mitsui Mining and Denka operated Daiichi and Claude. A few employees from Daiichi, including Isobe Tatsuji, also cooperated with the Mitsui men, although many engineers refused to work for

Mitsui and transferred to Sumitomo's electrochemical fertilizer works.[67] Scientists were heavily represented in the management group. During their first six months of work, they experimented with the Claude method, but, like their predecessors at the Suzuki plants, were largely unsuccessful in obtaining satisfactory results. Frustrated, many suggested that Mitsui Mining reopen negotiations for the TIEL method even though they knew Shōwa Fertilizer was soon to begin working the license. Many Mitsui scientists had lost confidence in the feasibility of the Claude method. Makita went one step farther and criticized the scientists at Suzuki Shōten for having even considered acquiring it. Indeed, he called the engineers hired by Suzuki Shōten incompetent "scatterbrains" (*awatemono*) and "thieves" (*dorobō*) for attempting to use the Claude method.[68]

Before abandoning that process, Makita made one last attempt to perfect it. In June 1928, he sent a study group from the Claude plants to France to investigate the method. Not surprisingly, the French developers at Aire Liquide were themselves unsatisfied with the process and were unable to offer clear answers to the questions asked by the visiting Japanese group. The Japanese were disconsolate and wired Makita that Mitsui Mining might best cut its losses by dropping the Claude patent in favor of the TIEL patent. Appropriately, a TIEL scientist had been accompanying the Mitsui group at the insistence of TIEL Director Kodera Fusajirō.[69] Although highly dissatisfied with the Claude method, Makita was unwilling at that point to support an alternate method of ammonia production. Enjoining the scientists to decide for themselves, he wired, "Act only after carefully considering whether you made the trip [to Europe] to establish negotiations or to break them."[70]

While the crew of engineers investigated the Aire Liquide operations, Isobe Fusanobu was lining up help for license negotiation. Sent separately to Europe, Isobe arrived first in London, where he solicited aid from Mitsui Bussan's European office in negotiating with Aire Liquide. Isobe succeeded in lowering the purchase price of the Claude method by one-third, to a level where further research and development to rectify problems in the technology would be economically feasible. In January, three months after Isobe and the

Japanese group returned to Japan, Mitsui Mining executives decided it would be worthwhile to use a modified Claude method. They decided to assume full responsibility for the Suzuki plants they had been holding temporarily, and purchased them and their technology cheaply, for 1.5 million yen.[71]

Mitsui Mining's deliberate care in the initial license acquisition was consistent with the company's somewhat more conservative style in strategic decision-making. Considerable research was still necessary to make the patent work, but Mitsui Mining had a license in hand by 1929. Throughout 1928, Mitsui Mining continued to carry out research, sent a scientific team to Europe, and negotiated lower fees and royalties. Though their process of acquiring the license had not demanded that Mitsui and Mitsubishi assume much risk, implementation of the license and diversification in new products in the next decade did require management to take many risks similar to those borne by the pioneering companies like Nitchitsu.

The first step Mitsui Mining took toward establishing its own role in ammonia production was to rename the Hikoshima plants purchased from Suzuki Shōten. In January 1929, they became the Nitrogen Plant of Hikoshima Refining, Inc. (KK Hikoshima Seirenjo Chisso Kōjō).[72] Having earlier decided that the Claude method for hydrogen production was dangerous because of the amount of pressure involved in the process, Mitsui substituted hydrogen produced in a coke oven. The new Special Nitrogen Research Section (Rinji Chisso Kōgyō Shikenbu) set up by Mitsui Mining in April 1929 to investigate the use of hydrogen from coke ovens confirmed the practicality of modifying the Claude process. Efficiency dictated that Mitsui try to find a nearby source of the necessary hydrogen; among the numerous coke ovens at Mitsui Mining locations, those at Miike Dyes appeared most attractive. Miike Dyes generated an enormous amount of hydrogen gas. In 1929, the gas had been used as an energy source for generation of electricity, but frequent equipment failures drove the price of a kilowatt-hour by gas-fueled generation to exorbitant levels. It was clear that using the hydrogen in ammonia production would make better use of it.[73] Greatest efficiency would be achieved by building the ammonia operation as close as possible to the hydrogen source. Sulfuric acid, which when combined with ammonia produced ammonium sulfate, was fortunately produced at Mitsui Refining, a neighbor of Miike Dyes. Therefore Miike was chosen as the site of Mitsui Mining's new modified-Claude-method plant because of its proximity to the raw materials for fertilizer manufacture. Makita was about to realize his dream.

The ground was broken for the ammonia plant at Miike in February 1930, and construction continued through the year. In August 1931, Miike Nitrogen Industries, Inc. (Miike Chisso Kōgyō KK) was founded. Capable of producing 36,000 tons of ammonium sulfate annually, Miike Nitrogen was controlled by Mitsui Mining, despite its status as an independent company. Mitsui Mining held 110,000 of the 200,000 shares issued (capitalization 10 million yen), Denka held 40,000 (a token for Denka's contributions during the exploratory year in 1928), and the remaining 50,000 were made available to the public.[74] This constituted a significant degree of public ownership, especially when compared with Nitchitsu, whose shares were not publicly offered. But this did not imply public influence over the company. In fact, Miike Mining controlled Miike Nitrogen, through domination of both management and capital. Miike Nitrogen had been established as an independent company only for tax purposes. Miike enjoyed an extensive line of credit from Mitsui Mining and shared its research and technology. Its product was first marketed in Japan as "Mitsui Ammonium Sulfate." And Miike merged with Oriental High Pressure, another subsidiary of Mitsui Mining, in February 1937 as soon as the company exhausted the 5-year business-tax exemption granted independent new companies.[75]

Miike Nitrogen grew rapidly. Employment jumped from 129 in 1932 to 653 in 1937. When it merged with Oriental High Pressure, the number of employees in the combined company reached 1,367. Chemicals were clearly a significant subsidiary of Mitsui Mining.[76] Having ventured into ammonia production, Mitsui Mining was finally ready to invest energetically in fertilizers and related chemicals. The new methods of production had given Mitsui access to resources required in chemical manufacture. Moreover, the recent turnabout in fertilizer imports had opened the market to domestic producers. In addition, access to capital for development was assured; Mitsui Mining and Denka held 75 percent of all shares, and loans up to 2.5 million yen were also available from Mitsui Mining. The professional managers at Miike Nitrogen were skilled engineers who understood new technology, including that which was developed while the Miike plant was under construction. In 1929, scientists at DuPont improved the efficiency of the Claude technology. The DuPont method recovered 99 percent of the generated hydrogen, compared to 94 percent by other methods.[77] In January 1932, therefore, Mitsui Mining purchased the license for DuPont's modification of the Claude method for 300,000 dollars, and began making plans for building a new plant to use the new process.

The new plant was built right next to Miike Nitrogen but was established as an independent corporation, Oriental High Pressure (Tōyō Kōatsu KK). Construction began in 1932, but the company was not officially established until April 1933, a strategy that maximized the length of time in which Mitsui Mining could take advantage of the law permitting tax exemptions for new corporations during their first 5 years. Oriental High Pressure, capitalized at 20 million yen, was twice as large as Miike Nitrogen. Of the 400,000 shares issued, 300,000 were held by Mitsui Mining, 40,000 by Miike Nitrogen, and the remaining 60,000 were offered to the public.[78] As in the case of the Nitchitsu Konzern, most (85 percent) of the shares were held by the parent company or another subsidiary, permitting flexibility in accounting. Tax benefits were apparently of greater concern to Mitsui Mining than flexible accounting, however, because the merger of Miike Nitrogen and Oriental High Pressure in 1937, when Miike's tax benefits ran out, obviated this flexibility.

As valuable as ammonia was to Mitsui Mining at this time, related chemicals were potentially as important. The first to be produced at a Mitsui facility was methanol, the high-powered fuel used in a variety of engines, including airplane engines. Methanol technology paralleled ammonia technology: Indeed, other producers of ammonia products, like Nitchitsu and DuPont, also manufactured methanol. An ammonia plant with technology based on coal gas was an appropriate location for methanol manufacture. The gas produced by coke ovens, which contained the hydrogen used in ammonia, was 35 percent methane.

Mitsui Mining moved into methanol production in 1928, when the company joined four other investigators in setting up the Methanol Research and Development Association (Metanoru Shisei Kumiai). Three years earlier, Shibata Katsutarō, the TIEL researcher instrumental in developing the TIEL method of ammonia synthesis, discovered a method for producing methanol. Impressed with Shibata's results, the newly formed Association studied the process over a two-year period, concluding its investigations in 1930.[79]

Mitsui Mining was then ready to begin production of methanol. The onset of the Depression may have impeded construction of a new plant, as building began only in February 1932, one month after Miike Nitrogen began producing ammonia.[80] Mitsui Mining named its methanol plant Synthetic Industries, Inc. (Gōsei Kōgyō KK); offically incorporated in July 1932, it used the coke-oven gas produced by the Claude-Method plant at Hikoshima as a basic raw material. Synthetic Industries produced Japan's first synthetic methanol on 28 December 1932.[81]

Shibata Katsutarō, Japan's pioneer in synthetic methanol production, soon quit his job at TIEL and joined the Mitsui operation. In February 1933, Shibata became a director of Mitsui Mining and head of the Hikoshima plant, following a pattern of *amakudari*, the practice of government employees' joining private enterprises in high managerial positions.[82] Shibata contributed his considerable technical skills to Mitsui Mining. By November 1933, the old Claude Method plant was producing methanol by the new DuPont method, and, in March 1935, the Claude facility was merged with Synthetic Industries. Two years later, technology developed at Synthetic Industries led to the production of carbolic acid and to Japan's first urea. (Sumitomo authorities debate the latter point, however, claiming Sumitomo's urea antedated Synthetic Industries'.) By 1938, Synthetic Industries merged with Oriental High Pressure, making Oriental a well-diversified chemical operation for Mitsui Mining.

The role of chemicals in the overall Mitsui zaibatsu increased rapidly during the next seven years, since chemicals were used as munitions. In September 1945, 14.1 percent (worth 132 million yen) of all Mitsui zaibatsu holdings were in chemicals. Only machinery with 26.3 percent (246 million yen) and mining with 24.1 percent (226 million yen) were larger investments. Furthermore, Mitsui's investment in chemicals jumped from 84 to 132 million yen in the last eighteen months of the war.[83]

Thus, as Mitsui Mining developed its chemical subsidiaries, particularly Oriental High Pressure, it became, like Nitchitsu, a modern vertically integrated company with technological interrelationships among the legally independent but closely connected corporations that were its production facilities. (See Figure 7.) It differed from Nitchitsu only in that certain key functions, like marketing, were handled outside the *konzern.*

SUMITOMO

Sumitomo's leaders first became interested in chemical manufacture as a way to deal with the pollution produced by their copper-smelting plant at Besshi. At the beginning of the twentieth century, farmers near the plant complained about the sulfurous gases poured into the air by the refinery. Sumitomo officials saw the problem as a social one and tried, in vain, to persuade the farmers to stop their demonstrations against the company for polluting their fields. Finally, they recognized the problem as environmental and concluded, in 1907, that the only solution was to restrict smelting activity. Of course, management did not think restricting output was a long-term solution. Sumitomo President Suzuki Masaya began searching for a method of controlling pollution while continuing to produce copper. He decided to use the effluent to produce sulfuric acid and then combine that acid with phosphatic rock to make phosphate fertilizer. Sumitomo's mining operation, like Mitsui's, then, spawned its chemical production.

In September 1913, after investigating phosphate-production methods throughout Europe and the United States, Suzuki settled on a sophisticated method pioneered by the Ernst Hartmann Company of Germany.[84] The Sumitomo Fertilizer Plant (Sumitomo Hiryō Seizōjo) was established under the direct management of

FIGURE 7 Integration at Ōmuta, 1941

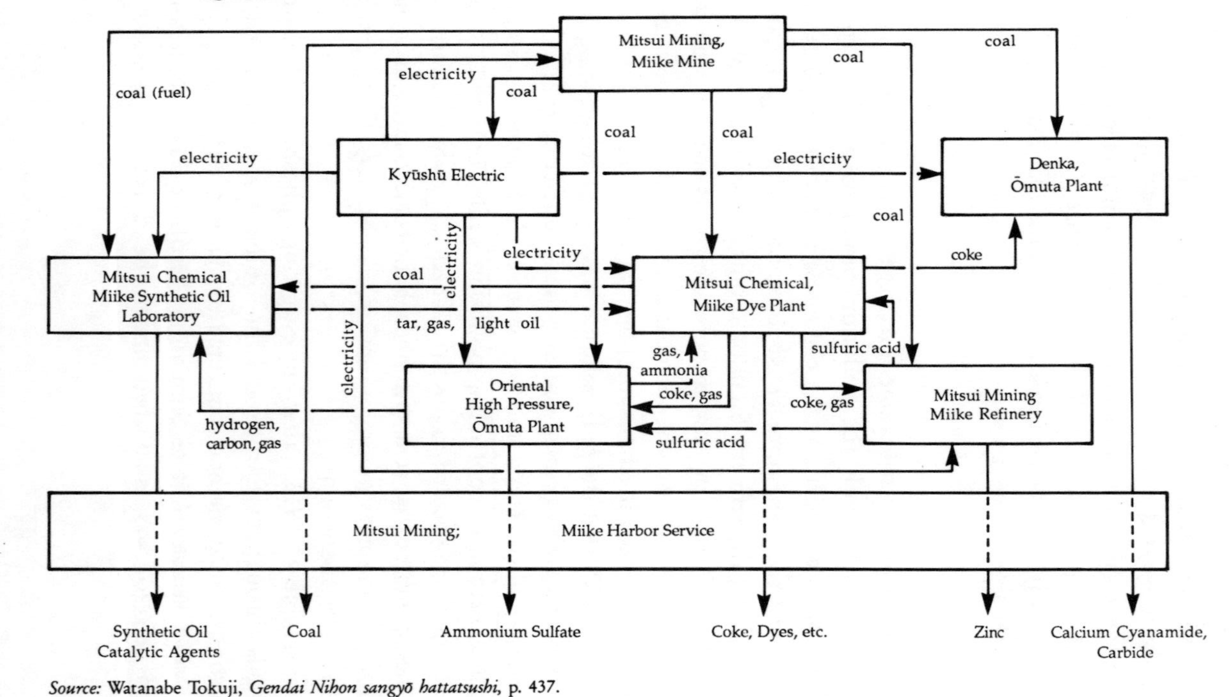

Source: Watanabe Tokuji, *Gendai Nihon sangyō hattatsushi*, p. 437.

Sumitomo Sōhonten (General Headquarters) on 22 September 1913, and the plant received prefectural permission to produce fertilizer in November. Construction commenced immediately on a facility to make one half ton daily of sulfuric acid and 75,000 tons annually of superphosphate and compound fertilizers.[85] Kajiura Kamejirō guided the plant's early development as its manager.[86] Sumitomo Fertilizer encountered difficulties almost from the start, however, when the onset of World War I prevented Hartmann engineers from remaining in Japan to supervise the application of the new technology. When the Germans returned home, Sumitomo was forced to replace its original plans with a more conventional process for superphosphate production.[87] Another problem was finding skilled technicians. At first the plant had just 9 management and technical employees and about 20 production workers. These 9 managers were divided in 1913 into just two sections, Production and Management. Moreover, most of the technical employees were transferred from Besshi Mines and knew nothing about chemicals. Three new experts had to be hired from outside the company.[88] Despite setbacks in installing technology and in locating knowledgeable managers, the company eventually came on stream in August 1915.[89]

Just as superphosphate production was starting up, Sumitomo President Suzuki began considering other products that would both be profitable and help him solve his pollution problem. He reasoned that new methods for synthesizing ammonium sulfate were potentially more effective than superphospate manufacture in controlling pollution, because ammonium sulfate would use up twice as much of the effluent sulfuric acid. But he could not get access to the ammonia process he wanted, the Haber method, during the first years of the war. Other methods were available, however, and Suzuki was open to suggestions. The scientist Takamine Jōkichi encouraged him to consider the recently developed General Chemical method. Sumitomo leaders expressed a genuine interest in the American method, but the Americans were less enthusiastic about Sumitomo's suggestion for an international effort with American, German, and Japanese input.[90]

President Suzuki maintained his interest in synthesis of ammonia throughout the war,[91] but postwar declines in the profitability of his chemical plant made large-scale investment in new processes impossible.[92] Thirty-three of the plant's 251 workers were let go in 1921. It appeared that the company was facing bleak times. But Suzuki's belief that production of ammonia was important to the nation (because of its role in the manufacture of explosives) impelled him to urge his fellow businessmen to pool their funds in the Eastern Nitrogen Association. Suzuki continued to make plans to produce ammonia using the expensive Haber-Bosch method long after his colleagues had abandoned such expectations. He even made a personal trip to BASF in Germany in December 1920, but returned home empty-handed when Bosch demanded an additional 3,000,000 yen for the method.[93]

Suzuki's efforts coincided with organizational changes in Sumitomo that later facilitated investment. Sumitomo Sōhonten, under which many separate undertakings had been managed directly, was reorganized on 26 February 1921 as Sumitomo Gōshi (Sumitomo Ltd.). The new company owned many of its subsidiaries' assets but did not manage them directly.[94] To be sure, top managers at Sumitomo Gōshi exercised considerable control over the subsidiaries; they even appointed subsidiaries' leading managers. But, for the chemical plant, as for other component enterprises of Sumitomo, independent corporate status meant that company managers carried out plans consonant with Sumitomo Fertilizer's own business strategies. This section will not focus on control, however, because a preoccupation with the issue of control by Sumitomo Gōshi in Ōsaka obscures the important role of the Besshi mine environment in the growth of chemical production. Like Mitsui Mining, Besshi was the origin of chemical manufacture in Sumitomo.

The fertilizer plant of Sumitomo was given an independent existence as Sumitomo Fertilizer Company, Inc. (KK Sumitomo Hiryō Seizōjo) on 1 June 1925. Established with a capitalization of 3 million yen (1.8 million paid up), it was Sumitomo Gōshi's sixth subsidiary.[95] On 3 June, its headquarters were moved from the plant at Niihama to the building that housed both Sumitomo Gōshi and Sumitomo Bank in Ōsaka.[96]

Both Sumitomo Gōshi and Sumitomo Fertilizer benefited from

the independent status of the latter. Although Sumitomo Fertilizer was almost wholly owned by Sumitomo Gōshi, the new corporate structure offered tax breaks to Sumitomo and independence to Sumitomo Fertilizer. That is, the new Sumitomo Fertilizer could avoid many of the corporate taxes previously assessed on the fertilizer plant within Sumitomo Sōhonten.[97]

The reorganization of Sumitomo Fertilizer in 1925 expedited the firm's move into the manufacture of electrochemicals. But the reorganization lagged three years behind Gōshi's reorganization. Then, in November 1922, Gōshi President Suzuki Masaya—a former bureaucrat from the Ministry of Agriculture and Commerce, a professional manager, and a champion of high-technology industry—having become ill, was replaced by Tanaka Kanekichi. Few top managers at Sumitomo Gōshi had matched Suzuki's enthusiasm for chemical production. Although new directions in fertilizer and other chemical manufactures at Gōshi increasingly came from on-site managers at Besshi like Kajiura Kamejirō, until 1925 Sumitomo Fertilizer was still managed by the main office. Without Suzuki, the impetus for developing chemicals disappeared. But the incorporation of the fertilizer company in 1925 led to its reorganization and the appointment of Ogura Masatsune, a Director of Sumitomo Gōshi, as President of Sumitomo Fertilizer, and of Kajiura Kamejirō, the manager who had been with the company from the beginning, as Managing Director.[98] They proceeded to hire scientists and engineers from Suzuki Shōten, the bankrupt company that spawned Mitsui Mining's chemical operations several years earlier. These were talented men, Makita Tamaki's opinion notwithstanding.[99] Four months after the independent incorporation of Sumitomo Fertilizer, Kajiura was replaced by Hidaka Naoji as Managing Director.

One of Hidaka's first decisions—after having to deal with a strike during his third month as manager[100]—was to resume preparations for synthesizing ammonia. Much of Sumitomo's production was marketed by designated retail shops and agricultural associations. The shops were bound contractually to sell only Sumitomo products, a distinct liability when Sumitomo made no nitrogenous fertilizer and farmers demanded it. Hidaka's concern that farmers would patronize other dealers led him to advocate ammonium sulfate manufacture. The potential profitability of ammonium sulfate also helped sway his decision.[101] But a method of production had still to be chosen.

President Ogura and his staff investigated various possibilities. Hidaka inspected Nitchitsu's and Suzuki Shōten's facilities. The head of the Niihama plant learned of a new method developed by America's Nitrogen Engineering Corporation (N.E.C.) while visiting with a friend at Mitsui Bussan in New York. This method held promise, since Sumitomo Fertilizer managers knew the American company from earlier negotiations. The Nitrogen Engineering Corporation had adopted many aspects of the General-Chemical method, itself basically a derivative of the Haber method. The N.E.C. process recommended itself to Ogura's team, who felt the N.E.C. engineers had sufficient experience to help Sumitomo install the plant in Japan. The N.E.C. method was also a coal-based process, rather than an electrochemical one like the Casale or the TIEL.[102] This was helpful because electrical capacity at Sumitomo Fertilizer's Niihama site was low. As a final test, Sumitomo had a member of the research staff at Niihama investigate the N.E.C. method; his report helped persuade the company that it was effective.[103]

Sumitomo purchased the N.E.C. license in 1928 and sent an inspection team to N.E.C.'s United States plant in March of that year. Construction of a plant, under the supervision of the N.E.C. engineers, began soon thereafter. A small pilot plant was completed by November 1928, and a large plant (15 tons per day capacity) was completed by November 1929. Construction costs reached 3,950,000 yen, of which 2,100,000 yen went for the purchase of foreign machinery and materials, and 1,850,000 for domestic materials.[104] But it took over a year for the first product to come off the line. Ammonia was first produced on 9 December 1930 and ammonium sulfate in April 1931.[105] During that time, Sumitomo Fertilizer had several changes in top management; these were filled by Sumitomo Gōshi appointments. Capitalization had been increased to 10 million yen when construction began, and subsequently increased with a series of plant expansions in the 1930s.

Sumitomo's product reached the Japanese market at a bad time. In

1931, European dumping had made competition for the market particularly intense. In 1928, costs per ton were estimated at 75 yen, while prices were 100 yen. But, by 1931, costs exceeded prices. Sumitomo Fertilizer leaders reacted to the poor market by increasing their outputs of liquid ammonia and finished fertilizer.[106] This strategy paid off, because expanded plant capacity came on stream in February 1933 as prices were rising to 90 yen. Although Sumitomo Fertilizer ended up by earning healthy profits, its tactic was perceived by contemporaries as contrary to Sumitomo Gōshi's policy of scaling down investments and production during the Depression.

Like other ammonia manufacturers, Sumitomo soon began to diversify. Nitric acid was the first new product they attempted. This was stimulated in 1932 by a request from the head of the Explosives Section of the Army's Ōshi Arsenal that Sumitomo Fertilizer work on nitric acid. To produce high-quality nitric acid for explosives, high-quality sulfuric acid was necessary.[107] This, in turn, led to the fertilizer company's contracting with America's Chemical Construction Corporation in 1933 for aid in construction of a new sulfuric acid plant using C.C.C.'s contact process.[108] The same year, scientists at Sumitomo Fertilizer decided to turn the carbolic acid gas produced during the manufacture of ammonia to good use and began planning a facility to make urea. These were profitable diversifications for Sumitomo Fertilizer. In 1934, the company produced 18.2 percent of the national output of nitric acid, which increased to 26.2 percent in 1936.[109]

Diversification made Sumitomo Fertilizer more than a fertilizer company, and, commensurate with its growth, the company decided to expand and change its name to Sumitomo Chemical Industries, Inc. (Sumitomo Kagaku Kōgyō KK) on 15 February 1934. Capitalization was raised to 20,000,000 yen, and shares were offered publicly to help pay for the expansion. Public sale of shares meant that "outsiders" like O.S.K.'s Murata Shōzō gained positions on the Board of Directors.[110]

Several months later, in June 1934, Sumitomo Chemical created Sumitomo Aluminum Refining, Inc. (Sumitomo Aruminiumu Seiren KK) as a subsidiary. Production began in February 1936, but the company posted no profits until 1939, after three years of trial and error in developing refining methods.[111] Assistance from the military and the Commerce and Industry Ministry facilitated the costly venture into aluminum refining. As the war effort expanded and orders for aircraft-quality aluminium rolled in, the aluminum plant began to earn hefty profits.[112] In 1935, the Navy urged Sumitomo Chemical to make formalin and methanol for fuel, which Sumitomo management agreed to do in 1938, because of the similarity of the technology for formalin, methanol, and ammonia.[113]

The next major innovation at Sumitomo was the construction in 1936 of the company's own coke ovens. Before 1936, the gas for ammonia manufacture had been purchased from the Ōsaka and Tokyo Gas Companies, but large increases in fertilizer output as well as the desire to liquify coal to make synthetic fuel encouraged the Sumitomo Chemical Company to run its own coke ovens.[114] These coke ovens permitted Sumitomo to begin selling heavy sulfuric acid and nitric acid to Japan Dyes, Inc. (Nihon Senryō Seizō KK); the two companies subsequently cooperated during World War II in supplying coal-tar products for military use.[115]

The war's expansion and the rise in military orders generally produced greater diversification at Sumitomo Chemical (see Table 43). Not only did the company become one of the largest producers of methanol and formalin, but it also started manufacturing isobutanol in 1938 for aviation fuel for the Army and began research in synthetic oil in cooperation with the Manchukuo authorities. In 1941, the Army pushed Sumitomo to use the German I.G. Farben method to increase ammonium sulfate output in Manchukuo and, when the company found implementation of that method problematic, the Army designated the firm a National-Policy Company in Manchukuo and gave it aid as an incentive to increase its output.[116]

Sumitomo Chemical's ties with the military intensified during the early 1940s. In 1941, the Army ordered hexogen and nitric acid for explosives. (Chōsen Chisso was the other company with similar orders.) Sumitomo Chemical attempted to manufacture synthetic rubber under contract, but production was not spectacular—a total of 18.07 tons between 1943 and 1945.[117] In January 1944, Sumitomo

TABLE 43 Diversified Sales in Sumitomo Chemical, 1929–1938 (10,000 yen)

Product	*1929*	*1930*	*1931*	*1932*	*1933*	*1934*	*1935*	*1936*	*1937*	*1938*
Total sales	621	542	468	629	1027	1389	1657	2024	3099	3729
Sulfuric acid	57	59	66	73	81	122	180	178	309	443
Superphosphate	385	358	230	274	257	274	316	302	458	456
Compound Fertilizer	179	125	94	99	234	214	264	304	359	541
Ammonium Sulfate			71	163	396	588	513	765	1213	1121
Ammonia			5	14	37	65	103	98	108	96
Ammonium Carbonate					7	4	6	8	9	8
Caustic Fertilizer					14	38	134	151	257	272
Nitric Acid						20	28	32	52	118
Potassium Carbonate							3	3	8	9
Sodium Phosphate							0.5	0.7	2	1
Sodium Nitrate							0.7	6	9	15
Pure Soda							0.07	8	19	7
Coke									15	79
Methanol										47
Formalin										33
Urea										11

Source: Mikami Atsufumi, "Sumitomo," p. 131.

Chemical was designated by the Munitions Ministry as one of 150 Munitions Companies (*gunju kaisha*). Its plants were, accordingly, renamed and some of its managers were designated as "responsible" for reporting to the Ministry.[118] Thus, the war eroded managerial independence for Sumitomo Chemical as it had for Nitchitsu.

While Sumitomo Chemical's managers still determined company policy, they molded their company's structure to respond to production and technological requirements. In 1934, the company that had been simply organized—at that time, it had 3 sections and 1 factory—expanded to 2 divisions (General Affairs and Sales) plus 1 factory, the Niihama plant, which itself had 8 sections and 15 subsections organized by product type. The following year, this structure was

Table 44 Subsidiaries and Investments of Sumitomo Chemical, 15 August 1945

Company	*Total Capital (1,000 yen)*	*Sumitomo Chemical Share (%)*	*Sumitomo Zaibatsu Share (%)*
Subsidiaries			
Sumitomo Taki Chemical	12,000	50.8	50.8
Shinto Paint	3,750	40.0	40.0
Special Glass Light	500	67.0	97.0
Investments			
Nihon Carbide	13,000	50.7	50.7
Manchuria Shinto Paint	2,500	40.0	40.0
Eastern Nitrogen Industries	5,000	58.0	70.5
Taiwan Organic Synthetic	5,000	10.0	10.0
Nihon Sulfuric and Nitric Acid Control Inc.	5,000	11.5	15.9
Imperial Compressed Glass	7,500	12.0	61.5
Sumitomo Associated Electric	20,000	18.8	97.0
Andō Light Metals	200,000	15.0	50.0
Sumitomo Aluminum Refining	20,000	17.5	87.5
Korean Sumitomo Light Metals	80,000	10.0	50.0
Atsukawa Coal Mines	2,000	50.0	100.0

Source: Sumitomo Kagaku, p. 137.

again changed: 1 division and 1 section were added and the Niihama plant ceased to be an organizational level within the company structure. That is, functionally defined divisions completely replaced the factory as part of the management reporting structure. In 1935, the Sales Division and the General Affairs Section were located at Ōsaka, and the Production Division and Business Division at Niihama.[119]

Ownership of the company also changed with expansion as additional capital had to be brought in from the outside. While Sumitomo Gōshi and Sumitomo family members held 98.17 percent of Sumitomo Fertilizer shares in 1928, their ownership of Sumitomo

Chemical dropped to 33.2 percent in 1937 and 24.8 percent in 1945 (see Table 44).[120]

Chemical production was generally quite profitable within the Sumitomo zaibatsu. Chemical profits rates were consistently higher than average profits for all types of investment within Sumitomo. Only machinery manufacturing usually showed a higher rate of profit.[121] Interestingly, Sumitomo Chemical, which made a wide variety of electrical products, earned profits for some products that were losers for Nitchitsu. Indeed, with few exceptions, Sumitomo Chemical had no products that failed to make a profit.

MITSUBISHI

Investment in chemicals in the Mitsubishi zaibatsu, like that in Mitsui and Sumitomo, grew from the mining industry. As in the other zaibatsu, scientists and engineers in Mitsubishi Mining spearheaded investment; in Mitsubishi, they were joined by scientists at Asahi Glass, the Mitsubishi-related glass company. But Mitsubishi investors approached investment in chemicals in a manner different from those in Mitsui Mining or Sumitomo Mining. Mitsui and Sumitomo first established basic production of a few chemical products; later, a wide variety of new products were derived from the related technologies. Mitsubishi planners, on the other hand, wished to begin all aspects of related production at once and to establish a coordinated, diversified chemical company. They took as their model the mature I.G. Farben. Carefully planning all aspects of their chemical operations, the Mitsubishi investors left little to chance in their diversification process. Thus, through more cautious investment, they estabished the last of the third-wave companies. Once they began production, they no longer lagged behind the first-, second-, or even third-wave firms in sophistication or diversity.

The development of Mitsubishi's chemical industry had its roots in the incorporation of Mitsubishi Mining as a separate company following the reorganization of Mitsubishi Limited Partnership (Mitsubishi Gōshi Kaisha) in 1917. Tax-law changes in 1913 made large partnerships (*gōshi* and *gōmei kaisha*) uneconomical. By taxing joint-stock companies at much lower rates, the law encouraged companies to offer their shares publicly. But, rather than do that, Mitsubishi used a different method of decreasing its tax liability. It decreased its size by establishing Mitsubishi Mining and other companies as independent firms.[122] Making Mitsubishi Mining independent aided chemical development—although this had not been a reason for making the company independent—because the smaller Mitsubishi Mining, more focused on certain types of production, was more likely to advocate chemical manufacture than the larger Mitsubishi Gōshi.

The other principal force in establishing a chemical industry related to Mitsubishi was Asahi Glass. Founded in 1907 by Iwasaki Toshiya, brother of Mitsubishi Gōshi President Iwasaki Koyata and nephew of the zaibatsu's founder, Asahi Glass relied on imported sources of soda ash for glass manufacture. Elimination of soda ash imports during World War I persuaded Toshiya that domestic sources were needed. This led Asahi Glass to use its own coking ovens to produce soda ash. Soda manufacture also required ammonia, and therefore Toshiya pressed for additional sources of ammonia as well.[123] But, though Toshiya tried to get Mitsubishi to manufacture ammonia, Koyata did not agree to new investment at that time. In fact, Koyata was not convinced that Mitsubishi had sufficient uses for ammonia to achieve economies of scale in production until the 1930s. Koyata changed his mind in the 1930s because, by then, Mitsubishi Mining was producing coal tar, which was used to make dyes, and dyes required ammonia. But, even during the long period before coal tar was produced, Toshiya did not abandon his quest for innovation. He continued to subsidize chemical research at Asahi Glass. In 1919, he set up the Asahi Glass Research Laboratory, where research on a wide variety of chemicals, not merely glass or soda ash, was conducted.[124] Clearly, Iwasaki Toshiya was strongly interested in chemicals.

It is also clear that Iwasaki Koyata continued to play an important role in approving the investment in new technologies even after the reorganization of Mitsubishi in 1917. It was Koyata who expressed an interest in the electrification of the Yalu River in 1919 and the Changjin in 1923, hoping to produce aluminum and electrochemicals.[125] And it was a scientist at Asahi Glass, Nishikawa Torakichi, who conducted negotations with I.G. Farben to develop the rivers jointly.[126]

Although the joint venture collapsed because of logistical problems of international cooperation and adverse economic conditions in both countries at the end of the 1920s, Koyata remained interested in chemical manufacture. Indeed, his enthusiasm was heightened after he read a 1926 report by Ikeda Kamesaburō, General Affairs Director for Mitsubishi Mining, on conditions in the chemical industries of the United States and Europe.[127] Ikeda continued to argue in favor of chemical manufacture, as did Nishikawa. Nishikawa had exceptional influence among researchers at Mitsubishi Mining and Asahi Glass; Koyata therefore listened attentively to Nishikawa's explanation of the benefits of a coordinated chemical industry. But the cautious behavior of many of Mitsubishi's top managers, together with the company's shortage of capital during the Taishō period due to its establishment of steel production and its extensive loans to newly independent subsidiaries, kept Mitsubishi from setting up chemical production in the 1920s. (The total value of Mitsubishi's loans and negotiable securities reached 34,100,000 yen in 1919.)[128] Improved exogenous conditions—a better market—as well as improved company conditions—an end to the severe capital shortage—in the 1930s presented Mitsubishi with different circumstances for investment.

When the depressed Japanese dye industry began to recover by the early 1930s, Koyata paid close heed to a 1933 report calling for a major effort in chemicals, issued in 1933 by top managers in Mitsubishi subsidiaries. An official of the Commerce and Industry Ministry lent his support to these tentative plans. Koyata then convened the Temporary Tar Industry Investigation Committee (Rinji Taru Kōgyō Chōsakai), made up of managers from Asahi Glass and Mitsubishi Mining.[129] Meeting in April 1934, these managers laid the groundwork for Japan Tar, Inc. (Nihon Taru Kōgyō KK), founded on 1 August 1934 to make dyes, tar, benzol, alcohol, and tar products from the residue of coke manufacture.

Mitsubishi Mining and Asahi Glass each invested 5,000,000 yen in the new company; Mitsubishi Gōshi did not invest but Iwasaki Koyata endorsed the venture.[130] Mitsubishi Mining also contributed its machinery for making coke, which it had previously used for steel manufacture. With the establishment of the coke plant, the new chemical company had its own supply of intermediate industrial chemicals for dye manufacture. Dye production began in April 1935. Dyes also required nitric acid, which was supplied by Sumitomo Chemical. At the July 1935 meeting of the Board of Directors, it was decided to begin ammonia production to make nitric acid. The Board selected I.G. Farben's method, and contracted for the license in February 1936.[131] Mitsubishi's Japan Tar, which had been founded on 1 August 1934, began production with this license only in December 1937. Unlike Mitsui, Mitsubishi founded its chemical company long before actual production of chemicals began. Mitsubishi managers wanted to do more studies first. One such study concerned additional uses for the ammonia the company would produce, because the planned nitric acid manufacture used relatively little ammonia by itself. This prevented economies of scale unless other uses for ammonia were found. Koyata was persuaded to use the rest of the ammonia for ammonium sulfate.[132]

In addition, new types of machinery were tested by Mitsubishi, especially the Winckler oven, an oven superior to the older coke ovens (used by Sumitomo and Mitsui) because it produced more gas and could be operated continously. As Mitsubishi had longstanding ties with I.G. Farben dating from their attempted collaboration in Korea, much new German technology was available to the company.[133]

By September 1936, most of the equipment was in place. The effort had taken two years and the energies of men from Mitsubishi Mining and Asahi Glass. Ammonia was to be just as important as tar to Japan Tar, so, in honor of the firm's completion, the Board of Directors decided at their October 1936 meeting to rename the company Japan Chemical (Nihon Kasei) and raised its capitalization to 30 million yen.[134] Yamada Saburō became President of Japan Chemical, and most of the new company's engineers moved from Asahi Glass.[135] By December 1937, the ammonia plant had started production. It attained 76 percent of its 21,000-ton capacity by 1939 and 87 percent by 1941. Ammonium sulfate capacity was 80,000 tons; 63 percent of that was produced in 1939 and 75 percent in 1941.[136]

The move into ammonia was highly profitable; profits jumped

from 8,000 yen in the second half of 1936 to 1,490,000 yen in the first half of 1939. Japan Chemical more than tripled its employment in the same period, from 1,017 to 3,546.[137]

Fertilizers were an important part of this growth, but they were not the whole story. Nitric acid for explosives was also a significant component in the company's expansion. The Navy requested explosives in 1937. Construction of production facilities was begun in November, and, by February 1938, Japan Chemical had started to supply explosives to the Navy. In 1940, ammonium sulfate production as a percentage of total ammonia use at Japan Chemical peaked at 93.7 percent; thereafter, explosives claimed an ever greater share, reaching 89.5 percent in 1945.[138] The government began to exert increasing influence on management decisions concerning ammonia soon after production began. Often, various government agencies worked at cross purposes; at times even the same agency had contradictory policies. For example, the Commerce and Industry Ministry attempted to stabilize fertilizer prices through subsidies to encourage fertilizer production under the Foodstuffs Control Law of 1942 and other earlier legislation. Subsidies rose from 14.5 percent in 1940 to 67.1 percent in 1945. To increase output, Japan Chemical installed the sulfurous gas method. But in 1942, just as the new technology was completed, the Commerce and Industry Ministry ordered doubled production of nitric acid for munitions, requiring installation of yet a different new method. Accelerated production of nitric acid began after February 1943.[139]

Another important change in Japan Chemical's fertilizer operations occurred in marketing. Eighty percent of all fertilizer producers were organized in the Ammonium Sulfate Producers' Association during the late 1930s. This helped market fertilizers, including those of Japan Chemical. Mitsubishi Trading continued marketing fertilizers, but its role was vestigial, as it handled just 3 to 5 percent of Japan Chemical's output at that time.[140]

Japan Chemical, carefully planned as a coordinated chemical company, produced more than dye and ammonia. It became a major producer of explosives, naphthalene, benzol, organic chemicals, and other products which soon came under government controls for mil-

TABLE 45 Mitsubishi Kasei Structure, 1944

President
Managing Director
- A. Planning Division
- B. Management
- C. Office of Oversight
- D. Head Office
  1. General Affairs Division
  2. Personnel Division
  3. Management Division
  4. Chemical Division
  5. Development Division
  6. Glass Division
  7. Soda Division
  8. Research Division
  9. Seoul Office
- E. Ōsaka Office
- F. Kurosaki Factory
- G. Coke Plant
- H–R. 11 additional plants

Source: Mitsubishi Kasei, p. 114.

Note: The Chemical Division produced coke and tar products, rayon, dyes, ammonia, explosives (both sales and manufacturing), carbide, fertilizers, inorganic acetic acid products. The Development Division handled magnesium, synthetic rubber, resins, liquid fuels. The Glass Division handled glass and plexiglass products. And the Soda Division handled various soda and soda-ash products.

itary reasons.[141] Though controls were imposed on specific products, the company was relatively independent of control over its energy sources because it had installed, between 1934 and 1936, its own coal-fired generators for electricity. Coal was supplied by the nearby Mitsubishi Mining.

Coal was also used for Japan Chemical's unprofitable attempt at low-temperature hydrogenation. In August 1937, Mitsubishi Mining and Mitsubishi Gōshi each invested 10,000,000 yen in Coal Olefaction Industries, Inc. (Sekitan Yuka Kōgyō KK). The following year, Imperial Fuels acquired Mitsubishi Gōshi's half of the shares, and, in August 1944, Coal Olefaction merged with Imperial Fuel.[142]

Mitsubishi's chemical operations were indeed multifaceted. In May

TABLE 46 Subsidiaries of Mitsubishi Chemical, August 1945

Company	*Capitalization (1,000 yen)*	*Mitsubishi Chemical Share (%)*
Tōyō Carbon	7,500	52.0
Japan Carbolic Acid	500	50.0
Takeda Chemical	9,000	50.0
Lion Fats and Oils	3,500	50.0
Mitsubishi Magnesium	5,000	50.0
Mitsubishi Kwantung Magnesium	15,000	50.0
South Manchuria Chemical	10,000	50.0
Kyōei Industries	150	66.7
Manchurian Soybean Chemical	30,000	20.0
Shōei Glass	4,500	100.0
Manchurian Shōei Glass	3,000	100.0
Shoka Glass	10,000	33.0
Tōkai Industries	1,000	100.0
Eastern Grinder Grindstone	1,000	50.0

Source: Mitsubishi Kasei, p. 115.

1935, 7 subsidiaries of Mitsubishi pooled resources and technology to establish Chemical and Industrial Machinery Production, Inc. (Kakōki Seisaku KK), thereby bringing the necessary machine-tool production into the industry.[143] (This was not, however, a subsidiary of Japan Chemical.) Japan Chemical moved into light metals and magnesium in Korea in 1943, and expanded into Manchuria and China as well.[144] When the Commerce and Industry Ministry ordered the merger of Asahi Glass and Japan Chemical on 1 April 1944 to form Mitsubishi Chemical (Mitsubishi Kasei), the truly coordinated chemical company Koyata had envisioned, a *konzern* with 14 subsidiaries, was created.

All three established zaibatsu eventually developed electrochemicals, and all three diversified in similar ways. Each entered the new field later than the new *konzern* chemical firms, but all entered when certain conditions of capital, resources, and managerial staff had been met. The independent incorporation of the component parts of the zaibatsu was the most important first step permitting this expansion into more risky areas of production, and the employment in decision-making positions of managers with professional skills in engineering, a result of the expansion of firms under zaibatsu jurisdiction, preceded entry into chemicals in all three cases. Finally, development of coal-based technology, which freed the zaibatsu from potential dependence on government-controlled sources of hydroelectricity, permitted their development of electrochemicals.

SEVEN

Conclusion

In capitalist economies, investment in high-technology industries occurs when conditions favorable to investors are created. (To be sure, the determination of the quality of the investors' environment is somewhat subjective.) If an economy is sufficiently advanced, the environment is responsive to manipulation by investors. That is, entrepreneurs shape the environment for successful development of technology by creating their own reserves of capital or obtaining necessary resources. But the environment must be fairly sophisticated before investors can begin to consider ways of affecting it to facilitate the industrial development of technology. Sophistication develops incrementally. Technological breakthroughs build on previous scientific discoveries, and their successful application assumes earlier incremental advances in conditions beyond the investors' control. New developments that appear to break with past practices are, in fact, the culmination of evolutionary change.

Investors can and do modify their environment by their investment decisions, but the decision to refine the environment presupposes that the infrastructure for technological development exists. Investors must take for granted that the political economy permits their capitalist activities. In advanced industrial economies, potential investors may lobby for political changes in the nature of government intervention in the economy, but these changes generally are refinements of established political economies. In less-developed countries, investment in higher levels of technology will occur only if investors are confident that a fundamental restructuring of the environment has been promoted. They must be able to assume the existence of an infrastructure for acquisition of capital, for delivery of resources to the work site, or for educating consumers to try new products and scientists to produce them. In most developing countries, potential investors cannot make such assumptions about the structure of the political economy.

Japan was different. Japanese investors in the late nineteenth and early twentieth centuries were able to make entrepreneurial decisions about investments assuming a supportive political economy. The Meiji government actively intervened in the economy to create new political, economic, and social institutions to foster the development of industrial capitalism. Like leaders in another late-nineteenth-century developer—Germany—Japanese leaders chose to accelerate the process of development experienced under conditions of laissez-faire in England. The shift from laissez-faire for the earliest developer, England, toward moderate government intervention for a later developer, the United States (as in construction of the Erie Canal, for example), to active involvement by the German and Japanese governments, coincided with the timing of industrialization in each society.

Decades of gradual growth did not have to pass before Japanese scientists could contemplate investment in technology. The Japanese government had created the schools to train scientists, the roads to move products, a banking system to assure retention of capital, the infrastructure for communication and dissemination of ideas through the postal and telegraph systems, and societal stability. Of particular importance to the chemical industry, the land tax reform of 1873 produced changes in ownership of land and use of its product which stimulated increased demand for fertilizers to raise productivity. Most of these aspects of government intervention eventually became standard in industrial economies, but Japan, unwilling to await the uncertain movement of an Invisible Hand in guiding development of the infrastructure, was an innovator among both developed and developing nations in stressing the contributions of the public sector.

The government, having guided the creation of Japan's economic infrastructure, then shifted course and proceeded to disengage itself from direct production well before the end of the nineteenth century. As in other industrial nations, entrepreneurs in Japan felt obliged to request that the government alter policies to facilitate investment, but entrepreneurs basically had to solve their own problems. They had to establish reliable access to capital and resources. They had to hire managers and workers able to adapt complicated technology. They had to find a steady market for their products and continue to develop new customers. And, above all, investors in high technology had to be willing to take risks with untested methods and to advance diversification of products spun off from previously developed technology. All these requirements for investment were within the control of potential developers.

The first large-scale investors in consumer chemical products were firms manufacturing superphosphate fertilizer. But superphosphate technology was both simple to master and difficult to modify to create a variety of related products. Until the 1920s, managers at Dai Nihon, Japan's most important superphosphate company, remained wedded to its simple technology. Company expansion was horizontal. Until after World War I, Dai Nihon acquired competing firms rather than spinning off new technology. Thus, it remained for the electrochemical producers—whose ranks Dai Nihon enthusiastically joined during the second wave of electrochemical investment during the 1920s—to invest aggressively in diverse products and to form modern vertically integrated firms.

The highest level of development in chemical technology before World War II was in the electrochemical industry. The first important developer in this field was Noguchi Jun, a pioneer not only in new technologies but also in the industrial structure for their commercial application. Noguchi was a product of the new educational system, which trained him to understand modern technology. Taking for granted the desirability of producing electrochemical fertilizers—not so obvious in turn-of-the-century Japan, which lagged behind Western nations in technology—Noguchi proceeded to create ways of gaining resources and capital. His main resource requirement was electricity, which led him to pursue good relations with officials able to grant the rights to generation of hydroelectricity. Capital was initially raised through stock issues and bank loans, but, as the capital markets matured in Japan, other sources, such as bonds, became more common. Furthermore, profits from sales were used to create new subsidiaries in related fields.

Once Noguchi proved that electrochemicals were a sound investment, other producers were encouraged to enter the field. The second wave of investors included firms like Dai Nihon and Shōwa Fertilizer. Like Noguchi's company, Nitchitsu, these firms came to be known as aggressive investors, with managers interested in diversifying their production in areas technologically related to their original products. The third wave came in the 1930s. This last group of investors were firms like Mitsui Mining, Mitsubishi Mining, and Sumitomo Mining. This wave differed from the previous two in that manufacture of chemicals was seen as an appropriate way for the mining companies to diversify; diversification started from mining, not from chemicals. Not only did the mining companies perceive an ever expanding market for chemical products, but the technology for manufacture of electrochemicals developed in the 1920s and 1930s was related to their mining operations.

During each of these three waves, firms commenced chemical manufacture when certain conditions had been met. All had access to resources, capital, and technology; all operated in a political climate that did not impede their investment, at least not initially; all had a market for their products; and all had managers who were committed to research, development, and commercial production of chemicals.

The companies of the three waves of investment shared other characteristics as well. Most important, all developed into vertically integrated firms with numerous subsidiaries. This type of company structure has been referred to here as the Konzern. The Konzern structure most readily emerged in the chemical industry because product diversification occurred by simple modification of established chemical processes. Mitsubishi Mining's venture in chemical production was the only one that deviated a bit from the pattern of gradual diversification in areas of related technology. Managers at Mitsubishi Mining (and at the Mitsubishi-related Asahi Glass) decided to found a chemical company already diversified. That company, Japan Tar (later Japan Chemical and later still Mitsubishi Chemical), manufactured a variety of products from the outset, but it, too, diversified in new areas as new technologies were developed and as military orders inspired new types of production. Thus, while Mitsubishi's investment in chemicals started differently from the others', by the end of the 1930s, all chemical companies were alike in expanding and diversifying rapidly because of Navy and Army orders for specific new products.

After 1937, these military orders had a result that was probably unanticipated and probably unwelcome among the entrepreneurs of the electrochemical companies: They deprived company managers of some of their independence in formulating business strategy. Products were developed less because they were closely related to other products than because the military needed them. In the case of Nitchitsu, this meant that the company's role in controlling its profitability by choosing to manufacture certain products was diminished. Although the military compromised Nitchitsu's ability to determine which products to make, the company did not lose its freedom to create and finance subsidiaries. Because creating subsidiaries and other structural modifications did affect the parent company's profitability, such types of changes became an increasingly important area for the company's initiative in decision-making. The company began to focus more attention on profiting from developing as a Konzern with numerous subsidiaries. By manipulating finances among the subsidiaries, possible only within a *konzern,* Nitchitsu managers retained some of the ability, which had eroded as military influence increased, to make strategic decisions. Thus, Nitchitsu's development as a Konzern was abetted by the growing influence of the military in its production decisions.

To greater and lesser degrees, the other chemical companies encountered similar problems of military influence. At the same time, though, all benefited from military contracts. Growth in the market for consumer goods slowed after 1937, when controls on marketing and distribution were instituted, making military contracts, especially those for technologically simple products, appear particularly welcome. In most cases, however, diversification into manufacture of innovative products for use as munitions proved unprofitable during World War II. Even Noguchi Jun, an aggressive investor, objected to some Navy orders.

What was the legacy of the prewar chemical industry? First, some of the areas of new investment during the World War II era—including plastics, resins, acetates, and petrochemicals—were important products in the immediate postwar years, even though these industries are somewhat depressed at present. Second, the postwar dissolution of the zaibatsu continued, in effect, the process of separation of their subsidiaries' management from the parent companies; as we have seen, this helped foster an environment more encouraging to innovation. Third, many of the subsidiaries of high technology companies became major postwar corporations. To be sure, companies like Nitchitsu were tremendously affected by Japan's postwar loss of its empire because a large part of Nitchitsu's (and Japan's) chemical output had been manufactured in the colonies, particularly in Korea and Manchuria. (Related to this was the loss of the major sources of hydroelectricity used by Nitchitsu's and Japan's chemical industry.) But, despite Nitchitsu's significant losses and its dissolution, some of its parts became major corporations, including Asahi Chemical, Sekisui Chemical, Sekisui House, and Nippon Kōei. Moreover, some of the managers who operated in the colonies before their independence continued their prewar projects after the war. Continuities may be seen in these companies and projects.

Fourth, there is an interesting continuity in the importance of technically trained managers in successful high-technology firms. As these types of firms mature, managers with backgrounds in finance or management have begun to rival in number those with backgrounds in science, but they have never dominated completely, as they have in trading companies or older, less technologically advanced manufacturing companies.

Fifth, the emphasis on manufacturing and marketing the company's own products has been characteristic both of the pre- and postwar high-technology firms. Sixth, the internationalization of technology acquisition and sale of finished products has been common both to pre- and postwar firms.

And finally, prewar high-tech chemical firms like Nitchitsu have had an important "demonstration effect" in showing the feasibility of scientific entrepreneurship. For example, Honda or Matsushita are often held up as representative of a type of entrepreneurship that many claim was not evident in the prewar period. Moreover, the zaibatsu firms were long seen as the originators and developers of the most sophisticated technology in the prewar period, although they relinquished that role to other companies like Honda or Matsushita after the war. The zaibatsus' ample supplies of capital and resources, it is usually asserted, gave them superior access to technology. Yet, it was companies like Nitchitsu, with its scientist-entrepreneurs, that led the development of the modern high-technology firm. It was their scientifically trained entrepreneurs who, in the early twentieth century, went out, found backers, and wagered those backers' funds to promote their planned production. This was an incentive to later producers able to follow these pioneers in private-sector technology acquisition and development.

But the postwar firms for which the demonstration effect was meaningful were largely in fields other than chemicals. The status of chemical firms in the Japanese economy has changed dramatically. Before World War II, the Japanese viewed their chemical industry as the developer of the most advanced technology available. Since the war, the chemical industry has failed to live up to its prewar promise, and high-technology industry today has come to mean the electronic, semiconductor, fiberoptic, biotechnology, and other industries. Moroever, much chemical production in areas like fertilizers has moved offshore—as Nitchitsu did by moving to the colonies—in search of new markets and cheaper resources and labor. But, whatever its current position, the chemical industry in the prewar years laid the groundwork for Japan's successful high-technology industries in the postwar decades.

Notes

Bibliography

Index

Notes

Introduction

1. Some informative studies representative of this approach are Chalmers Johnson, *MITI and the Japanese Miracle: The Growth of Industrial Policy, 1925–1975*; Edward J. Lincoln, *Japan's Industrial Policies*; Daniel I. Okimoto, "Political Context," and "Conclusions," in Daniel I. Okimoto, Takuo Sugano, and Franklin B. Weinstein, eds., *Competitive Edge: The Semiconductor Industry in the U.S. and Japan;* Daniel I. Okimoto, "Regime Characteristics of Japanese Industrial Policy"; and Gary R. Saxonhouse, "Industrial Policy and Factor Markets: Biotechnology in Japan and the United States."

2. See, for example, Toshio Shishido, "Japanese Technological Development"; and Sheridan Tatsuno, *The Technopolis Strategy: Japan, High Technology, and the Control of the Twenty-first Century.*

3. Almost every author discussing technology-intensive industries in Japan notes the significance of Japan's easy and inexpensive acquisition of imported technology and the skill of Japanese technicians in adapting basic science for commercial purposes. These studies are premised on the ostensible opposition of innovation and importation of technology. Western authors usually state that cheap technology was one major reason for Japan's successes in recent decades. Some who fear America may fall behind in technology take comfort in the belief that, although Americans "let the Japanese steal the store," as they might put it, the era of rapid growth in Japan will be coming to an end either because Americans, wise to Japanese methods, will sell less technology or because the Japanese have caught up with the Americans and will have to invest in the more expensive course of development of basic science. At the same time, Japanese are often sensitive to charges of imitation or adaptation of imported technology. Ryuzo Sato quotes what he finds to be a typical Japanese view that "the 'imitation' strategy is a Japanese strategy and . . . we feel ashamed of this deep down in our thinking." Sato refutes the idea that innovation and imitation are polar opposites. Indeed, in a case similar to the focus of this book, he notes that American chemical firms lagged behind their European counterparts in initial development of

technology and therefore made extensive use of European innovations. In time, however, they were not only "imitators" but also "creators." Ryuzo Sato, "Nothing New? An Historical Perspective on Japanese Technology Policy," pp. 299–302. The concepts of technology acquisition and commercial development as I use them in this book parallel Sato's. That is, being able to understand and to adopt technology from abroad—to imitate it, as it were—is the first step toward being able to create it. Imitation and innovation are two ends of a continuum and have not historically been opposites. An excellent historical example of this pattern occurred in the United States in the nineteenth and early twentieth centuries; American technicians, using "Yankee ingenuity," as it was called, took European basic technology and skillfully applied it. In time, their understanding of that technology was strong enough to use as the foundation for advanced research in basic science. Individual Japanese investors in technology showed the same pattern in the prewar period. In the chemical industry, mastery of technical application was soon followed by innovation.

4. Studies taking this approach include William Ouchi, *Theory Z: How American Business Can Meet the Japanese Challenge*; and Richard Tanner Pascale and Anthony G. Athos, *The Art of Japanese Management*.

5. An excellent historical treatment of this subject is in Andrew Gordon, *The Evolution of Labor Relations in Japan: Heavy Industry, 1853–1955*.

6. Several authors have found the greatest dynamism among Japanese companies not related to "groups," some of which were prewar zaibatsu. Well-known and successful firms independent of groups include Toyota, Honda, Shiseido, Sony, Hitachi, Matsushita, Canon, and Seiko. Many of these new or newly rebuilt companies developed new products and processes. See, for example, James C. Abegglen and George Stalk, Jr., *Kaisha, the Japanese Corporation*, pp. 189–190; and Sheridan Tatsuno, p. 58.

7. Edwin O. Reischauer and Albert M. Craig, *Japan: Tradition and Transformation*, p. 199. The role of the zaibatsu in the development of new technology may be commonly (and understandably) overstated because of the zaibatsus' apparent dominance of a concentrated product market. For example, Kozo Yamamura states: "In the ammonium sulphate industry, the Mitsui and Mitsubishi firms by 1930 virtually divided the total output between them." Kozo Yamamura, "The Japanese Economy, 1911–1930: Concentration, Conflicts, and Crises," p. 312. But Mitsui was just beginning to invest in ammonium sulfate production, and Mitsubishi had not yet begun in 1930, so they certainly did not dominate the market, although the banks of these two zaibatsu had indeed lent funds to companies that did dominate it. To be sure, there were disagreements between lenders and borrowers about the profitability of the timing of certain investments—there is a particularly important case, discussed in Chapter 4, in which Mitsubishi Bank, concerned about profits, withdrew support from Noguchi Jun—but the borrowers were generally independent in developing new technology. It should

be noted that Yamamura himself does not imply in this article that a concentration of market share necessarily produced a concentration of technology.

8. Alfred D. Chandler, *The Visible Hand: The Managerial Revolution in American Business*, p. 6.

9. Hugh Patrick, "Japanese High Technology Policy in Comparative Context," pp. 4–5.

10. Postwar industrial policy in Japan generally took the form of supporting the "supply side"; that is, the government tended to facilitate investment and production in areas seen as important. The market, whether domestic or international, was supposed to contribute to this investment by absorbing its products. The government itself created no extensive demand. During the war, the thrust of policy was quite different; support was given mainly on the "demand side," although the government did also try to increase supply. As in other contemporary belligerent nations, as well as in the postwar United States, public purchases of military supplies encouraged private investment in high-technology research in wartime Japanese companies.

11. Merton J. Peck, Richard C. Levin, and Akira Goto, "Picking Losers: Public Policy Toward Declining Industries in Japan," p. 116. The authors note that the National Federation of Agricultural Cooperatives and similar organizations purchase fertilizer for resale to Japanese consumers; much of this may be imported fertilizer. Abegglen and Stalk, pp. 23–24, emphasize as reasons for the decline of the Japanese electrochemical industry high energy costs, excessive pollution, high real-estate costs, and the shared benefit to resource-rich countries and to Japanese firms of the former's producing finished fertilizers with purchased Japanese technology.

12. A British researcher, William Henry Perkin, mistook aniline dyes for quinine in 1856, since the two have similar formulae. L.F. Haber, *The Chemical Industry During the Nineteenth Century*, p. 80.

13. William D. Wray, *Mitsubishi and the N.Y.K., 1870–1914: Business Strategy in the Japanese Shipping Industry*, p. 8.

14. Ibid., p. 508.

15. Christopher Freeman, *The Economics of Industrial Innovation*, pp. 46–47, 92.

16. Yasuo Okamoto, "The Grand Strategy of Japanese Business," pp. 278–279.

ONE *Early Development of the Chemical Industry in Japan*

1. Arisawa Hiromi, ed., *Nihon sangyō hyakunenshi*, I, 83; Watanabe Tokuji and Hayashi Yūjirō, *Nihon no kagaku kōgyō*, pp. 71–72; Watanabe Tokuji, *Gendai Nihon sangyō kōza, Kagaku kōgyō*, IV, 31.

2. Haber, *The Chemical Industry During the Nineteenth Century*, p. 144. While

Japanese combined production of caustic soda and soda ash reached approximately 3,000 to 4,000 tons annually in the period 1905–1910, imports during the same period hovered around 20,000 to 30,000 tons per year, mostly from Great Britain. Watanabe Tokuji, *Gendai Nihon sangyō hattatsushi:* Vol. XIII, *Kagaku kōgyō*, pp. 27–29 of appendix (repaginated), Tables III and V.

3. Kobayashi Masaaki, "Kangyō to sono haraisage," p. 41.

4. Koyama Hirotake, *Nihon gunji kōgyō no shiteki bunseki*, p. 67. For an interesting study of the role of arsenals in technology development, see Kozo Yamamura, "Success Illgotten? The Role of Meiji Militarism in Japan's Technological Progress."

5. There are some notable exceptions, though. The foremost family in the American chemical industry, the DuPonts of Delaware, owe their fortune to their pivotal position among suppliers of munitions to federal arsenals. Beginning with the Tripoli campaign of 1810 and the War of 1812, DuPont explosives have played a large role in every American war. The family's work with government authorities and researchers has been close and profitable. Haber, *Nineteenth Century*, p. 53.

6. Nakamura Chūichi, *Nihon sangyō no kigyōshiteki kenkyū*, pp. 10–11.

7. Watanabe Tokuji, *Gendai Nihon sangyō hattatsushi*, p. 116.

8. Ibid., p. 116.

9. Ibid., pp. 120–125; Nakamura Chūichi, *Nihon sangyō*, p. 18.

10. By 1903, there had apparently been some rationalization, even in the match industry. Just seven years earlier, in 1896, there had been 307 factories producing matches; the 239 figure for 1903 represents a 20% decrease. The labor force required to make those matches dropped more than 50% in those seven years, from over 46,000 to just above 20,000. The number of hours worked per worker is, however, unclear, and "rationalization" may not have increased labor productivity. In 1896, the match industry dragged down the entire chemical industry in average capital investment per worker.

11. Repeated references to explosions and other industrial accidents throughout the 133 contributions by former Nitchitsu employees collected in Kamata Shōji, ed., *Nippon Chisso shi e no shōgen*, Vols. I–XXX, indicate that the special problems of work conditions in the chemical industry were an important concern of Nitchitsu employees.

12. Takekazu Ogura, ed. *Agricultural Development in Modern Japan*, p. 13.

13. Organic fertilizers are defined here as those that derive from waste (excrement, compost, etc.), vegetable matter (soy cakes, grasses, etc.), or animal matter (fish cakes, oil). Inorganic or chemical fertilizers include those directly produced from minerals. Note the difference in definition from the chemical definition of organic compounds; the use of *organic* in discussing fertilizers has no relation to the question of the presence of carbon.

14. Haber, *Nineteenth Century*, p. 50.

15. Morikawa Hidemasa, "Shibusawa Eiichi, Nihon kabushiki kaisha no sōritsusha," pp.60–61; Yamashita Yukio, "Kagaku kōgyō no kusawaketachi," p. 91; Shiobara Matasaku, *Takamine Hakushi,* pp. 28–29.

16. Expansion of the superphosphate industry during the Sino-Japanese War, however, encouraged the Osaka sulfuric acid firm's management to resume superphosphate production, and, in 1896, after changing its name to Osaka Alkali, the company became one of Japan's largest manufacturers of fertilizers. *Meiji kōgyōshi: Kagaku kōgyō,* p. 731; Watanabe Tokuji, *Gendai Nihon sangyō hattatsushi,* p. 87; Nissan Kagaku Shashi Hensan Iinkai, *Hachijūnenshi,* pp. 35, 45.

17. Shiobara, p. 31; *Meiji kōgyōshi: Kagaku kōgyō,* p. 730; Nissan Kagaku, p. 35. Note that the Ministry of Technology was absorbed into other ministries in 1885.

18. Watanabe Tokuji, *Gendai Nihon sangyō hattatsushi,* p. 85.

19. Morikawa, "Shibusawa Eiichi, Nihon kabushiki kaisha no sōritsusha," pp. 60–61; Nissan Kagaku, p. 35. The superphosphate firm was, however, independent of Mitsui, as is evident from its rejection of Mitsui's role in sulfuric acid transactions by 1889. Other investors were Kashiwamura Shin, Nishimura Torashirō, Kawamura Den'ei, Tsuji Morinari, Kawasaki Hachiyuemon, Kimura Masami, Taneda Yoshikazu, Saionji Kōsei, Wada Tadatarō, Yonekura Ippei, Sudō Tokiichirō, Sasaki Yūnosuke, Sueda Yukio, Satō Hikosuke, Tsuda Tsukane, Okabe Fumio, Miyamoto Shinyuemon, Furuya Ryūzō, and logically, Takamine Jōkichi. Dai Nihon Jinzō Hiryō Kabushiki Kaisha, *Dai Nihon Jinzō Hiryō Kabushiki Kaisha gojūnenshi,* pp. 24–27.

20. Because of farmers' reluctance to use the new fertilizer, success in marketing was uneven in Japan. Most fertilizer was sold in the more advanced areas, and relatively little was sold in the backward areas. According to the records of one major fertilizer merchant, Ida of Nihonbashi, over two-thirds of its sales in 1887 were in five prefectures plus Ōsaka; farmers in the northeast and southwest shied away from chemical fertilizers at that time. Watanabe Tokuji, *Gendai Nihon no sangyō hattatsushi,* p. 86.

21. For a list of publications and organizations promoting use of superphosphates, see Nissan Kagaku, p. 39. Itinerant students and graduates taught modern farming methods from 1877 to 1893. Ogura, pp. 158–159.

22. Nissan Kagaku, p. 43.

23. The compilers of *Meiji kōgyōshi: Kagaku kōgyō* note that producer Taki Kumejirō received this type of aid, thereby helping him establish efficient production. Watanabe Tokuji states, on the other hand, that Taki's request was ignored, setting him back a few years. *Gendai Nihon sangyō hattatsushi,* p. 92.

24. Ogura, pp. 162–163.

25. Watanabe Tokuji, *Gendai Nihon sangyō hattatsushi,* pp. 89–90; Nissan Kagaku, p. 37. Furthermore, Tokyo Artificial Fertilizer's management preferred to sell its product as part of a compound, as it could receive a higher markup on

the compound than on plain superphosphate. Shimotani Masahiro, "Dai Nihon Jinzō Hiryō torasuto to karinsan sekkai kōgyō," p. 36.

26. For a description of the process, see: Nakamura Chūichi, *Nihon kagaku kōgyōshi,* p. 9.

$$Ca_3(PO_4)_2 + 2H_2SO_4 + 4H_2O \rightarrow CaH_4(PO_4)_2 + 2(CaSO_4 + 2H_2O)$$

tricalcium phospate	+	sulfuric acid	+	water	=	monocalcium phosphate	+	gypsum

The mixture of monocalcium phosphate and gypsum is superphosphate of lime fertilizer.

27. Shimotani Masahiro, "Dai Nihon Jinzō Hiryō," p. 36; Nakamura Chūichi, *Nihon sangyō no kigyōshiteki kenkyū,* p. 14. Although soybean cakes failed to supply the majority of all types of nutrients, they did supply most of the nitrogenous component of fertilizers after the turn of the century, until overtaken by synthetic fertilizers a few decades later. *Chōki keizai tōkei: nōringyō,* Vol. IX, Table 20, pp. 196–197. Japanese demand for soybean fertilizers had an enormous effect on Chinese exports as well; soybeans, soy oil, and soy cakes accounted for 0.3% of China's exports in 1880, climbing to 14% by 1910. Albert Feuerwerker, *The Chinese Economy, ca. 1870–1911,* p. 53.

28. L. F. Haber, *The Chemical Industry 1900–1930,* p. 98.

29. Nakamura Chūichi, *Nihon sangyō no kigyōshiteki kenkyū,* p. 43.

30. World production of superphosphate increased dramatically on the eve of World War I. In 1913, the United States was producing 3.75 million tons annually, and France and Germany each produced 2 million tons. Italy's production jumped from 146,000 tons in 1895 to 1 million tons in 1910. All this superphosphate production meant that the fertilizer industry became the world's major user of sulfuric acid. Indeed, countries that had not produced sulfuric acid by the Leblanc process before the end of the nineteenth century—since the world market had been dominated by the British—were forced to establish sulfuric acid industries to supply their superphosphate manufacturers. Superphosphate production alone required 3.5 million tons of sulfuric acid (as expressed in 100% H_2SO_4) in 1913 throughout the world. As a weaker solution was generally used, more was actually consumed (1 ton of 62.5% sulfuric acid was used to make 2 tons of superphosphate). Haber, *The Chemical Industry 1900–1930,* pp. 104–105.

31. The process whereby sulfuric acid manufacturers took up fertilizer production occurred in Japan as well as in Europe. Sumitomo, for instance, had excess sulfuric acid production and decided to use it for fertilizer manufacture in 1915. Nakamura Chuichi, *Nihon sangyō no kigyōshiteki kenkyū,* p. 57.

32. Shōji Tsutomu, *Nihon soda kōgyōshi,* pp. 206–207.

33. Ibid., p. 203.

34. Shimotani Masahiro, "Dai Nihon Jinzō Hiryō," p. 37. For a list of the new companies and new facilities producing superphosphate immediately after the

Russo-Japanese War, see Nissan Kagaku, p. 49; Shimotani Masahiro, "Dai Nihon Jinzō Hiryō," p. 38.

35. For a list of the major cartels and industrial organizations in mining and manufacturing from 1880 to 1932, see Miwa Ryōichi, "Nihon no karuteru," pp. 171–173.

36. Nissan Kagaku, p. 52. The initial attempt to organize in 1904, undertaken by 3 firms (Ōsaka Alcali, Ōsaka Sulfuric Acid and Soda, and Nihon Sulfuric Acid and Soda), failed because organizing appeared to make little sense in that war-boom year. By August 1907, however, the situation had changed dramatically, and 6 firms in the Kansai area agreed to try to regulate prices. Nevertheless, the group of 6 failed to overcome their rivalries. Shimotani Masahiro, "Dai Nihon Jinzō Hiryō," p. 39.

37. Nissan Kagaku. p. 53.

38. The change of name in 1910 recognized Dai Nihon's growth from a regional to a national firm. Later, the firm changed its name again, to Nissan Chemical.

39. Shimotani Masahiro, "Dai Nihon Jinzō Hiryō," p. 58. See *Meiji kōgyōshi: Kagaku kōgyō*, pp. 732–733, for a list of factories and their dates of initial production and of absorption into Dai Nihon.

40. Shimotani Masahiro, "Dai Nihon Jinzō Hiryō," p. 48; Nissan Kagaku, pp. 62–69; Nihon Kasei Hiryō Kyōkai, *Rinsan hiryō kōgyō no ayumi*, pp. 50–52.

41. Ogura, pp. 222, 163, 308; Nihon Kasei Hiryō Kyōkai, p. 49. Nissan Kagaku, p. 61. Production cutbacks were 20% from March 1922 to 31 May 1922 and 30% from 1 June 1922 to 15 November 1922. Suspended until 1 March 1923, they were reimposed, at 30%, until December 1925. Nihon Kasei Hiryō Kyōkai, p. 50.

42. Approximately 70% of the cost of making superphosphate was raw materials cost, and 19% was cost of packaging the product. Labor and management costs accounted for significantly less than 20% of the final value. Nakamura Chūichi, *Nihon sangyō no kigyōshiteki kenkyū*, p. 43. Shimotani Masahiro, "Dai Nihon Jinzō Hiryō," p. 44. By the end of the 1920s, capital investment in superphosphate plant capable of producing 100,000 tons annually would be about 2 million yen, while it would cost about 6 times that amount to set up a comparably sized ammonium sulfate plant. Shimotani Masahiro, "Dai Nihon Jinzō Hiryō," p. 59, footnote 46.

43. By-product ammonium sulfate is made by mixing sulfuric acid with the ammonia wastes, or by-products, of steel-refinery or natural-gas production. Its production is, of course, limited by the amount of ammonia wastes generated.

44. This advance in technology will be discussed in Chapter 2.

45. Advances in scientific education and research during the Meiji period have been treated by a number of authors. Watanabe Tokuji, *Gendai Nihon sangyō hattatsushi*, pp. 182–189, and Nakamura Chūichi, *Nihon sangyō no kigyōshiteki*

kenkyū, Chapters 1 and 2, analyze the significance of technological and educational developments. *Meiji kōgyōshi: Kagaku kōgyō*, pp. 1117–1137, offers the most comprehensive discussion of chemical education of the general histories of the industry. Nihon Kagakukai, *Nihon no kagaku hyakunenshi*, treats technical advancement implicitly throughout its 1,300 pages, and Meiji schooling explicitly, pp. 92–113. Development of technical education, among other topics, is dealt with in Nihon Kagakushi Gakkai, *Nihon kagaku gijutsu shi taikei*, Vol. XXI, *Kagaku gijutsu*. Yuasa Mitsutomo, *Kagakushi*, has perhaps the best analysis of education and research and their roles in the history of science. Miyoshi Nobuhiro, *Nihon kōgyō kyōiku seiritsu no kenkyū*, discusses the founding of the College of Technology, pp. 269–270. Technicians a step below the professionals of the College of Technology and the Imperial Universities are treated by Ryōichi Iwauchi, "Institutionalizing the Technical Manpower Formation in Meiji Japan." Finally, an interesting series of articles in English appears in the *Journal of World History* 9 (1965).

46. Iwauchi, p. 422.

47. Miyoshi, pp. 361–364.

48. Foreign teachers of chemistry at various schools are listed in Yuasa, pp. 106–107; Watanabe Minoru, "Japanese Students Abroad and the Acquisition of Scientific and Technical Knowledge," pp. 270–271.

49. *Meiji kōgyōshi: kagaku kōgyō*, p. 1122; Watanabe Minoru, pp. 271–272.

50. Nihon Kagakukai, p. 94.

51. Miyoshi, p. 269; Yuasa, p. 107; Nakamura Takeshi, "The Contribution of Foreigners," p. 301. Yuasa notes that Divers remained in Japan from 1873 until 1899, while Nakamura has him leaving the College of Technology in 1885. As Divers taught at Tokyo Imperial University, according to Yuasa, after leaving the College, it is possible that the Irishman's stay was extended until 1899.

52. *Meiji kōgyōshi: kagaku kōgyō*, p. 1122.

53. Nihon Kagakukai, p. 101; Watanabe Minoru, p. 271.

54. Fertilizers were not his only claim to fame. Takamine had leadership qualities, being a prime mover behind the effort to establish a national research laboratory. He also had high scholarly ability, discovering the substances adrenalin and Takadiastase.

55. Nihon Kagakukai, p. 99.

56. German scientific education created a standard for Europe. Many of England's best chemists received their education from German instructors, but the lack of sustained chemical education in Britain meant that British products appeared embarrassingly poor next to German products at the Paris Exhibition of 1867. Even less chemistry was being studied in the United States at the time. Both countries eventually considered the types of programs Germany had instituted; although the United States and England did not copy most of the German

system's characteristics, they did greatly expand their scientific educational establishments at the end of the nineteenth century.

The German system was essentially multi-tiered and recognized a relationship between industry and science. These two characteristics produced a university system free in spirit and inexpensive in cost. Students could move from institution to institution, acquiring whatever knowledge they felt important. Furthermore, science was necessary for the state, and so education was seen as a right and a duty. A university education was not for all, but some form of learning was. Students at technical colleges (technische Hochschulen), although possessing little of the academic freedom to move about enjoyed by their peers at the universities, nonetheless received subsidization in high-level education that prepared them well for entry into leadership positions in business. University enrollment of Germans surpassed technical-college enrollments. By the end of the nineteenth century, foreign admiration for the German system meant that British, American, and Russian students flocked to the inexpensive German colleges. The lowest level of technical education were the trade schools (Fachschulen), which trained teenagers in practical skills.

The relationship between industry and science was another important consideration in Germany. University laboratories prepared experiments to benefit private producers, and technical colleges educated an elite class of technically trained business leaders capable of understanding the technical aspects of their firms' operations.

Development of the best principles at each level of education and the offer of education at low cost to the student allowed any student to pursue a scientific education. This meant that Germany would be able to develop more intellectual talent, even if the cost of educating all those children and young people was high. Expenses invested were repaid handsomely by Germany's technological lead in the early twentieth century; investments in technology and knowledge repay exponentially. Education was not the only reason for Germany's intellectual leadership. Promotion of research was helpful as well. Haber, *Nineteenth Century,* pp. 74–78; Haber, *The Chemical Industry 1900–1930,* pp. 43–47.

57. Watanabe Minoru, pp. 278–282.

58. Nihon Kagakukai, p. 99.

59. Indeed, the need for better technical training and scientific studies was one important reason for government support of Kyoto University. As one Diet member stressed during debate on an 1890 bill to establish the university:

> There is only the Imperial University in Tokyo, and, because of the lack of competition, the professors have ceased to discover new scientific theories, and the students have ceased to pursue their scientific objectives. . . . The establishment of an imperial university in the Kansai district is indispensible to the development of education in Japan.

Quoted in Nakayama Shigeru, "The Role Played by Universities in Scientific and Technological Development in Japan," p. 348.

60. Ibid., pp. 349–352.

61. Watanabe Tokuji, *Gendai Nihon sangyō hattatsushi,* pp. 182–183.

62. Iwauchi, pp. 426–428.

63. Nihon Kagakukai, p. 98.

64. Iwauchi, pp. 428–432.

65. *Meiji kōgyōshi: Kagaku kōgyō,* pp. 1125–1128.

66. Ibid., pp. 1129–1137; Iwauchi, pp. 431–433.

67. *Meiji kōgyōshi: Kagaku kōgyō,* p. 1074.

68. Haber, *The Chemical Industry 1900–1930,* Chapter 3. Government scientific laboratories existed in several countries before World War I, but none was specifically responsible for chemical analysis. The National Bureau of Standards in the United States, founded in 1901, mainly benefited physicists and engineers, as did Britain's National Physical Laboratory. Britain had had a Chemical Inspectorate of the War Office since 1854, but it was minuscule. Haber, *The Chemical Industry 1900–1930,* pp. 220–221.

69. Chikayoshi Kamatani, "The Role Played by the Industrial World in the Progress of Japanese Science and Technology," pp. 401–402.

70. Asai Yoshio, "Nōshōkō kōtō kaigi," p. 144.

71. The statement issued by the third conference meeting of the group of officials, businessmen, and scholars calling for the establishment of research laboratories is excerpted in Watanabe Tokuji, *Gendai Nihon sangyō hattatsushi,* pp. 184–185.

72. Asai, p. 144.

73. *Tōkyō Kōgyō Shikenjo rokujūnenshi,* p. 1.

74. Watanabe Tokuji, *Gendai Nihon sangyō hattatsushi,* p. 186.

75. *Meiji kōgyōshi: Kagaku kōgyō,* p. 1063.

76. Watanabe Tokuji, *Gendai Nihon sangyō hattatsushi,* pp. 185, 187–188; Nihon Kagakushi Gakkai, p. 136.

TWO *Pioneers in Electrochemicals: The First Wave*

1. To be sure, there were other more creative minds in Japan's scientific community, such as Takamine Jōkichi. But Takamine found the United States a much more congenial place to work, because scientific facilities were more readily available, so he set up his own research laboratories in Chicago and New York after 1890. Noguchi and Fujiyama are noteworthy for their commercial use in Japan of processes either purchased or developed.

2. Haber, *The Chemical Industry 1900–1930,* p. 39.

3. Ibid., p. 32.

4. There was a 250% rise in consumption of artificial fertilizers between 1905 and 1913. Uchisaburo Kobayashi, *The Basic Industries and Social History of Japan, 1914–1918,* p. 83.

5. Nobufumi Kayō, "The Characteristics of Heavy Application of Fertilizers in Japanese Agriculture," pp. 378–389. Overfertilization could cause many varieties of plants to topple over; government researchers, therefore, developed stronger strains of rice.

6. G. Daikuhara and T. Imaseki, "On the Behavior of Nitrate in Paddy Soils"; S. Uchiyama, "On the Manurial Effect of Calcium Cyanamide under Different Conditions." The work of these scientists is erudite and carefully researched.

7. Daikuhara and Imaseki, p. 35.

8. Haber, *The Chemical Industry 1900–1930,* p. 32.

9. Ibid., pp. 85–86. After Cavendish, scientists, beginning with Mrs. Lefebre (personal name unknown) in 1859, worked on methods for fixing atmospheric nitrogen. Nitrogen research occupied the attention of the leading chemists in the United States, Germany, and Britain during the last half of the nineteenth century.

10. Sir William Crookes, quoted in Haber, *The Chemical Industry 1900–1930,* p. 85.

11. Haber, *The Chemical Industry During the Nineteenth Century,* p. 105.

12. Haber, *The Chemical Industry 1900–1930,* p. 101; Haber, *Nineteenth Century,* p. 106. Ammonium sulfate was produced not only by gasworks operators, but also by shale-oil refiners (in Scotland) and by others. Though the percentage declined, most of the fertilizer came from the by-product ovens of the gasworks during the nineteenth century.

13. Haber, *The Chemical Industry 1900–1930,* p. 144.

14. Nihon Ryūan Kōgyō Kyōkai, *Nihon ryūan kōgyōshi,* pp. 44–46. Nations producing the largest amounts of all types of ammonium sulfate used in Japan were: until 1914, England; 1919–1925, the United States and England; 1926–1931, Germany; after 1932, Japan.

15. For instance, Tokyo Gas Company's Ōshima plant began production of sulfuric acid in 1922 for use by Tokyo Gas. Kantō Taru Seihin Kabushiki Kaisha Sōritsu Jūshūnen Kinen Jigyō Iinkai, *Taru kōgyōshi,* p. 191.

16. Nakamura Chūichi, *Nihon kagaku kōgyōshi,* pp. 25–26; Nihon Ryūan Kōgyō Kyōkai, p. 47.

17. Nakamura Chūichi, *Nihon sangyō no kigyōshiteki kenkyū,* p. 53. Scholars, including Shimomura Kotarō, Takahashi Toyokichi, Nagai Nagayoshi, Hiraga Toshitomi, and Toshitake Einoyuki, did examine European processes, however. In referring to coal-tar derivatives, the term *organic* is used in the standard way; that is, organic chemicals are those with carbon.

18. Nakamura Chūichi, *Nihon sangyō no kigyōshiteki kenkyū,* p. 53.

19. Kondō Yasuo, *Ryūan: Nihon shihonshugi to hiryō kōgyō,* p. 73. Output of

by-product ammonium sulfate, though small throughout the 1920s, increased its market share in the 1930s as mining companies invested more strongly in recovery operations. Several factors produced this increase in the 1930s. One was an increase in coal-tar production as the economy geared up for war and electricity costs fluctuated.

20. Yokohama was the first city in Japan to have gas service. Initially, no Japanese in Yokohama saw the need for a municipal gasworks. The foreign community there petitioned the Governor of Kanagawa for permission to start a gas company, and the Governor, wary of foreign investment in an enterprise essentially requiring a local monopoly, pressed a group of Japanese merchants to supersede the foreign petition by filing their own request in 1870. Though the foreigners complained, the gas company stayed in Japanese hands and began serving customers in September 1872, setting the stage for further development of the gas industry. Watanabe Tokuji, *Gendai Nihon sangyō hattatsushi*, p. 152; Kantō Taru Seihin KK, p. 15.

21. Run for about a year and a half by the Tokyo Chamber of Commerce, the gasworks passed into the hands of the municipal government in May 1876. Little changed as far as management was concerned; Shibusawa Eiichi, who as head of the Chamber of Commerce had directed development of the gasworks, stayed on as head of Tokyo's Gas Bureau. Members of the Tokyo Municipal Assembly decided in June 1885 to discontinue support of the Gas Bureau and to allow the gasworks to be managed by private entrepreneurs. Tōkyō Gasu Kabushiki Kaisha, *Tōkyō Gasu shichijūnenshi*, pp. 7–13, 23.

22. Ibid., insert after p. 498, p. 37

23. Ibid., insert after p. 498. Starting with 61 employees in 1885 (the year private management began), Tokyo Gas employed 564 in 1900 and 1,265 in 1906. Between 1901 and 1905, capitalization doubled. Although ammonium sulfate did not account for the entire increase in Tokyo Gas, it was one factor contributing to the company's growth. Most important, gas output increased and with it capacity for output of nitrogenous fertilizer.

24. Tōkyō Gasu KK, p. 41; Kantō Taru Seihin KK, pp. 26–28; *Meiji kōgyōshi: Kagaku kōgyō*, pp. 173–175.

25. Kakujiro Yamasaki and Gataro Ogawa, *The Effects of the World War upon the Commerce and Industry of Japan*, p. 14.

26. Shōji Tsutomu, *Jinzō hiryō kōgyō*, p. 133.

27. Kantō Taru Seihin KK, p. 193.

28. Noguchi is generally known as Jun. He is said to have despised his childhood name of Shitagau, which his wet nurse continued to call him. To appease his son, the father, Yukinobu, changed the boy's name to Juntarō, shortened by the child's school chums to Junta-san. Hoshino Yoshiro, "Noguchi Jun to gijutsu no kakushin," p. 360. The short piece on Noguchi is part of a special series honoring 10 people whose contributions to the Japanese economy the editors deemed

sufficiently noteworthy to include among the "100 people who made Japan."

29. Ibid., p. 360. The Faculty of Engineering at Tokyo Imperial University graduated just 80 students that year; after 1885 and its absorption of the College of Technology, Tokyo University offered the only university-level engineering course in Japan.

30. Kamata Shōji, "Waga kuni saisho no kābaido jigyō—Kōriyama Kābaido seizōjo," pp. 75–76.

31. Matsushita Denkichi, *Kagaku kōgyō zaibatsu no shinkenkyū*, p. 152. Fujiyama's educational background and motivation are discussed in Shibamura Yōgo, *Kigyō no hito Noguchi Jun den: Denryoku, kagaku kōgyō no paionia*, pp. 38–39.

32. Fujiyama's genius appears to be downplayed in most of the biographies lauding Noguchi. Because the two split up acrimoniously, Noguchi's biographers seem to "take sides" subtly by such tactics as accrediting him with Fujiyama's discoveries which were, as most industrial discoveries, patented in the name of the discoverer's firm.

33. Hoshino Yoshiro, pp. 360–361; Denka Rokujūnenshi Hensan Iinkai, *Denka rokujūnenshi*, p. 87.

34. Kamata, "Waga kuni saisho no kābaido kōgyō," p. 76.

35. Denka, p. 87.

$$CaO + 3C \rightarrow CaC_2 + CO$$

quicklime + coke = calcium carbide + carbon monoxide

Nakamura Chūichi, *Nihon kagaku kōgyōshi*, p. 24. Later commercial production followed similar lines, only the process began with calcium carbonate (limestone) instead of quicklime. ($CaCO_3 = CaO + CO_2$)

36. Shimotani Masahiro, "Hensei ryūan, sekkai chisso kōgyō to Denki Kagaku Kōgyō Kabushiki Kaisha no seiritsu: Waga kuni kagaku kōgyō ni okeru dokusen keieishi," p. 30. *Meiji kōgyōshi: Kagaku kōgyō* differentiates between the two earliest producers of calcium carbide, noting that the chemical was first discovered at Sankyozawa and first produced commercially at Kōriyama. Both began producing in 1901, though Fujiyama first made carbide in his laboratory in 1900. See *Nihon Kagakukai*, p. 141, and Ōshio Takeshi, "Nippon Chisso Hiryō Kabushiki Kaisha ni yoru hensei ryūan seizo kigyōka no katei," p. 255. Furthermore, real production had to await Noguchi's participation, since he brought Siemens machinery with him. Watanabe Tokuji, *Gendai Nihon sangyō hattatsushi*, p. 168. The basic difference between the two producers was that Miyagi Spinning produced carbide as a side operation at Sankyozawa whereas the Kōriyama Carbide Company had the first independent facility. Nakamura Chūichi, *Nihon sangyō no kigyōshiteki kenkyū*, p. 46.

37. Kābaidō Kōgyō no Ayumi Hensan Iinkai, *Kābaido kōgyō no ayumi*, p. 31; Kamata Shōji, "Waga kuni saisho no kābaido jigyō," p. 74.

38. Kagaku Keizai Kenkyūjo, *Kagaku kōgyō no jissai chishiki*, p. 14.

39. For a discussion of the importance of hydroelectric power, see Kuriyama Toyo, *Gendai Nihon sangyō hattatsushi: Denryoku*, pp. 54–62.

40. Although the percentage of capital represented by fixed capital decreased as the electric industry matured, it remained quite high.

Fixed Capital in the Hydroelectric Industry

Year	*Amount (million yen)*	*% of Total Capital*
1914	556	99
1920	1,042	91
1925	2,567	88
1929	4,478	86

Source: Adapted from Kondō Yasuo, *Ryūan*, p. 132, Table 73.

41. While in America he, rather prosaically, assisted his brother who was an importer there. Shibamura Yōgo, *Noguchi Jun den*, p. 40.

42. Hoshino, p. 361; Yoshioka Kiichi, *Noguchi Jun*, Chapter 2. Yoshioka takes an entirely different view of Noguchi's early adult years. Despite the hard work, assiduous study, and demands on his time that forced him to go from company to company helping to set up electrical plants all over the country ("almost single-handedly" according to Yoshioka, p. 46), Noguchi remained in good spirits. Yoshioka was, at the time he wrote the book, President of Chisso, successor firm to Noguchi's, and had worked under Noguchi throughout his adult life. Yoshioka's adulatory tones are not unique, however. Accounts of the man's life written during World War II, such as Kamoi Yū's *Noguchi Jun*, and Matsushita Denkichi's *Kagaku kōgyō zaibatsu no shinkenkyū*, similarly paint Noguchi in rather grand terms. To them, the chemical entrepreneur's patriotism and readiness to adjust his production to military needs made him a praiseworthy leader. Because his biographers make Noguchi larger than life, it is important to place his contributions in proper perspective. More objective measures of his importance and that of his company in the Japanese and colonial economies do, nonetheless, substantiate the claim that Noguchi played a significant role in the economy.

43. Yoshioka, p. 58; Fukumoto Kunio, ed. *Noguchi wa ikite iru: Jigyō supiritto to sono tenkai*, pp. 15–18.

44. While several authors state that each man's contribution was 1,000 yen, Ōshio Takeshi claims the contribution was 2,000 yen per investor. "Nippon Chisso Hiryō Kabushiki Kaisha," p. 255.

45. Watanabe Tokuji, *Gendai Nihon sangyō hattatsushi*, p. 168. Herman Kessler, Siemens representative in Japan after 1887, was active in promoting use of Siemens machinery. The Siemens firm wisely sent representatives to countries in which their products were sold to assure proper installation and maintenance.

See Sigfrid von Weiher, "The Rise and Development of Electrical Engineering and Industry in Germany in the Nineteenth Century: A Case Study—Siemens and Halske," p. 39.

46. Kābaido Kōgyō no Ayumi Hensan Iinkai, p. 31. As in Japan, Western commercial production of calcium carbide followed soon after its discovery. In 1894, a Frenchman, Moissan, continued Willson's researches, and, in the following year, Willson set up a plant in South Carolina. Almost immediately, Japan began importing calcium carbide, and it was just four years until Fujiyama applied the scientific information published in journals to a manufacturing process in Japan.

47. Kābaido Kōgyō no Ayumi Hensan Iinkai, p. 44. The firm was Nihon Kābaido KK, not to be confused with the later firm at Minamata with a similar but not identical name.

48. Watanabe Tokuji, *Gendai Nihon sangyō hattatsushi,* p. 174; Denka, p. 88n. Sankyozawa, again independent after dissolution of its merger, was left in the hands of Itō and his colleagues and renamed Yamamitsu Carbide.

49. Later, Nagasato's son and Hino became directors at Chōsen Chisso and Nitchitsu, respectively. Ōshio Takeshi, "Nippon Chisso Hiryō Kabushiki Kaisha, p. 258. Hoshino, p. 361; Watanabe Tokuji, *Gendai Nihon sangyō hattatsushi,* p. 175; Nakamura Chūichi, *Nihon sangyō no kigyōshiteki kenkyū,* p. 46; Matsushita, p. 65; Ōshio, "Nippon Chisso Hiryō Kabushiki Kaisha," p. 258.

50. Ōshio, "Nippon Chisso Hiryō Kabushiki Kaisha," p. 259.

51. Yoshioka, p. 67. For an analysis of why Minamata was chosen, see Shibamura, *Noguchi Jun den,* pp. 47–48.

52. Yoshioka, p. 59.

53. Nihon Ryūan Kōgyō Kyōkai, pp. 51–52; Denka, p. 89. This firm was Nihon Kābaido Shōkai.

54. Watanabe Tokuji, *Gendai Nihon sangyō hattatsushi,* p. 176, including a list of these men, whom Noguchi discovered in Tokyo, Kōriyama, and other places.

55. Ichikawa, cited in Shibamura, *Noguchi Jun den,* p. 17.

56. Haber, *The Chemical Industry 1900–1930,* pp. 89–90.

57. Calcium carbide was placed in a large tub and heated to 1100 degrees Centigrade; nitrogen was then introduced.

$CaC_2 + N_2 \rightarrow CaCN_2 + C$

Nitrogen, which under normal atmospheric conditions does not readily combine with other elements, reacts efficiently with calcium carbide when heated, thereby being "fixed" so that it can be used. The remaining carbon was reused in the production of additional calcium carbide from lime. Nihon no Kagakukai, pp. 640–641.

58. Denka, p. 91; Nihon Ryūan Kōgyō Kyōkai, p. 50.

59. Katagiri Ryūkichi, *Hantō no jigyō-ō Noguchi Jun,* pp. 184–185.

60. Ōshio, "Nippon Chisso Hiryō Kabushiki Kaisha," p. 260. Together with the Bank of Germany, the firm held over 75% of all shares.

61. Haber, *The Chemical Industry 1900–1930,* p. 88; Watanabe Tokuji, *Gendai Nihon sangyō hattatsushi,* p. 178. Shibamura, *Noguchi Jun den,* also credits Noguchi's brother, who was in Paris studying painting, with helping him. Ōshio, "Nippon Chisso Hiryō Kabushiki Kaisha," p. 260.

62. Yoshioka, pp. 70–71. The tiny firm's acquisition of the license caused quite a stir in financial circles in Japan.

63. Shibamura, *Noguchi Jun den,* p. 54.

64. Udagawa Masaru, *Shōwashi to shinkō zaibatsu,* p. 260.

65. It has been suggested elsewhere that family ties among women are instrumental in interfirm ties. Matthews M. Hamabata, "Love and Work in Japanese Society: The Role of Women in Large-Scale Family Enterprise." Further study would discover whether this pattern was typical in the prewar period as well.

66. Nakahashi Tokugorō, one of the most influential men of affairs in Osaka, probably viewed Noguchi Jun as a protégé. Born in Noguchi's birthplace, Ishikawa prefecture, Nakahashi had graduated from Tokyo University ten years before Noguchi. Birthplace ties were important to the O.S.K. President; his beneficiaries, including men he recruited to work for Noguchi, often hailed from Ishikawa. Nakahashi was as conversant in the world of politics as he was in commercial circles, having been a member of the Lower House of the Diet, a councillor in the Communications Ministry and head of the Railroad Board. He later became a major force in the Seiyūkai, one of the established political parties, and held various cabinet portfolios. Both his political and financial ties were of help to Noguchi. Ōshio, "Nippon Chisso Hiryō Kabushiki Kaisha," p. 260; Katagiri Ryūkichi, p. 31–33; Fukumoto, p. 24. Miyamoto Mataji, *Kansai zaikai gaishi,* p. 76; Ichikawa Homei, *Ichikawa Seiji den,* p. 59.

67. Ōshio, "Nippon Chisso Hiryō Kabushiki Kaisha," p. 262; Shibamura, *Noguchi Jun den,* p. 57; Nagasawa Kosho, "Nichitsu Kasare-hō kōgyōka ni tsuite no ichikōsatsu: Denki Kagaku Kōgyō (KK) to no hikaku ni oite."

68. Yamamoto Tomio, *Nippon Chisso Hiryō jigyō taikan,* pp. 438–439; Matsushita, p. 65.

69. Shimotani Masahiro, "Nippon Chisso Hiryō KK to takakuka no tenkai: Waga kuni kagaku kōgyō ni okeru dokusen keiseishi," p. 98; Katagiri Ryūkichi, pp. 26, 248; Ōshio, "Nippon Chisso Hiryō Kakushiki Kaisha," p. 261; Kawamura Kazuo, "Nippon Chisso kosho zakkan, sono shi: Sōgyō jidai no hitotachi, kābaido o chūshin ni," p. 113. With the exception of Shiraishi, who had been an engineer at Mitsubishi, the directors were all professional managers.

70. Ōshio Takeshi, "Chōshinkō kaihatsu o meguru Nitchitsu to Mitsubishi no tairitsu ni tsuite," p. 167.

71. Takahashi Kamekichi, *Nihon zaibatsu no kaibō,* pp. 108–109.

72. Matsushita, pp. 65, 70.

73. Haber, *The Chemical Industry 1900–1930,* p. 111; Matsushita, p. 65.

74. Denka, p. 92; Nakamura Chūichi, *Nihon sangyō no kigyōshiteki kenkyū,*

p. 48; Ichikawa, p. 60; "Nippon Chisso sanjūnen kinen zadankai," p. 17.

75. Denka, p. 90; Watanabe Tokuji, *Gendai Nihon sangyō kōza: Kagaku kōgyō,* p. 129.

76. This problem was not unique to Japan. When forced to use cyanamide during World War I instead of other fertilizers, German farmers balked. Haber, *The Chemical Industry 1900–1930,* p. 200.

77. Nihon no kagakukai, p. 92.

$$CaCN_2 + 3H_2O \rightarrow 2NH_3 + CaCO_3$$

calcium cyanamide + steam = ammonia + limestone

$$2NH_3 + H_2SO_4 \rightarrow (NH_4)_2SO_4$$

ammonia + sulfuric acid = ammonium sulfate

78. The cyanamide method for production of ammonium sulfate is referred to as the metamorphosis method in Japanese, and the product is called metamorphosized ammonium sulfate (*hensei ryūan*).

79. Shimotani Masahiro, "Nippon Chisso Hiryō KK," p. 98; "Nippon Chisso sanjūnen kinen zadankai," p. 17.

80. Udagawa Masaru, *Shōwashi to shinkō zaibatsu,* p. 52; "Zadankai," pp. 17–18.

81. Shibamura, *Noguchi Jun den,* pp. 63–64; Ichikawa, pp. 68–69.

82. Nihon Ryūan Kōgyō Kyōkai, p. 53.

83. Yoshioka, p. 67.

84. Denka, pp. 91–92; Watanabe Tokuji, *Gendai Nihon sangyō kōza,* p. 129.

85. Yoshioka, p. 75; Udagawa, *Shōwashi to shinkō zaibatsu,* p. 50.

86. Ōshio, "Nippon Chisso Hiryō Kabushiki Kaishao," pp. 263–265.

87. "Nippon Chissō sanjūnen kinen zadankai," pp. 18–19.

88. Denka, p. 92. Not only was the process more appropriate to the raw materials available in Japan, but it was also able to take advantage of the low cost of Japanese labor, according to the aged Albert Frank in a 1956 commemorative speech at Tokyo University.

89. Shimotani, "Nippon Chisso Hiryō," p. 98.

90. "Nippon Chissō sanjūnen kinen zadankai," pp. 32–34.

91. Denka, pp. 92–93.

92. Watanabe Tokuji, *Gendai Nihon sangyō hattatsushi,* p. 279.

THREE *Nitchitsu's Profitability and Expansion: 1912–1922*

1. Yamamoto Tomio, p. 445; Ōshio Takeshi, "Nitchitsu kontsuerun no seiritsu to kigyō kin'yū,"pp. 120, 123; Ichikawa, p. 69; Udagawa, *Shōwashi to shinkō zaibatsu,* p. 39. Hypothec Bank President Shimura Gentarō was the brother of Noguchi's longtime friend Watanabe Yoshirō of Aichi Bank. Also, persuading

the stockholders would have been easy, as Noguchi held 27% of all shares and the largest 10 holders held 71%.

2. Yamamoto Tomio, p. 446.

3. Yoshioka, pp. 82–83; Yamamoto Tomio, p. 445; Ichikawa, pp. 69–70. While passing through tunnels in the mountains of Kyūshū, railroad passengers were covered in soot. Railroad Board President Gotō Shinpei (Sengoku's predecessor) decided that electrification was the way to remove this problem. "Nippon Chisso sanjūnen kinen zadankai," p. 18.

4. Yoshioka, pp. 82–83; Matushita, pp. 70–71; Ōshio Takeshi, "Nitchitsu kontsuerun no seiritsu to kigyō kin'yū," p. 63.

5. "Nippon Chisso sanjūnen kinen zadankai," pp. 18–19.

6. Kawamura Kazuo, "Nippon Chisso kosho zakkan, sono go: Taishō shōki no Minamata no geppō to Kagami no jūyō nikki," p. 78.

7. The return of the water rights for the Sogi and Minamata plants in 1915 was similarly political. Ichikawa, pp. 74–75.

8. Miyamoto, p. 76; Yoshioka, p. 101. Sengoku was a close personal friend of Shiraishi Naojirō.

9. Yamamoto Tomio, p. 446.

10. Yoshioka, p. 84.

11. Ōshio Takeshi, "Nippon Chisso Hiryō Kabushiki Kaisha," p. 268. The ammonium sulfate plant at Kagami was completed six months before the Shirakawa electric plant in November 1914.

12. Morikawa Hidemasa, "The Increasing Power of Salaried Managers in Japan's Large Corporations," p. 46.

13. Chandler, p. 10.

14. Hoshino, p. 362.

15. Shimotani Masahiro, "Nippon Chisso Hiryō KK," p. 98. Not one of the important factories in Noguchi's operations, the Ōita plant did not generate its own electricity—it bought electricity from Ōita Electric—and was sold to Kyūshū Hydroelectric in 1921.

16. Yoshioka, p. 97; Shibamura, *Noguchi Jun den*, p. 32. The Siemens Affair was a major political scandal leading to the forced resignation of Prime Minister Yamamoto Gonnohyōe. Siemens had bribed naval officers during contract negotiations. Protests erupted, the House of Peers rejected Yamamoto's budget, and the Cabinet fell.

17. Ōshio Takeshi, "Nippon Chisso Hiryō Kabushiki Kaisha," pp. 267–268.

18. Kondō, p. 105.

19. Uchisaburo Kobayashi, pp. 96–97.

20. Ōshio Takeshi, "Nippon Chisso Hiryō Kabushiki Kaisha," p. 268.

21. Yamamoto Tomio, p. 445. It was said that "all Nitchitsu employees had gold watches dangling from gold chains," according to Ichikawa, p. 78.

22. Ichikawa, p. 78.

23. Kawamura, "Nippon Chisso kosho zakkan, sono go," pp. 84–85.
24. Uchisaburo Kobayashi, p. 86.
25. Yoshioka, p. 102.
26. Yamamoto Tomio, p. 451.
27. The "new zaibatsu" or "Konzern" will be discussed in Chapter 5.
28. Ōshio Takeshi, "Nitchitsu Kontsuerun no kin'yū kōzō," p. 115.
29. For further information on cement, see Kawamura Kazuo, "Taishō chūki no hitotachi—Noguchi san no omokage o motometsutsu," pp. 100–104.
30. Shimotani Masahiro, "Nitchitsu kontsuerun to gōsei ryūan kōgyō: Waga kuni kagaku kōgyō ni okeru dokusen keiseishi," p. 70. The expansion and conversion of the Minamata plant in 1915 is discussed in the company documents reprinted in Kawamura, "Nippon Chisso kosho zakkan, sono go," pp. 56–94.
31. Shimotani, "Nippon Chisso Hiryō KK," pp. 102–103; Shimotani Masahiro, *Nihon kagaku kōgyō shiron*, p. 118.
32. Shimotani, "Nitchitsu kontsuerun," p. 65.
33. Katagiri Ryūkichi, p. 246.
34. Chandler, p. 11.
35. Yamamoto Tomio, p. 518.
36. Gordon, p. 423, suggests that it was fairly standard for firms in the Meiji period to offer benefits including job security and good wages to white-collar employees, evidences of these employees' membership in their firms. In other words, company identification, which certainly decreases the wish to leave, was fostered by benefits and career opportunities. Apparently, Nitchitsu was not unique in offering benefits to encourage loyalty.
37. Kawamura Kazuo, "Nippon Chisso kosho zakkan, sono ni: Himekawa Hatsudenjo no koto," pp. 86–87; Kawamura, "Taishō chūki no hitotachi," pp. 79–86.
38. Kawamura, "Taisho chūki no hitotachi," pp. 77–79.
39. Ashimura Masando, "Kagami, Nobeoka, Kōnan, Eian no omoide," p. 28.
40. Katagiri Ryūkichi, pp. 245–247; Matsushita, pp. 155–158.
41. Yamamoto Tomio, p. 452; Matsushita, p. 159; Katagiri Ryūkichi, pp. 248–249.
42. Transfer of management trainees among departments and sections of a corporation is typical for young managers on the "fast track" in Japan today. It is assumed that a rising manager will have experienced several departments before taking on top managerial responsibilities. This permits an understanding of all important aspects of a company and fosters a spirit of integration.
43. Yamada Yutaka, "Chisso seikatsu 23 nen no omoide," pp. 54–57.
44. Kubota Masao, "Furui hanashi to atarashii hanashi," p. 9; Tashiro Saburō, "Noguchi Kenkyūjo sōritsu gakuya hanashi," pp. 5–13.
45. Shimotani Masahiro, "Nitchitsu kontsuerun to gōsei ryūan kōgyō," pp. 64–65.
46. Yoshioka, p. 68.

47. Kondō, pp. 128–130.

48. Haber, *The Chemical Industry 1900–1930,* p. 103. Some British manufacturers miscalculated high costs for synthetic ammonia and therefore believed the cyanamide process to be cheaper.

49. Nakamura Chūichi, *Nihon sangyō no kigyōshiteki kenkyū,* pp. 60–61. He indicates that, although the number of workers in the chemical industry increased by 119% during World War I, the number of plants increased by 207%. Throughout industry in Japan, the increase in employment surpassed the increase in the number of plants (78% to 57%), indicating either that the number of tiny workshops counted as "plants" was higher than in other industries or that some of the new chemical plants were less labor-intensive and thus more modern. Use of electric prime movers was highest in the chemical industry, and that is a common standard for assessing modernity. Matsushima Harumi, "Jūkagaku kōgyōka no katei," pp. 594–595.

50. Shimotani, *Nihon kagaku kōgyō shiron,* p. 118.

51. Shimotani, "Nippon Chisso Hiryō KK," pp. 102–103; Watanabe Tokuji, *Gendai Nihon sangyō hattatsushi,* p. 298.

52. Shimotani, *Nihon kagaku kōgyō shiron,* pp. 118–119, notes that it was either the first or the second company to use the 8-hour work day.

53. The famous good relations enjoyed by higher-level technical staff are noted in several reminiscences by company executives. Yoshioka, p. 106; Yoshida Mitsukuni, *Zaikaijin no gijutsuken,* p. 118.

54. Kawamura Kazuo, "Nippon Chisso kosho zakkan, sono go," pp. 86–87. The number of employees per subsection is listed, by gender.

55. Takeoka Koji, "Kagami kōjō to Shin'etsu Chisso no kaisō," p. 55.

56. Katagiri Yasuo, "Kagami kōjo no omoide," pp. 71–77.

57. Takeoka, p. 55.

58. Shimotani, *Nihon kagaku kōgyō shiron,* pp. 118–119; Katagiri Yasuo, pp. 75–76; Takeoka, p. 55.

59. Ebihara Yoshio, "Chisso ni nyūsha shite 50 nen," p. 116.

60. Yamasaki and Ogawa.

61. Ryoshin Minami, *Power Revolution in the Industrialization of Japan: 1885–1940,* p. 159.

62. Shimotani, *Nihon kagaku kōgyō shiron,* p. 118.

63. Yamamoto Tomio, p. 453.

64. Shimotani, "Nitchitsu kontsuerun to gōsei ryūan kōgyō," p. 75.

65. Ron Wells Napier, "Prometheus Absorbed: The Industrialization of the Japanese Economy, 1905–1937," also notes that productivity does not tend to increase when an industry is doing well. That is, productivity can be raised by technological innovation; investment in innovation appears superfluous when it is not needed. Napier's data, pp. 144–145, indicate that productivity growth for fertilizers was high—40%—between 1919-1923 and 1923–1929, a period corre-

sponding to high innovation and competitive struggle against market saturation by imports.

66. Noguchi's article, published in 1914 as part of the series *Denki hyoron* was entitled "Kōgyōjo yori mita kūchū chisso kotei hō" (An industrial view of methods of fixation of atmospheric nitrogen). It is cited in several sources on ammonium sulfate or on Nitchitsu. The longest quotation from the article is in the official company history. Yamamoto Tomio, p. 457.

67. Haber, *The Chemical Industry 1900–1930,* Chapter 2.

68. Haber devised a way in which steam would be blown over hot coal, producing a mixture of hydrogen and carbon monoxide; and air would be blown over coal, producing nitrogen and carbon monoxide. These two mixtures would be put together so that the hydrogen and nitrogen were in a 3 to 1 ratio and then passed through a catalyst to form hydrogen, nitrogen, and carbon dioxide. At this point, high pressure was needed to scrub out the carbon dioxide. The gases could also be obtained by methods already in use, such as the Linde method for nitrogen used in the cyanamide process, but the lime process as well as methods for hydrogen was not satisfactory for large-scale production, because of high cost. Bosch and William Wild developed the method for water-gas described here (the Bamag process). Haber, *The Chemical Industry 1900–1930,* pp. 94–95.

69. Haber, *The Chemical Industry 1900–1930,* p. 94; Nihon Ryūan Kōgyō Kyōkai, pp. 55–56.

70. Nihon Ryūan Kōgyō Kyōkai, pp. 123–124.

71. Ibid., p. 124.

72. Ibid., p. 60.

73. Ibid., pp. 125–126.

74. Ibid., p. 127.

75. William J. Reader, *Imperial Chemical Industries: A History–The Forerunners 1870–1926,* p. 350.

76. Haber, *The Chemical Industry 1900–1930,* pp. 198–202.

77. Ibid., pp. 205–206.

78. Ibid., p. 208.

79. Watanabe Tokuji, *Gendai Nihon sangyō hattatsushi,* pp. 258, 301; Nihon Ryūan Kōgyō Kyōkai, p. 128. See also Michael A. Cusumano, "Scientific Industry: Strategy, Technology, and Entrepreneurship in Prewar Japan."

80. Nihon Kagakukai, p. 115. Signers of the petition included Takamatsu Toyokichi, Takamine Jōkichi, Nakano Takeyoshi, Ikeda Kikunae, Tahara Ryōjun, Sakurai Jōji, and Suzuki Umetarō.

81. Watanabe Tokuji, *Gendai Nihon sangyō hattatushi,* p. 252; Nippon Kagakushi Gakkai, p. 188.

82. Watanabe Tokuji, *Gendai Nihon sangyō hattatushi,* p. 242.

83. Ibid., p. 242.

84. Ibid., p. 242.

85. The Commission's investigations led to the Law to Encourage Production

of Dyes and Drugs, passed 29 May 1915, and the subsequent establishment of Nippon Dyes under that Law. Watanabe Tokuji, *Gendai Nihon sangyō hattatsushi,* pp. 258–261.

86. Nihon Kagakukai, pp. 116, 146–147.

87. Ibid., p. 147.

88. Nihon Ryūan Kōgyō Kyōkai, pp. 60–62.

89. Nippon Kagakushi Gakkai, p. 215.

90. Nihon Ryūan Kōgyō Kyōkai, p. 129. Shibata was assisted by Shōji Nobumori, and Yokoyama by Nakamura Kenjirō and Tokuoka Matsuo.

91. Nihon Ryūan Kōgyō Kyōkai, p. 130.

92. Ibid., pp. 131–132.

93. Ibid., pp. 130–131; Nihon Kasei Hiryō Kyōkai, p. 50.

94. Nihon Ryūan Kōgyō Kyōkai, pp. 139–140.

95. Watanabe Tokuji, *Gendai Nihon sangyō kōza,* p. 147.

96. Hashimoto Jurō, "1920 nendai no ryūan shijo," p. 56.

97. Yoshioka, pp. 125–126.

98. Yamamoto Tomio, p. 458.

99. Ibid., p. 458.

100. Yoshioka, p. 129.

101. Kitayama Hisashi, "Kasare hō ni yoru anmonia gōsei hō," p. 22.

102. "Nippon Chisso sanjūnen kinen zadankai," p. 54.

103. Yoshioka, p. 132.

104. "Nippon Chisso sanjūnen kinen zadankai," p. 56.

105. Yoshioka, pp. 133–134.

106. Ibid., p. 135.

107. "Nippon Chisso sanjūnen kinen zadankai," p. 57

108. Yamamoto Tomio, p. 459; Yoshioka, p. 136.

109. Yoshioka, p. 137; Shiraishi Muneshiro Kankōkai, ed., *Shiraishi Muneshiro,* pp. 60–66.

110. Yoshida, p. 208.

111. Yoshioka, p. 137. Kagami, he said, was surrounded by farmers and fishermen who objected to industrial expansion. The area around Nobeoka, on the other hand, was underpopulated.

112. Yoshioka, p. 138.

113. Yamamoto Tomio, p. 460.

114. For a detailed description of the politics of acquiring water rights for the Gokasegawa, see Kubota Yutaka, "Nippon Chisso jidai no kaikō: Noguchi san no omoide o chūshin ni," p. 7.

115. Ōshio, "Nitchitsu Kontsuerun no kin'yū kōzō," pp. 115–116.

116. Ibid., pp. 214–215.

117. Ibid., Table 47-1, Nitchitsu balance sheet, 1920–1944, pp. 216–220.

118. Yamamoto Tomio, p. 465.

119. Katagiri Ryūkichi, p. 248; Yamamoto Tomio, p. 518.

120. Yoshioka, p. 143.

121. Ibid., p. 143; Matsushita, p. 36.

122. Yamamoto Tomio, p. 464.

123. Shimotani Masahiro, "Nippon Chisso Hiryō KK," p. 107.

124. Yoshioka, p. 147. Kudō was brought back from Germany, where he had been studying.

125. Yoshioka, pp. 127–128.

126. Yamamoto Tomio, p. 465; Shimotani, "Nippon Chisso Hiryō KK," p. 109.

127. Takeoka, p. 68.

128. Shimotani Masahiro, "Nitchitsu kontsuerun to gōsei ryūan kōgyō," p. 69; Matsushita, p. 73.

129. Shifts were from 7 a.m. to 3:15 p.m.; 3 p.m. to 11:15 p.m.; and 11 p.m. to 7:15 a.m. Shimotani, "Nippon Chisso Hiryō KK," p. 108.

130. Differentials on treatment of the various levels of employees were quite sharp. For example, bonuses were distributed in an extremely inequitable manner, with section heads (*kachō*) receiving 4 months' pay, regular white-collar employees 3 months, foremen 15 days, and workers 10 days. Shimotani, "Nippon Chisso Hiryō KK," p. 108.

131. Sugimoto Toshio, "Ōtsu, Nobeoka kōjo no kaiko: Noguchi san no jinken kōgyō kaihatsu e no kaiko," pp. 5–99.

132. Fukumoto, p. 46.

133. Kondō, pp. 77–78; Shibamura, *Noguchi Jun den*, pp. 77–78.

134. Freeman, p. 256.

135. Shibamura, *Noguchi Jun den*, p. 78.

136. Denka, pp. 93–95; Shimotani Masahiro, "Hensei ryūan, sekkai chisso kōgyō to Denki Kagaku Kōgyō KK no seiritsu"; Fukumoto, p. 31.

137. Frank commented, on learning of Fujiyama's new process, "I was surprised that he discovered the continuous method in Japan, where labor costs were low compared to Europe at that time." Cited in Nagasawa, p. 72.

138. Watanabe Tokuji, *Gendai Nihon sangyō hattatsushi*, p. 295; Shibamura Yōgo, *Nihon kagaku kōgyōshi*, pp. 80–81.

139. Denka, pp. 50–51; Kābaido, pp. 134–136, has a good description of the various methods pioneered by Fujiyama.

140. Watanabe Tokuji, *Gendai Nihon sangyō hattatsushi*, p. 285, lists the 22; they include many of the same men who invested in other fertilizer companies, such as Dai Nihon, a quarter of a century earlier. Most backed their intentions with money and action.

141. Shimotani, "Hensei," p. 34; Nagasawa, p. 71. All 11 members of the Board of Directors of Denka had been signers of the declaration of support for Denka. Of the 19 largest investors, excluding Mitsui Hachirōemon, all but 6 had signed the document (3 of the 6 appear to have been family members of Fujiyama, and

2 were women). Only 5 signers were neither on the Board nor among the firm's 19 largest shareholders (of 267 shareholders). And 7 signers were not only members of the Board but also among the largest shareholders.

142. Denka, pp. 99, 100–101; Shimotani, "Hensei," p. 34.

143. Shimotani, "Hensei," p. 35; Denka, p. 99.

144. Denka, p. 100.

145. Shimotani, "Hensei," pp. 37–38; Denka, p. 101.

146. Watanabe Tokuji, *Gendai Nihon sangyo hattatsushi,* p. 287. Hibi was subsequently used to organize other Denka plants as well.

147. Shimotani, "Hensei," p. 37; Denka, p. 101.

148. Shimotani, "Hensei," p. 39.

149. Denka, p. 105.

Sales by Denka
(in tons)

Year	*Calcium Cyanamide*	*Cyanamide-Method Ammonium Sulfate*
1915	428	2,990
1916	249	5,823
1917	1,281	9,608
1918	1,985	11,233
1919	2,636	18,078
1920	1,600	9,479
1921	424	25,247
1922	4,699	18,343
1923	5,693	46,583
1924	2,943	21,276
1925	10,336	44,725
1926	15,918	33,519

Source: Denka, p. 105.

150. Denka, p. 110; Watanabe Tokuji, *Gendai Nihon sangyō hattatsushi,* pp. 288.

151. Ibid.

152. Haber, *The Chemical Industry 1900–1930,* pp. 86–87.

153. Watanabe Tokuji, *Gendai Nihon sangyō hattatsushi,* p. 289.

FOUR *Nitchitsu's Diversification and Search for Resources: 1924–1933*

1. Watanabe Tokuji, *Gendai Nihon sangyō hattatsushi,* p. 368. Nissan's Aikawa used his relationship with administrators to invest in Manchuria; Noguchi invested in Korea.

2. Nihon Ryūan Kōgyō Kyōkai, pp. 92–93.
3. Ibid., p. 64.
4. Shimotani, *Nihon kagaku kōgyō shiron*, p. 158.
5. Nihon Ryūan Kōgyō Kyōkai, p. 64.
6. Graham D. Taylor and Patricia E. Sudnik, *Du Pont and the International Chemical Industry*, pp. 76–79.
7. Ibid., pp. 80–87.
8. Albert Frank, "Noguchi Jun shi no tsuisō," p. 188.
9. Yoshioka, p. 151.
10. Ibid., p. 151.
11. Ichikawa, p. 120; Ōshio Takeshi, "Nitchitsu Kontsuerun no seiritsu to kigyō kin'yū," p. 74.
12. Ichikawa, p. 122.
13. Ibid., pp. 122–123.
14. Ōshio Takeshi, "Nitchitsu Kontsuerun no kin'yū kōzō," pp. 143–144.
15. Yoshioka, p. 205.
16. Kariya Susumu, "Nippon Chisso no kayaku jigyō," pp. 6–7.
17. Fukumoto, pp. 46–47.
18. Kariya, "Nippon Chisso," p. 8.
19. Yoshioka, p. 206.
20. Kariya, "Nippon Chisso," p. 9.
21. Ibid.
22. Ibid., pp. 9–10.
23. Ibid., pp. 10–13.
24. Ibid., pp. 11–12.
25. Ibid., p. 15.
26. Kamoi, pp. 216–217.
27. Kobayashi Hideo, "1930 nendai Nihon Chisso Hiryō Kabushiki Kaisha no Chōsen e no shinshitsu ni tsuite," p. 144.
28. Yoshioka, pp. 124–125.
29. Sang-chul Suh, *Growth and Structural Changes in the Korean Economy*, pp. 9–10.
30. Han-kyo Kim, "The Colonial Administration in Korea."
31. Samuel Pao-San Ho, "Colonialism and Development: Korea, Taiwan, and Kwantung," pp. 362–363.
32. Kobayashi Hideo, p. 141.
33. Andrew J. Grajdanzev, *Modern Korea*, pp. 133–134. Chōsen Denki Jigyō Shi Henshū Iinkai, *Chōsen denki jigyō shi*, pp. 116–117.
34. Ibid., p. 139.
35. Hatade Isao, *Nihon no zaibatsu to Mitsubishi*, pp. 228–229; Nihon Ryūan Kōgyō Kyōkai, p. 153.
36. Nagatsuka Riichi, *Kubota Yutaka den*, p. 116; Ichikawa, p. 95.

37. Chōsen Denki, pp. 558–559.
38. Nagatsuka, p. 100.
39. Chōsen Denki, p. 249.
40. Morita Kazuo, "Noguchi Jun no Fusenkō kaihatsu," pp. 383–389; Yoshioka, pp. 153–154; Hoshino, p. 363.
41. Nagatsuka, p. 105; Kubota Yutaka, p. 8; Chōsen Denki, p. 250.
42. Kubota Yutaka, p. 10.
43. Nagatsuka, pp. 109–110.
44. Ibid., pp. 110–112.
45. Yoshioka, pp. 155–156; Fukumoto, p. 60; Nakamura Seishi, "Kyōdai denryoku kagaku konbināto no kensetsu," p. 64.
46. Nagatsuka, p. 116; Kubota Yutaka, pp. 10–11.
47. Kubota Yutaka, pp. 10–11.
48. Nagatsuka, p. 118.
49. Yoshioka, p. 158.
50. Shiraishi Muneshiro Kankōkai, *Shiraishi Muneshiro*, p. 77.
51. Noguchi Jun, "Nihon Kai ni kiriotoshita Fusenkō no suiden jigyō," p. 116.
52. Shibamura, *Noguchi Jun den*, p. 139.
53. Fukumoto, p. 63.
54. Chōsen Denki, p. 252; Kubota Yutaka, p. 12.
55. Kamoi, p. 264.
56. Noguchi Jun, p. 117; Yoshioka, pp. 159–160.
57. Nagatsuka, p. 120.
58. Kubota Yutaka, p. 12.
59. Yoshioka, pp. 164–165.
60. Nagatsuka, p. 130; p. 175.
61. Chōsen Denki, p. 133.
62. Suzuki Tsuneo, "Nihon ryūan kōgyō no jiritsuka katei," p. 74.
63. Ōshio Takeshi, "Nitchitsu Kontsuerun to Chōsen Chisso Hiryō," pp. 68–69.
64. Ōshio Takeshi, "Chōsen Chisso Hiryō Kabushiki Kaisha no shūeki ni kansuru ichikōsatsu," pp. 117–123.
65. Kobayashi Hideo, p. 150.
66. Shimotani Masahiro, *Nihon kagaku kōgyō shiron*, p. 178.
67. Suzuki Otokichi, "Kyūnenkan no Kōnan seikatsu dampen," p. 32. This represents a large pay raise since 1935, when university graduates earned 60 yen per month plus a 27-yen allowance, compared to starting salaries of 50 yen at other chemical companies. Ebihara, p. 113.
68. Suzuki Otokichi, pp. 32–33.
69. Kobayashi Hideo, pp. 171, 180.
70. Kamata Shōji, *Hoku-sen no Nihonjin kunan ki: Nitchitsu Kōnan kōjō no saigo*, pp. 15–19.

71. Ho, p. 364.
72. Noguchi Jun, p. 116.
73. Katagiri Ryūkichi, pp. 129–150, discusses ramifications of Noguchi's spirit.
74. Kamoi, p. 262.
75. Shimotani Masahiro, "Nippon Chisso Hiryō," p. 118.
76. Noguchi Jun, p. 117.
77. Nagatsuka, pp. 121–126; p. 52.
78. Shimotani, "Nippon Chisso Hiryō," p. 114.
79. Yoshioka, p. 162; Chōsen Denki, p. 253.
80. Nagatsuka, pp. 139–140.
81. Noguchi Jun, p. 117; Yamamoto Tomio, p. 522.
82. Noguchi Jun, p. 117; Yoshioka, p. 167.
83. Nakamura Seishi, p. 69.
84. Grajdanzev, p. 136.
85. Nagatsuka, pp. 134–139.
86. Yoshioka, pp. 167–168.
87. Ibid., p. 168; Shimotani, "Nippon Chisso Hiryō," p. 107.
88. Kamoi, pp. 252–254.
89. Shiraishi Muneshiro Kankōkai, p. 79.
90. Yoshioka, p. 162; p. 170. Leading managers from Kyūshū included Kudō Kōki, Nagasata Takao, Ōishi Takeo, and Sakaguchi Tokuzō. Kamoi, pp. 249, 252; Yoshioka, p. 171.
91. Nihon Ryūan Kōgyō Kyōkai, p. 99.
92. Japan received 20% of all German synthetic ammonia exports in 1928–1929. Nihon Ryūan Kōgyō Kyōkai, pp. 94, 99–100.
93. Suzuki Tsuneo, "Nihon ryūan kōgyō no jiritsuka katei," p. 68.
94. For more on the *sangyō kumiai* and other farmers' organizations, see Ogura, pp. 247–258.
95. Nihon Ryūan Kōgyō Kyōkai, pp. 75–77.
96. Ibid., p. 80.
97. Ibid., pp. 71–72.
98. Ogura, p. 224.
99. Satō Kanji, *Hiryō mondai kenkyū*, pp. 122–126.
100. Ibid., p. 122.
101. Nihon Ryūan Kogyō Kyōkai, p. 73.
102. Hashimoto, "1920 nendai no ryūan shijo," p. 59.
103. Suzuki Tsuneo, "Nihon ryūan kōgyōno jiritsuka katei," p. 75.
104. Yoshioka, p. 188.
105. Suzuki Tsuneo, "Nihon ryūan kōgyō no jiritsuka katei," p. 73; Hashimoto, "1920 nendai," p. 52. Mitsubishi Trading Company had exclusive rights to import American ammonium sulfate from U.S. Steel and Barrett. While imports by Mitsui of German fertilizer increased, there was little increase in imports by

Mitsubishi of American fertilizer. Because the trade in American fertilizer was limited to Mitsubishi on the Japanese side and was dominated by U.S. Steel on the American, we have only a limited sample from which to extrapolate conclusions. Mitsubishi claimed to be losing 40 yen per ton on its contracted sales of U.S. Steel's ammonium sulfate after 1922, but this may be insufficient explanation for the loss of the American product's market share. Mitsubishi Shōji Kabushiki Kaisha, eds., *Ritsugyō bōekiroku*, p. 470.

106. Nihon Ryūan Kōgyō Kyōkai, pp. 85–86.

107. These regulations were decided by the Cabinet and implemented by the bureaucracy. No political fights had to occur on the floor of the Diet, and this facilitated the promulgation of the regulations.

108. Ogura, p. 224.

109. Hashimoto Jūrō, "Ryūan dokusentai no seiritsu," pp. 55–59.

110. Nihon Ryūan Kōgyō Kyōkai, p. 100.

111. Permissable exports were: England 50%; Poland 40%; Germany, Belgium, Holland, 30%; France 10%; average 30%. Suzuki Tsuneo, "Nihon ryūan kōgyō no jiritsuka katei," p. 76 fn.

112. Nihon Ryūan Kōgyō Kyōkai, p. 103.

113. Kondō, p. 154.

114. Ibid., p. 154.

115. Yoshioka, pp.189–190.

116. Fukumoto, p. 75.

117. Ōshio Takeshi, "Fujiwara-Bosch kyōteian to Nihon no ryūan kōgyō," p. 48.

118. Ishiguro Tadayuki, "Ryūan mondai to Noguchi-san," pp. 72–79; Yoshioka, p. 193.

119. See Nihon Ryūan Kōgyō Kyōkai, pp. 113–116, for Cabinets' proposals to stabilize imports.

120. Kondō, pp. 156–157.

121. Ōshio, "Fujiwara-Bosch," p. 47. See also Barbara Molony, "Noguchi Jun and Nitchitsu: Colonial Investment Strategy of a High Technology Enterprise," pp. 245–248.

122. Nihon Ryūan Kōgyō Kyōkai, pp. 107–108.

123. Kondō Yasuo and Nihon Ryūan Kōgyō Kyōkai both take this position, which overlooks some of the complex aspects of ammonia imports as well as the attempts by firms related to Mitsui and Mitsubishi to import production technology.

124. Kondō, p. 157.

125. Noguchi could avoid such pressure, as most of his output was marketed directly by the Nitchitsu subsidiary, Nitrogenous Fertilizer Sales (Chisso Hiryō Hanbai KK), founded in Ōsaka in 1926. Matsushita, p. 89.

126. Ōshio, "Fujiwara-Bosch," p. 50.

127. Ibid., p. 52.

128. Morikawa Hidemasa, *Zaibatsu no keiei shiteki kenkyū*, pp. 173, 206–209; Mikami Atsufumi, "Kyūzaibatsu to shinkō zaibatsu no kagaku kōgyō: Sumitomo Kagaku to Shōwa Denkō o chūshin to shite," pp. 148–158; Watanabe Tokuji, *Gendai Nihon sangyō hattatsushi*, pp. 448–453.

129. Hashimoto Jurō, "Ryūan dokusen tai no seiritsu," p. 62.

130. Ibid., p. 61.

131. Nihon Ryūan Kōgyō Kyōkai, pp. 118–120.

132. Shibamura Yōgo, *Kagaku hiryō*, pp. 48–49.

133. Shimotani Masahiro, "Nitchitsu kontsuerun to gōsei ryūan kōgyō," pp. 84–86.

134. Nihon Ryūan Kōgyō Kyōkai, p. 120.

135. Shibamura, *Kagaku hiryō*, p. 49.

136. Nihon Ryuan Kōgyō Kyōkai, p. 199.

137. Kobayashi Hideo, p. 163.

138. Shimotani, "Nitchitsu kontsuerun to gōsei ryūan kōgyō," p. 65.

139. Yoshioka, p. 181.

140. Ōshio, "Nitchitsu kontsuerun no kin'yū kōzō," pp. 116–117.

141. Ōshio Takeshi, "Nitchitsu kontsuerun no kigyō haichi," pp. 66–67.

142. Ashimura, pp. 40–41.

143. This structural evolution at Nitchitsu in many ways resembled the reorganization of another chemical giant, DuPont. World War I had induced the American company to diversify in new areas, but the old company structure, with centralized direction of important functions like marketing, manufacturing, and purchasing, was unsuccessful in the new environment. In the early 1920s, the company was forced to decentralize day-to-day operations, giving its branches and subsidiaries greater autonomy by making "product rather than function the basis of the organization." As at Nitchitsu, DuPont senior managers reserved to themselves larger decisions involving company direction. Taylor and Sudnik, pp. 77–79.

144. Ōshio, "Nitchitsu Kontsuerun no kin'yū kōzō," p. 128.

145. Ōshio, "Chōsen Chisso Hiryō Kabushiki Kaisha," p. 113.

146. Yoshioka, p. 181.

147. Ichikawa, p. 100.

148. Shiraishi Muneshiro kankōkai, p. 82.

149. Katagiri Ryūkichi, p. 65. Kushida has been characterized as a dull "typical banker" and "English gentleman" without the wisdom to understand Noguchi's investment decisions.

150. Ibid., p. 74.

151. Nagatsuka, pp. 157–159.

152. Kubota Yutaka, p. 21.

153. Nagatsuka, p. 160.
154. Ogawa Masao, "Noguchi-san to yushi kōgyō," pp. 31–32.
155. Tamaki Shōji, former President Japan Consulting Engineers Association and long-time Nitchitsu employee, corroborated this view of Noguchi's attitude toward money in an interview, 10 August 1983. Tamaki noted that "Noguchi knew how to spend money; Ichikawa knew how to make it."
156. Kubota Yutaka, p. 23.
157. Ōshio, "Chōshinkō," p. 176.
158. Ibid., p. 176.
159. Kobayashi Hideo, p. 139.
160. James B. Crowley, *Japan's Quest for Autonomy*, pp. 87–89, notes that Ugaki was a noted supporter of military modernization and economic rationalization.
161. Ugaki Kazushige, "Noguchi Jun-kun o omou," p. 39; Kobayashi Hideo, p. 174; Fukumoto, p. 77. Ugaki was temporarily Governor General (April-October 1927) when then-Governor General Saitō Makoto was called away to attend the Naval Reduction Conference. At that time, Ugaki was Army Minister. "Ugaki Kazushige nikki ni kakareta Noguchi-san," p. 67.
162. Yoshioka, p. 178; Matsushita, p. 152.
163. Ichikawa, p. 103.
164. Nakamura Seishi, "Kyōdai denryoku kagaku konbināto no kensetsu," pp. 70–71.
165. Miyake Haruteru, *Shinkō kontsuerun tokuhon*, p. 119.
166. Matsushita, pp. 150–151.
167. Ugaki, "Noguchi Jun-kun o omou," p. 42.
168. Yoshioka, p. 232–233.
169. Imai Raijirō, "Denki tōsei to denryokukai no genzai oyobi shōrai," pp. 328–331; Chosen Denki, pp. 156–180.
170. Grajdanzev, p. 135; Imai Raijirō, p. 320.
171. Imai, pp. 332–335.
172. Ugaki, p. 44; Yoshioka, pp. 233–234; Kubota Yutaka, p. 24.
173. Quoted in Yoshioka, p. 223.
174. "Ugaki Kazushige nikki," p. 70.
175. Nagatsuka, pp. 166–167. Final costs of construction were almost 60,000,000 yen.
176. Chōsen Denki, p. 256.
177. Nagatsuka, p. 169.
178. Ōshio, "Chōshinkō," p. 184.
179. Ōshio, "Nitchitsu Kontsuerun no kin'yū kōzō," pp. 142–143.
180. Ugaki Kazushige, "Noguchi Jun-kun o omou," pp. 45–46.
181. Yoshioka, p. 290.
182. "Bukachō ni yoru Nippon Chisso Hiryō 30 shūnen kinen zadankai (Shōwa jūsan nen shichigatsu)," p. 131.

183. Ichikawa, p. 104; Shiraishi Muneshiro Kankōkai, p. 83.
184. Sugita Motomu, "Nippon Chisso no omoide," p. 48.
185. Shimotani, "Nippon Chisso Hiryo KK," p. 123.
186. Katagiri Ryūkichi, pp. 84–85.
187. Ibid., p. 85.
188. Kamoi, p. 262.
189. For information about Katō's desire to assert his independence of the Ministry of Finance I am indebted to Karl Moskowitz, conversation 31 July 1981.
190. Ōshio, "Chōshinkō," p. 177.
191. Katagiri Ryūkichi, p. 98.
192. Following great growth of the Bank of Chōsen after 1937—it was the official bank for funding the Kwantung Army in the China War—Noguchi's percentage of total loans dropped as the bank expanded.
193. Ōshio, "Chōshinkō," p. 177.
194. Miyake, pp. 104–105.
195. Shimotani, "Nitchitsu kontsuerun," p. 63.
196. Ōshio, "Nitchitsu Kontsuerun no kin'yū kōzō," pp. 236, 141.
197. Ibid, pp. 173, 182–183.
198. Yoshioka, p. 240.
199. Chōsen Denki, p. 314.
200. Kubota Yutaka, p. 26.
201. Nagatsuka, pp. 172–174.
202. Yoshioka, p. 240.
203. Ibid., pp. 236–238; Chōsen Denki, p. 258.
204. Chōsen Denki, p. 258.
205. Nagatsuka, p. 177.
206. Ōshio, "Nitchitsu Kontsuerun no kin'yū kōzō," pp. 174–180.
207. Nagatsuka, p. 177.
208. Ibid., pp. 177–178.
209. Ibid., p. 177.
210. Wray, p. 357, defines economic expansion as imperialist when "first, an economic privilege [is] acquired by force, and second, its impact . . . contribute(s) to an erosion of the subject country's autonomy."
211. See, for example, Hilary Conroy, *The Japanese Seizure of Korea: 1868–1910*; Samuel Ho; Peter Duus, "Economic Dimensions of Meiji Imperialism: The Case of Korea, 1895–1910"; Mark R. Peattie, "Introduction"; Ramon H. Myers, "Post World War II Japanese Historiography of Japan's Formal Colonial Empire"; Akira Iriye, "The Failure of Economic Expansion: 1918–1931"; James I. Nakamura, "Incentives, Productivity Gaps, and Agricultural Growth Rates in Prewar Japan, Taiwan and Korea"; and Paul W. Kuznets, *Economic Growth and Structure in the Republic of Korea.*

212. Duus, p. 147.
213. Peattie, p. 12.
214. See, for example, Ho, p. 350; Duus, p. 141; and James Nakamura, p. 346.
215. Wray, pp. 359–360.

FIVE *Nitchitsu and the Japanese Empire*

1. Chandler, *The Visible Hand.* Although each subsidiary was legally a separate corporation, they functioned as if analogous to branches of a unified corporation.
2. Yoshioka, p. 303.
3. Yamamoto Tomio, p. 413.
4. Shibata Kenzō, "Suginishi 50 nenkan no hansei," p. 7.
5. Shimotani, "Nippon Chisso Hiryō," p. 67.
6. Reader, Part Five.
7. Matsushita, p. 84; Yoshioka, p. 229. Unfortunately, the catch began to diminish after 1937. Ogawa, p. 27.
8. Ōshio, "Chōsen Chisso Hiryō Kabushiki Kaisha," p. 99.
9. Sagami Teruo, "Chisso Sekken no omoide," pp. 76–79.
10. Yoshioka, p. 231; Kamoi, p. 210.
11. Itō Masafumi, "Kōnan no omoide," p. 112.
12. Yoshioka, p. 267; Kariya Susumu, "Kayaku kōjō," p. 80.
13. Kariya, "Kayaku kōjō," p. 80; Yoshioka, p. 267.
14. Kamoi, p. 214.
15. Kariya, "Kayaku kōjō," p. 80; Katagiri Ryūkichi, p. 225.
16. Iwama Shigenori, "Nitchitsu no yushi jigyō to watakushi," p. 51.
17. Kariya Susumu, "Nippon Chisso no kayaku jigyō," p. 18.
18. Yoshioka, pp. 268–269.
19. Kariya, "Kayaku kōjō," p. 81.
20. Ibid., p. 80.
21. Matsushita, p. 118.
22. Kariya, "Kayaku kōjō," p. 81.
23. Katagiri Ryūkichi, p. 223.
24. Kamoi, p. 30.
25. Watanabe Tokuji, *Gendai Nihon sangyō hattatsushi,* p. 394.
26. Ōshio, "Chōsen Chisso," pp. 99–104.
27. Ōshio, "Nitchitsu kontsuerun no kin'yū kōzō," pp. 160–161.
28. Kamoi, pp. 209–210.
29. Shimotani, "Nippon Chisso Hiryō," p. 117; Kamoi, p. 327; Shimotani, *Nihon kagaku kōgyō shiron,* p. 138.
30. Yoshioka, pp. 227–228.
31. Arnold Krammer, "Fueling the Third Reich," p. 395.

32. Hoshiko Toshiteru, "Chōsen ni okeru sekiyu jigyō," pp. 364–367; Watanabe Tokuji, *Gendai Nihon sangyō hattasushi*, p. 513; Kamoi, p. 327.

33. Munekata Eiji, "Noguchi-san hassō no sekitan ekika ni eikō are," p. 5.

34. Tashiro Saburō, "Kōnan kenkyūjo no koto," pp. 6–7.

35. Ōshio, "Nitchitsu Kontsuerun no kin'yū kōzō," p. 152.

36. Shinoda Keiji, "Idai naru rōhi: Kichirin Jinzō Sekiyu," p. 74.

37. Yoshioka, p. 274.

38. Ōshio, "Chōsen Chisso," pp. 104–107; Shimotani, *Nihon kagakui kōgyō shiron*, p. 139.

39. Ōshio, "Chōsen Chisso," p. 105.

40. Munekata, "Noguchi-san," p. 6.

41. Abe Kun'ichi, "Kichirin Jinzō no omoide," p. 64.

42. Munekata Eiji, "Agochi (Sekitan ekika) kōjō no omoide," pp. 29–31.

43. Krammer, p. 396; Shiraishi Muneshiro, "Kōnan kōjō no gaisetsu," pp. 36–37; Nenryō Konwakai, eds., *Nihon kaigun nenryōshi, Jō*, p. 76. The Navy wanted private interests to develop its patents.

44. Ōshio, "Chōsen Chisso," p. 106.

45. Even this amount, if maintained on an annual basis, would total just 36,500 tons as compared to Germany's 1.2 million tons in 1939. Krammer, pp. 403–404.

46. Munekata, "Noguchi-san," pp. 6–8.

47. Munekata, "Agochi," pp. 72–74, 80–81.

48. Ōshio, "Chōsen Chisso," p. 106.

49. Ōshio, "Nitchitsu Kontsuerun no kin'yū kōzō," pp. 169–173; Ōshima Mikiyoshi, "Nitchitsu Nenryō–Ryūkō kōjō," pp. 89–106.

50. Kodama Noritada, "Kaisō danpen," p. 47.

51. Sugita, p. 53.

52. Ōshio "Nitchitsu Kontsuerun no kin'yū kōzō," pp. 152–157.

53. Kudō Kōki, "Noguchi-san no omokage," p. 811.

54. Ichikawa, pp. 158–159.

55. Ōshio, "Nitchitsu Kontsuerun no kin'yū kōzō," pp. 166–168; Ichikawa, p. 160.

56. Tamaki Shōji, interview.

57. Yoshioka, p. 275.

58. Ibid., p. 276.

59. Nenryō Konwakai, p. 305.

60. Yoshioka, p. 257.

61. Watanabe Tokuji, *Gendai Nihon sangyō hattatsushi*, p. 391.

62. Oshio, "Nitchitsu Kontsuerun no kin'yū kōzō," pp. 158–159; Nakano Hitoshi, "Kōnan maguneshiumu kōjō ki," p. 138.

63. Fritz Hansgierg was brought to Korea to advise Nitchitsu on magnesium production, but he soon turned to other areas. He advised Nitchitsu to found Nitchitsu Gemstones (Nitchitsu Hōseki), a maker of munitions-use ball bear-

ings; he held 15% of this company's shares. He also helped to develop aluminum refining. Ōshio, "Nitchitsu Kontsuerun no kin'yū kōzō," pp. 164–166; Shiraishi Muneshiro Kankōkai, p. 101. Noguchi was not initially interested in aluminum refining; according to a researcher in Hungnam, he declared in 1934 or 1935 that, if he had to invest in metal refining, he preferred to emphasize magnesium. But he let interested researchers develop aluminum production because they showed they had already collected considerable data. Marui Ryōsei, "Denkai kōjō kinmu nijūnen," p. 80.

64. Nakano, pp. 136–138.

65. Yoshioka, p. 257.

66. Ibid., p. 248.

67. Ibid., pp. 248–251.

68. Kitayama, p. 23.

69. Albert Frank, originator of the Frank-Caro method of cyanamide production, likened Noguchi's production of cyanamide to a person's return to his "first love." Frank, pp. 186–191.

70. Shimotani, "Nippon Chisso Hiryō," p. 119; Ōshio, "Chōsen Chisso," p. 104.

71. Yoshioka, p. 246; see also Ōshio, "Chōsen Chisso," table inserted between pages 98 and 99.

72. Ōshio, "Chōsen Chisso," pp. 110–115.

73. Yamamoto Shigeru, *Dai Tō-A kagaku kōgyō ron*, p. 156.

74. Ibid., pp. 259–261.

75. Kubota Yutaka, pp. 29–30.

76. Ibid., p. 30; Nagatsuka, p. 190.

77. Nagatsuka, pp. 190–192; Kubota Yutaka, p. 31.

78. Nagatsuka, pp. 197–198.

79. Chōsen Denki, p. 292.

80. Tamaki Shōji, interview.

81. Chōsen Denki, p. 292; Nagatsuka, p. 201.

82. Ōshio, "Nitchitsu Kontsuerun no kigyō haichi," p. 72.

83. Nagatsuka, p. 202.

84. Chōsen Denki, p. 292; Kubota Yutaka, p. 37; Nagatsuka, pp. 216–217.

85. Kubota Yutaka, p. 35.

86. Chōsen Denki, p. 371.

87. Nagatsuka, p. 207.

88. Ibid., p. 211.

89. Kubota Yutaka, p. 38.

90. Ibid., p. 38.

91. Ibid., p. 38.

92. Nagatsuka, pp. 211–212.

93. Ibid., pp. 213–215.

94. Ibid., pp. 216–217; Kubota Yutaka, pp. 39–40.

95. Nagatsuka, p. 221; Chōsen Denki, p. 375.
96. Udagawa Masaru, *Shinkō zaibatsu*, p. 135.
97. Ibid., p. 136.
98. Ibid., p. 131.
99. Ōshio, "Nitchitsu Kontsuerun no kin'yū kōzō," insert between pages 132 and 133.
100. Udagawa, *Shinkō zaibatsu*, p. 126.
101. Ibid., p. 136.
102. Kamata Shōji, *Hoku-Sen no Nihonjin kunan ki: Nitchitsu Kōnan kōjō no saigo*, p. 13. Kamata (p. 435) also breaks down conscription rates by job classification; most draftees came from the ranks of "regular employees."
103. Ibid., pp. 15–19.
104. Ōshio, "Nitchitsu Kontsuerun no kin'yū kōzō," pp. 150–164.
105. Ibid., pp. 174–179.
106. Tamaki Shōji, interview.
107. Chōsen D nki, p. 299.
108. Ōshio, "Nitchitsu Kontsuerun no kin'yū kōzō," pp. 164–166.
109. Ibid., pp. 194–196.
110. Ibid., pp. 182–186, p. 192.
111. Ibid., p. 203.
112. Ibid., pp. 169–173.
113. Kubota Yutaka, pp. 42–43.
114. Ishiguro Ribee, "Ka-hoku Chisso no koto," pp. 41–42; Itō Bunkichi, "Taigen kōjō no omoide," p. 46.
115. Chihara Matsuo, "Ka-hoku Chisso no koto domo," pp. 36–38.
116. Ōshio, "Nitchitsu kankei kaisha no seiritsu to idō," pp. 66–67.
117. Yokota Shigeru, "Taiwan Chisso no omoide," pp. 27–34.
118. Egami Masao, "Shūsengo no Taiwan Chisso," pp. 42–43.
119. Nagatsuka, pp. 225–229.
120. Kubota Yutaka, p. 47.
121. Nagatsuka, pp. 230–231.
122. Udagawa, *Shōwashi to shinkō zaibatau*, pp. 192–193.
123. Nagatsuka, pp. 236–240.
124. Kubota Yutaka, p. 49.
125. Nagatsuka, p. 244.
126. Ibid., p. 239.
127. Nagatsuka, p. 245.
128. Kubota Yutaka, p. 54.
129. Tamaki Shōji, interview.
130. Sangyō Seisakushi Kenkyūjo, *Waga kuni dai kigyō no keisei hatten katei*, pp. 38-39.
131. Ōshio Takeshi, "Shinkō kontsuerun."

132. Udagawa, *Shinkō zaibatsu,* pp. 11–12.
133. Yasuoka Shigeaki, *Zaibatsu no keieishi,* p. 101.
134. Nakamura Seishi, *Shinkō zaibatsu no hatten,* p. 114.
135. Ōshio, "Shinkō Kontsuerun," p. 90.
136. Udagawa, *Shinkō zaibatsu,* passim.
137. Shimotani, *Nihon kagaku kōgyō shiron,* p. 173.
138. Udagawa Masaru, *Shinkō zaibatsu,* p. 222; p. 228.
139. Shimotani, *Nihon kagaku kōgyō shiron,* p. 181.
140. Ibid., p. 180.
141. Ōshio, "Nitchitsu Kontsuerun no kin'yū kōzō," p. 118.
142. Ibid., p. 123.
143. Ibid., p. 123.
144. Ishino Inosuke, "Omoide are kore: Shōwa jūichinen kara jūgonen made no honsha seikatsu," p. 24.
145. Ōshio, "Nitchitsu Kontsuerun no kin'yū kōzō," pp. 119–120.
146. Ibid., pp. 208–209.
147. Udagawa, *Shinkō zaibatsu,* p. 268.

SIX *The Second and Third Waves: Widening the Circle of Investors in High Technology*

1. Morikawa, *Zaibatsu no keiei shiteki kenkyū,* p. 205; Nihon Ryūan Kōgyō Kyōkai, p. 60.
2. Nihon Ryūan Kōgyō Kyōkai, p. 60; Atsufumi Mikami, " Old and New Zaibatsu in the History of Japan's Chemical Industry," pp. 202–204.
3. Nihon Ryūan Kōgyō Kyōkai, p. 61.
4. Nissan Kagaku, p. 97.
5. Nihon Ryūan Kōgyō Kyōkai, p. 62.
6. Ibid., p. 90.
7. Nissan Kagaku, p. 80.
8. Morikawa, *Zaibatsu no keiei shiteki kenkyū,* p. 205.
9. Nihon Ryūan Kōgyō Kyōkai, p. 63; Nissan Kagaku, p. 80.
10. "Caustic" or the trade-marked names Mizuho fertilizer and Tokiwa fertilizer were simply Dai Nihon's own names for certain compounds containing nitrogen, ammoniated nitrogen, hydrogenated potassium, phosphatic acid, and potassium, in specific proportions. Kondō, p. 140.
11. Nissan Kagaku, p. 75.
12. Ibid., p. 85.
13. Hashimoto, "Ryūan dokusentai no seiritsu," p. 53
14. Ibid., p. 86.
15. Nihon Ryūan Kōgyō Kyōkai, p. 141.

16. Nissan Kagaku, p. 88.
17. Ibid., p. 88.
18. Shimotani, "Dai Nihon Jinzō Hiryō," p. 49.
19. Nissan Kagaku, p. 70.
20. Ibid., p. 87.
21. Ibid., p. 92.
22. Kondō, p. 200; Nissan Kagaku, p. 92.
23. Nissan Kagaku, p. 92.
24. Ibid., p. 93.
25. Ibid., p. 96.
26. Wanatabe Tokuji, *Gendai Nihon sangyō hattatsushi,* pp. 411–412.
27. Hashimoto, "Ryūan dokusentai no seiritsu," p. 48, cites data from Minami Manshū Tetsudō Keizai Chōsakai (1935) and Nōmura Shōken Chōsabu (1932) to show that Shōwa Fertilizer's output (144,000 tons), though just half of Chōsen Chisso's (285,000 tons), was the second largest.
28. Shōwa Denkō Kabushiki Kaisha Shashi Hensanshitsu, *Shōwa Denkō 50nenshi,* pp. 1–10.
29. Ko-Suzuki Saburōsuke-kun Den Kiroku Hensankai, *Suzuki Saburōsuke den,* p. 83.
30. Shōwa Denkō, p. 10.
31. Ibid., p. 32.
32. Ibid., p. 27.
33. Watanabe Tokuji, *Gendai Nihon sangyō hattatushi,* p. 304.
34. Kondō, p. 141.
35. Watanabe Tokuji, *Gendai Nihon sangyō hattatsushi,* p. 304.
36. Nihon Ryūan Kōgyō Kyōkai, p. 143.
37. Shibamura Yōgo, *Nihon kagaku kōgyōshi,* p. 101.
38. Hashimoto, "Ryūan dokusentai no seiritsu," p. 54.
39. Shibamura, *Nihon kagaku kōgyōshi,* p. 101.
40. Shōwa Denkō, pp. 33–34; Hashimoto, p. 55, says that the loan was given in June 1941.
41. Watanabe Tokuji, *Gendai Nihon sangyō hattatsushi,* pp. 402–403.
42. Nihon Ryūan Kōgyō Kyōkai, p. 143.
43. Shōwa Denkō, p. 40.
44. Matsushita, p. 206; Ishikawa Teijirō, *Suzuki Saburōsuke den, Mori Nobuteru den.* p. 249.
45. Watanabe Tokuji, *Gendai Nihon sangyō hattatsushi,* p. 405.
46. Ibid., pp. 370, 405; Matsushita, p. 206.
47. Matsushita, pp. 252–256.
48. Ishikawa Teijirō, p. 234.
49. Watanabe Tokuji, *Gendai Nihon sangyō hattatsushi,* p. 403.
50. Katagiri Ryūkichi, pp. 154–155, 162–163.

51. Other sources place the firms in a slightly different hierarchical order. Mikami Atsufumi, for instance, eliminates Denki Kagaku and Ube Chisso and places Nissan (Dai Nihon) behind Nitchitsu, but generally agrees that Nitchitsu's domestic output was greatly exceeded by newer producers by 1935. Mikami Atsufumi, "Sumitomo Kagaku no keisei, hatten katei," p. 134.

52. Morikawa, "The Increasing Power of Salaried Managers in Japan's Large Corporations." pp. 61–64.

53. Morikawa, *Zaibatsu no keiei shiteki kenkyū,* pp. 61–64.

54. Mishima Yasuo, *Mitsubishi Zaibatsu,* p. 317; Mitsubishi Kasei Kōgyō Kabushiki Kaisha, *Mitsubishi Kasei shashi,* p. 35.

55. Shimotani, *Nihon kagaku kogyō shiron,* pp. 238–240.

56. Ibid., p. 240n.

57. Morikawa, *Zaibatsu,* p. 205.

58. Though the cost per year was less than the fees Noguchi paid, Mitsui would not have been relieved of payments after a single year, making it difficult to recoup its original investment.

59. See Chapters 3 and 4 above.

60. Morikawa, *Zaibatsu,* pp. 206, 172.

61. Ibid., p. 220.

62. Ibid., p. 220n.

63. Ibid., p. 220. Even after modification, the Hikoshima plants under Mitsui management operated just 200 days a year.

64. Ibid., p. 173; p. 184n. Tatsumi Eiichi headed the Meguro Laboratories, and Hibi Tatsuji headed the Industrial Affairs Section at Denka.

65. Ibid., p. 173. This auditor was Soeda Juichi.

66. Nihon Ryūan Kōgyō Kyōkai, p. 147.

67. Morikawa, *Zaibatsu,* p. 173. The management group at Daiichi and Claude included men from both Mitsui Mining and Denka: Fuwa Kumao of the Tagawa Mine, Tatsumi Eiichi of Meguro Laboratories, Ishige Ikuji of Miike Dyes, Hibi Tatsuji of Denka, and Furusaki Hidejirō of Denka.

68. Ibid., p. 207.

69. Nihon Ryūan Kōgyō Kyōkai, p. 147; Morikawa, *Zaibatsu,* p. 206.

70. Makita Tamaki, quoted in Morikawa, p. 206.

71. Morikawa, *Zaibatsu,* p. 173; Nihon Ryūan Kōgyō Kyōkai, p. 247.

72. Nihon Ryūan Kōgyō Kyōkai, p. 147.

73. Ibid.; Morikawa, *Zaibatsu,* pp. 207–208.

74. Shimotani, *Nihon kagaku,* p. 232.

75. Morikawa, *Zaibatsu,* pp. 208–209.

76. Shimotani, *Nihon kagaku,* p. 231.

77. Nihon Ryūan Kōgyō Kyōkai, p. 148.

78. Morikawa, *Zaibatsu,* pp. 208–209.

79. Ibid., p. 209.

80. Ibid.

81. Ibid.

82. Yasuo Horie, "The Tradition of the *Ie* (House) and the Industrialization of Japan," pp. 231–245.

83. Yasuda Shigeaki, *Mitsui Zaibatsu,* p. 287.

84. Mikami Atsufumi, "Kyūzaibatsu to shinkō zaibatsu no kagaku kōgyō: Sumitomo Kagaku to Shōwa Denkō o chūshin to shite," pp. 148–149.

85. Sumitomo Kagaku Kōgyō Kabushiki Kaisha, *Sumitomo Kagaku Kōgyō Kabushiki Kaishashi,* pp. 23–24.

86. Sakudō Yotarō, *Sumitomo Zaibatsu,* p. 185.

87. Mikami, "Sumitomo Kagaku," p. 125.

88. Sumitomo Kagaku, pp. 23–24.

89. Sakudō, p. 185.

90. Morikawa, *Zaibatsu,* p. 171; Watanabe Tokuji, *Gendai Nihon sangyō hattatsushi,* p. 441; Sumitomo Kagaku, p. 41.

91. Watanabe Tokuji, *Gendai Nihon sangyō hattatsushi,* p. 441.

92. Sumitomo Chemical profits plunged from 176,000 yen in 1918 and 559,000 yen in 1919 to -112,000 yen in 1920, -170,000 yen in 1921, and -200,000 yen in 1922. Profits recovered in 1923. Mikami, "Sumitomo Kagaku," p. 127.

93. Sumitomo Kagaku, p. 42.

94. Ibid., p. 30.

95. Asajima Shōichi, *Senkanki Sumitomo Zaibatsu keieishi* p. 198.

96. Mikami, "Sumitomo Kagaku," pp. 128–129.

97. Ibid., p. 129.

98. Watanabe Tokuji, *Gendai nihon sangyō hattatsushi,* p. 441.

99. Mikami, "Kyūzaibatsu," p. 157.

100. Sumitomo Kagaku, p. 35.

101. Watanabe Tokuji, *Gendai Nihon sangyō hattatsushi,* pp. 443–444, quotes the company's records on its decision to invest in ammonia production.

102. Mikami, "Kyūzaibatsu," p. 158.

103. Sumitomo Kagaku, p. 44.

104. Ibid., p. 47.

105. Mikami, "Kyūzaibatsu," p. 158; "Sumitomo Kagaku," p. 130.

106. Mikami, "Kyūzaibatsu," p. 159.

107. Watanabe Tokuji, *Gendai Nihon sangyō hattatsushi,* p. 445.

108. Mikami, "Kyūzaibatsu," p. 161.

109. Sumitomo Kagaku, p. 63.

110. Ibid., p. 64.

111. Asajima, p. 201.

112. Sumitomo Kagaku, p. 91; p. 108.

113. Asajima, p. 201.

114. Watanabe Tokuji, *Gendai Nihon sangyō hattatsushi*, p. 445; Mikami, "Kyū-zaibatsu," p. 161.
115. Watanabe Tokuji, *Gendai Nihon sangyō hattatsushi*, p. 446.
116. Sumitomo Kagaku, pp. 71–73.
117. Ibid., p. 106; p. 111.
118. Ibid., p. 101.
119. Ibid., p. 65.
120. Mikami, "Sumitomo Kagaku," p. 129; p. 133.
121. Asajima, p. 477.
122. Mishima, p. 85.
123. Watanabe Tokuji, *Gendai Nihon sangyō hattatsushi*, pp. 448–449.
124. Mitsubishi Kasei, p. 26.
125. Hatade, pp. 228–229.
126. Nihon Ryūan Kōgyō Kyōkai, p. 153.
127. Morikawa, *Zaibatsu*, p. 172.
128. Mishima, p. 317.
129. Mitsubishi Kasei, p. 35.
130. Mishima, p. 319.
131. Mitsubishi Kasei, pp. 52–53.
132. Nihon Ryūan Kōgyō Kyōkai, p. 153; Watanabe Tokuji, *Gendai Nihon sangyō hattatsushi*, p. 453.
133. Nihon Ryūan Kōgyō Kyōkai, p. 154.; Watanabe Tokuji, *Gendai Nihon sangyō hattatsushi*, p. 455.
134. Nihon Ryūan Kōgyō Kyōkai, p. 154. The term *kasei* for "chemical" was invented by Iwasaki Koyata.
135. Watanabe Tokuji, *Gendai Nihon sangyō hattatsushi*, p. 455.
136. Mitsubishi Kasei, pp. 70–72.
137. Ibid., p. 66.
138. Ibid., p. 85.
139. Ibid., p. 84.
140. Ibid., pp. 71–72.
141. Watanabe Tokuji, *Gendai Nihon sangyō hattatsushi*, pp. 456–457.
142. Mishima, p. 320.
143. Ibid., p. 320.
144. Mitsubishi Kasei Shashi, pp. 99–103.

Bibliography

Abe Kun'ichi. "Kichirin Jinzō no omoide" (Reminiscences of Jilin Artificial). In Kamata Shōji, ed., *Nippon Chisso shi e no shōgen* (Eye-witness accounts of Nitchitsu history), Vol. XI. Tokyo, Tokyo Shinku Sabisu-nai Nippon Chisso Shi e no Shōgen Iinkai, 1980.

Abegglen, James C., and George Stalk, Jr. *Kaisha, the Japanese Corporation.* New York, Basic Books, 1985.

Arisawa Hiromi, ed. *Nihon sangyō hyakunenshi* (One hundred-year history of Japanese industry). 2 vols. Tokyo, Nihon Keizai Shinbunsha, 1967.

Asai Yoshio. "Nōshōkō kōtō kaigi" (High-level conferences on agriculture, commerce, and industry). In Ōishi Kaichirō and Miyamoto Ken'ichi, eds., *Nihon shihonshugi no hattatsushi no kisō chishiki.* Tokyo, Yuhikaku, 1975.

Asajima Shoichi. *Senkanki Sumitomo Zaibatsu keieishi* (Business history of the wartime Sumitomo Zaibatsu). Tokyo, Tokyo Daigaku Shuppankai, 1983.

Ashimura Masando. "Kagami, Nobeoka, Kōnan, Eian no omoide" (Reminiscences of Kagami, Nobeoka, Hungnam, and Yong'an). In Kamata Shōji, ed., *Nippon Chisso shi e no shōgen,* Vol. VIII. Tokyo, 1979.

Ban, Sung Hwan, Pal Yong Moon, and Dwight Perkins, eds. *Rural Development.* Cambridge, Council on East Asian Studies, Harvard University, 1980.

"Bukachō ni yoru Nippon Chisso Hiryō 30 shūnen kinen zadankai [Shōwa 13 nen shichigatsu]" (Division and section chiefs' symposium for Nitchitsu's thirtieth anniversary [July 1938]). In Kamata Shōji, ed., *Nippon Chisso shi e no shōgen, Dokukan,* Vol. I. Tokyo, 1987.

Chandler, Alfred. *The Visible Hand: The Managerial Revolution in American Business.* Cambridge, Harvard University Press, 1977.

Chihara Matsuo. "Ka-hoku Chisso no koto domo" (About North China Nitrogenous). In Kamata Shōji, ed., *Nippon Chisso shi e no shōgen,* Vol. I. Tokyo, 1977.

Chōki keizai tōkei: nōringyō (Long-term economic statistics: agriculture and forestry industries), Vol. IX. Tokyo, Tōyō Keizai Shinposha, 1965.

Chōsen Denki Jigyō Shi Henshū Iinkai, eds. *Chōsen denki jigyō shi* (History of the electric industry in Korea). Tokyo, Shadan Hōjin Chūō Nikkan Kyōkai, 1981.

Chōsen Ginkō. *Shōkeisansho* (General accounts of the Bank of Chōsen). 1930–1943. Seoul, Chōsen Ginkō.

Conroy, Hilary. *The Japanese Seizure of Korea: 1868–1910.* Philadelphia, University of Pennsylvanina Press, 1960.

Crowley, James B. *Japan's Quest for Autonomy: National Security and Foreign Policy 1930–1938.* Princeton, Princeton University Press, 1966.

Cusumano, Michael A. "Scientific Industry: Strategy, Technology, and Entrepreneurship in Prewar Japan." In William D. Wray, ed., *Managing Industrial Enterprise: Cases in Japan's Prewar Experience.* Cambridge, Council on East Asian Studies, Harvard University, 1989.

Dai Nihon Jinzō Hiryō Kabushiki Kaisha, eds. *Dai Nihon Jinzō Hiryō Kabushiki Kaisha gojūnenshi* (Fifty-year history of Dai Nihon Artificial Fertilizers, Inc.). Tokyo, Dai Nihon Jinzō Hiryō, 1936.

Daikuhara, G., and T. Imaseki. "On the Behaviour of Nitrate in Paddy Soils," *Bulletin of the Imperial Central Agricultural Station* 1.2:7–36 (October 1907).

Denka Rokujūnenshi Hensan Iinkai, eds. *Denka rokujūnenshi* (Sixty-year history of Denka). Tokyo, Denka, 1977.

Duus, Peter. "Economic Dimensions of Meiji Imperialism: The Case of Korea, 1895–1910." In Ramon H. Myers and Mark R. Peattie, eds., *The Japanese Colonial Empire, 1895–1945.* Princeton, Princeton University Press, 1984.

Ebihara Yoshio. "Chisso ni nyūsha shite 50 nen" (Fifty years since I entered Nitchitsu). In Kamata Shōji, ed., *Nippon Chisso shi e no shōgen,* Vol. XXVII. Tokyo, 1986.

Egami Masao. "Shūsengo no Taiwan Chisso" (Taiwan Nitrogenous after the end of the war). In Kamata Shōji, ed., *Nippon Chisso shi e no shōgen,* Vol. II. Tokyo, 1977.

Feuerwerker, Albert. *The Chinese Economy, ca. 1870–1911.* Michigan Papers in Chinese Studies, no. 5, Ann Arbor, 1969.

Frank, Albert. "Noguchi Jun shi no tsuisō" (Reminiscences about Noguchi Jun). In Takanashi Kōji, ed., *Noguchi Jun no tsuikairoku.* Ōsaka, Noguchi Jun Tsuikairoku Hensankai, 1952.

Freeman, Christopher. *The Economics of Industrial Innovation.* Harmondsworth, Penguin Books, 1974.

Fukumoto Kunio, ed. *Noguchi wa ikite iru: Jigyō supiritto to sono tenkai* (Noguchi lives: The development of his entrepreneurial spirit). Tokyo, Fuji Intanashiyonaru Konsarutanto, 1964.

Gordon, Andrew. *The Evolution of Labor Relations in Japan: Heavy Industry, 1853–1955.* Cambridge, Council on East Asian Studies, Harvard University, 1985.

Grajdanzev, Andrew J. *Modern Korea.* New York, Institute of Pacific Relations, 1944.

Haber, L. F. *The Chemical Industry During the Nineteenth Century.* London and Oxford, Clarendon Press, 1958.

——, *The Chemical Industry 1900–1930: International Growth and Technological Change.* Oxford, Clarendon Press, 1971.

Hamabata, Matthews M. "Love and Work in Japanese Society: The Role of Women in Large-Scale Family Enterprise." Paper presented at Association for Asian Studies meeting, 1983.

Hashimoto Jurō. "1920 nendai no ryūan shijō" (Ammonium sulfate market in the 1920s), *Shakai keizai shigaku* 43.4:45–70 (1978).

——, "Ryūan dokusentai no seiritsu" (Establishment of the monopoly system for ammonium sulfate), *Keizaigaku ronshū* 45.4:44–68 (1980).

Hatade Isao. *Nihon no zaibatsu to Mitsubishi* (Japan's zaibatsu and Mitsubishi). Tokyo, Rakuyu Shobō, 1978.

Ho, Samuel Pao-San. "Colonialism and Development: Korea, Taiwan, and Kwantung." In Ramon H. Myers and Mark R. Peattie, eds., *The Japanese Colonial Empire, 1895–1945.* Princeton, Princeton University Press, 1984.

Horie, Yasuo. "The Tradition of the *Ie* (House) and the Industrialization of Japan." In Keiichiro Nakagawa, ed., *Social Order and Entrepreneurship.* International Conference on Business History, Vol. II. Tokyo, University of Tokyo Press, 1977.

Hoshiko Toshiteru. "Chōsen ni okeru sekiyu jigyō" (The oil industry in Korea). In Shibuya Reiji, ed., *Chōsen no kōgyō to sono shigen.* Seoul, Chōsen Kōgyō Kyōkai, 1937.

Hoshino Yoshiro. "Noguchi Jun to gijutsu no kakushin" (Noguchi Jun and technological reform), *Chūō kōron* 80.928:359–365 (February 1965).

Ichikawa Homei. *Ichikawa Seiji den* (Biography of Ichikawa Seiji). Ōsaka, Bunshindō, 1974.

Imai Raijirō. "Denki tōsei to denryokukai no genzai oyobi shōrai" (Electric controls and the present and future of the Electric Power Association). In Shibuya Reiji, ed., *Chōsen no kōgyō to sono shigen.* Seoul, Chosen Kōgyō Kyōkai, 1937.

Iriye, Akira. "The Failure of Economic Expansion: 1918–1931." In Bernard S. Silberman and H. D. Harootunian, eds., *Japan in Crisis: Essays on Taishō Democracy.* Princeton, Princeton University Press, 1974.

Ishiguro Ribee. "Ka-hoku Chisso no koto" (About North China Nitrogenous). In Kamata Shōji, ed., *Nippon Chisso shi e no shōgen,* Vol. I. Tokyo, 1977.

Ishiguro Tadayuki. "Ryūan mondai to Noguchi-san" (The ammonium sulfate problem and Mr. Noguchi). In Takanashi Kōji, ed., *Noguchi Jun o tsuikairoku.* Ōsaka, Noguchi Jun Tsuikairoku Hensankai, 1952.

Ishikawa Teijirō. *Suzuki Saburōsuke den, Mori Nobuteru den* (Biographies of Suzuki Saburōsuke and Mori Nobuteru). Tokyo, Tōyō Shokan, 1954.

Ishino Inosuke. "Omoide are kore: Shōwa jūichinen kara jūgonen made no honsha seikatsu" (Various memories: Life at company headquarters from 1936 to 1940). In Kamata Shōji, ed., *Nippon Chisso shi e no shōgen,* Vol. XVII. Tokyo, 1983.

Itō Bunkichi. "Taigen kōjō no omoide" (Memories of the Taiyuan plant). In Kamata Shōji, ed., *Nippon Chisso shi e no shōgen,* Vol. I. Tokyo, 1977.

Itō Masafumi. "Kōnan no omoide" (Memories of Hungnam). In Kamata Shōji, ed., *Nippon Chisso shi e no shōgen,* Vol. X. Tokyo, 1980.

Iwama Shigenori. "Nitchitsu no yushi jigyō to watakushi" (Nitchitsu's fats and oils industry and I). In Kamata Shōji, ed., *Nippon Chisso shi e no shōgen,* Vol. II. Tokyo, 1977.

Iwauchi, Ryōichi. "Institutionalizing the Technical Manpower Formation in Japan," *Developing Economies* 15.4:420–439 (December 1977).

Johnson, Chalmers. *MITI and the Japanese Miracle: The Growth of Industrial Policy, 1925–1975.* Stanford, Stanford University Press, 1982.

Kābaido Kōgyō no Ayumi Hensan Iinkai, eds. *Kābaido kōgyō no ayumi* (Development of the carbide industry). Tokyo, Kābaido Kōgyōkai, 1956.

Kagaku Keizai Kenkyūjo, eds. *Kagaku kōgyō no jissai chishiki* (Factual information about the carbide industry). Tokyo, Tōyō Keizai Shinposha, 1956.

Kajinishi Mitsuhaya, Katō Toshihiko, Ōshima Kiyoshi, and Ōuchi Tsutomu, eds. *Nihon ni okeru shihonshugi no hattatsushi* (History of the development of capitalism in Japan), Vol. III. Tokyo, Tokyo Daigaku Shuppansha, 1975.

Kamata Shōji. *Hoku-Sen no Nihonjin kunan ki: Nitchitsu Kōnan kōjō no saigo* (Record of the sufferings of the Japanese in North Korea: The last days of Nitchitsu's Hungnam plant). Tokyo, Jiji Tsūshinsha, 1970.

——. "Waga kuni saisho no kābaido jigyō—Kōriyama Kābaido Seizōjo" (Japan's first carbide industry—the Kōriyama carbide plant). In Kamata Shōji, ed., *Nippon Chisso shi e no shōgen,* Vol. XXIV. Tokyo, 1985.

Kamatani, Chikayoshi. "The Role Played by the Industrial World in the Progress of Japanese Science and Technology," *Journal of World History* 9 (1965).

Kameyama Naoto. *Kagaku kōgyō gairon* (Outline of the chemical industry). Tokyo, Hyōronsha, 1940.

Kamoi Yū. *Noguchi Jun.* Tokyo, Tōkōsha, 1943.

Kantō Taru Seihin Kabushiki Kaisha Sōritsu Jūshūnen Kinen Jigyō Iinkai, eds. *Taru kōgyōshi: Tōkyō Gasu o chūshin to shite* (History of the tar industry, with a focus on Tokyo Gas). Tokyo, Kantō Taru Seihin KK, 1960.

Kariya Susumu. "Kayaku kōjō" (The explosives plant), *Kagaku kōgyō*, January 1951: 80–81.

——. "Nippon Chisso no kayaku jigyō" (Nitchitsu's explosives industry). In Kamata Shōji, ed., *Nippon Chisso shi e no shōgen*, Vol. VIII. Tokyo, 1979.

Katagiri Ryūkichi. *Hantō no jigyō-ō Noguchi Jun* (Noguchi Jun, entrepreneurial king of the peninsula). Tokyo, Tōkai Shuppansha, 1939.

Katagiri Yasuo. "Kagami kōjō no omoide" (Memories of the Kagami plant). In Kamata Shōji, ed., *Nippon Chisso shi e no shōgen*, Vol. VIII. Tokyo, 1979.

Kawamura Kazuo. "Nippon Chisso kosho zakkan, sono ni (2): Himekawa Hatsudenjo no koto" (Nitchitsu's old documents, #2: Himekawa generating plant). In Kamata Shōji, ed., *Nippon Chisso shi e no shōgen*, Vol. XII. Tokyo, 1981.

——. "Nippon Chisso kosho zakkan, sono shi (4): Sōgyō jidai no hitotachi, kābaido o chūshin ni" (Nitchitsu's old documents, #4: The people of the early years, with a focus on carbide). In Kamata Shōji, ed., *Nippon Chisso shi e no shōgen*, Vol. XIII. Tokyo, 1981.

——. "Nippon Chisso kosho zakkan, sono go (5): Taishō shōki no Minamata no geppō to Kagami no jūyō nikki" (Nitchitsu's old documents, #3: Minamata's monthly reports and Kagami's important diaries in the early Taishō period). In Kamata Shōji, ed., *Nippon Chisso shi e no shōgen*, Vol. XXVI. Tokyo, 1985.

——. "Taishō chūki no hitotachi–Noguchi san no omokage o motometsutsu" (The people of mid-Taishō–the continuing search for people like Noguchi). In Kamata Shōji, ed., *Nippon Chisso shi e no shōgen*, Vol. XXVIII. Tokyo, 1986.

Kayō, Nobufumi. "The Characteristics of Heavy Application of Fertilizers in Japanese Agriculture," *The Developing Economies* 2.4:373–396 (1964).

Kim, Han-kyo. "The Colonial Administration in Korea." In Andrew Nahm, ed. *Korea under Japanese Colonial Rule*. Western Michigan University Press, 1973.

Kitayama Hisashi. "Kasare hō ni yoru anmonia gosei hō" (Casale-method ammonia synthesis), *Kagaku kōgyō* 2.5:22–25 (May 1951).

Ko-Suzuki Saburōsuke-kun Den Kiroku Hensankai, eds. *Suzuki Saburōsuke den* (Biography of Suzuki Saburōsuke). Tokyo, 1932.

Kobayashi Hideo. "1930 nendai Nippon Chisso Hiryō Kabushiki Kaisha no Chōsen no shinshutsu ni tsuite" (Nitchitsu's advance into Korea in the 1930s). In Yamada Hideo, ed., *Shokuminchi keizaishi no shōmondai*. Tokyo, Ajia Keizai Kenkyūjo, 1973.

Kobayashi Masaaki. "Kangyō to sono haraisage" (The sale of government industries). In Nakagawa Keiichirō, Morikawa Hidemasa, and Yui Tsunehiko, eds., *Kindai Nihon keieishi no kisō chishiki*. Tokyo, Yūhikaku, 1974.

Kobayashi, Uchisaburo. *The Basic Industries and Social History of Japan, 1914–1918*. New Haven, Yale University Press, 1930.

Kodama Noritada. "Kaisō danpen" (Fragmentary memories). In Kamata Shōji, ed., *Nippon Chisso shi e no shōgen*, Vol. XXII. Tokyo, 1985.

Kōgyō Kagakukai. *Saikin jūnen ni okeru honpō kagaku kōgyō no gaikan* (Outline of Japan's chemical industry in the last 10 years). Tokyo, Seibundo Shinkōsha, 1947.

Kondō Yasuo. *Ryūan: Nihon shihonshugi to hiryō kōgyō* (Ammonium sulfate: Japanese capitalism and the fertilizer industry). Tokyo, Nihon Hyōronsha, 1950.

Koyama Hirotake. *Nihon gunji kōgyō no shiteki bunseki* (Historical analysis of Japan's armaments industry). Tokyo, Ochanomizu Shobō, 1972.

Krammer, Arnold. "Fueling the Third Reich," *Technology and Culture* 19.3:394–422 (July 1978).

Kubota Masao. "Furui hanashi to atarashii hanashi" (Old stories and new stories). In Kamata Shōji, ed., *Nippon Chisso shi e no shōgen,* Vol. XI. Tokyo, 1980.

Kubota Yutaka. "Nippon Chisso jidai no kaiko: Noguchi san no omoide o chūshin ni" (Memories of Nitchitsu: With a focus on my memories of Mr. Noguchi). In Kamata Shōji, ed., *Nippon Chisso shi e no shōgen,* Vol. VII. Tokyo, 1979.

Kudō Kōki. "Noguchi-san no omokage" (Traces of Mr. Noguchi). In Takanashi Kōji, ed., *Noguchi Jun o tsuikairoku.* Ōsaka, Noguchi Jun Tsuikairoku Hensankai, 1952.

Kuriyama Toyo. *Gendai Nihon sangyō hattatsushi: Denryoku* (History of the industrial development of modern Japan: Electricity). Tokyo, Gendai Nihon sangyō hattatushi kenkyūkai, 1964.

Kuznets, Paul W. *Economic Growth and Structure in the Republic of Korea.* New Haven, Yale University Press, 1977.

Lincoln, Edward J. *Japan's Industrial Policies.* Washington, D.C., Japan Economic Institute of America, 1984.

Marui Ryōsei. "Denkai kōjō kinmu nijūnen" (Twenty years of service in the electrolysis plant). In Kamata Shōji, ed., *Nippon Chisso shi e no shōgen,* Vol. X. Tokyo, 1980.

Matsushima Harumi. "Jūkagaku kōgyōka no katei" (The process of industrialization in the heavy and chemical industries), *Shakai keizai shigaku* 33.6 (1968).

Matsushita Denkichi. *Kagaku kōgyō zaibatsu no shinkenkyū* (New research on zaibatsu in the chemical industry). Tokyo, Chūgai Sangyō Chōsakai, 1938.

Meiji kōgyōshi: Kagaku kōgyō (Meiji industrial history: Chemical industry). Tokyo, Meiji Kōgyōshi Hakkojo, 1925.

Mikami Atsufumi. "Sumitomo Kagaku no keisei, hatten katei" (The process of formation and development of Sumitomo Chemical), *Ōsaka Daigaku keizaigaku* 25.1:123–151 (June 1975).

———. "Kyūzaibatsu to shinkō zaibatsu no kagaku kōgyō: Sumitomo Kagaku to Shōwa Denkō o chūshin to shite" (The chemical industry in the old zaibatsu and new zaibatsu: With a focus on Sumitomo Chemical and Shōwa Denkō). In

Yasuoka Shigeaki, ed., *Zaibatsu shi kenkyū.* Tokyo, Nihon Keizai Shinbunsha, 1979.

——. "Old and New Zaibatsu in the History of Japan's Chemical Industry: With Special Reference to the Sumitomo Chemical Co. and the Showa Denko Co." In Akio Okochi and Hoshimi Uchida, eds., *Development and Diffusion of Technology: Electrical and Chemical Industries.* International Conference on Business History, Vol. VI. Tokyo, University of Tokyo Press, 1980.

Minami, Ryoshin. *Power Revolution in the Industrialization of Japan: 1885–1940.* Tokyo, Kinokuniya, 1987.

Mishima Yasuo. *Mitsubishi Zaibatsu* (Mitsubishi Zaibatsu). Tokyo, Nihon Keizai Shinbunsha, 1981.

Mitsubishi Kasei Kōgyō Kabushiki Kaisha, eds. *Mitsubishi Kasei shashi* (Company history of Mitsubishi Kasei). Tokyo, 1981.

Mitsubishi Shōji Kabushiki Kaisha, eds. *Ritsugyō bōekiroku* (Company trade records). Tokyo, 1958.

Miwa Ryōichi. "Nihon no karuteru" (Japanese cartels). In Morikawa Hidemasa, ed., *Nihon no kigyō to kokka.* Tokyo, Nihon Keizai Shinbunsha, 1976.

Miyakawa Saburō, ed. *Shōwa sangyōshi* (History of industry in the Shōwa era). 3 vols. Tokyo, Tōyō Keizai Shinposha, 1950.

Miyake Haruteru. *Shinkō kontsuerun tokuhon* (New zaibatsu reader). Tokyo, Shunjusha, 1937.

Miyamoto Mataji. *Kansai zaikai gaishi* (History of the Kansai business community). Ōsaka, Kansai Keizai Rengōkai, 1976.

Miyoshi Nobuhiro. *Nihon kōgyō kyōiku seiritsu no kenkyū* (Research on the establishment of industrial education in Japan). Tokyo, Kazama Shobō, 1979.

Molony, Barbara. "Noguchi Jun and Nitchitsu: Colonial Investment Strategy of a High Technology Enterprise." In William D. Wray, *Managing Industrial Enterprise: Cases from Japan's Prewar Experience.* Cambridge, Council on East Asian Studies, Harvard University, 1989.

Morikawa Hidemasa. "Shibusawa Eiichi, Nihon kabushiki kaisha no sōritsusha" (Shibusawa Eiichi, founder of the joint-stock company in Japan). In Morikawa Hidemasa, *Nihon keieishi kōza,* Vol. IV, *Nihon no kigyō to kokka.* Tokyo, Nihon Keizai Shinbunsha, 1976.

——. *Zaibatsu no keiei shiteki kenkyū* (Business historical research on zaibatsu). Tokyo, Tōyō Keizai Shinposha, 1980.

——. "The Increasing Power of Salaried Managers in Japan's Large Corporations." In William D. Wray, ed., *Managing Industrial Enterprise: Cases from Japan's Prewar Experience,* Cambridge, Council on East Asian Studies, Harvard University, 1989.

——, ed. *Nihon no kigyō to kokka, Nihon keieishi kōza,* Vol. IV (Enterprises and the state in Japan). Tokyo, Nih on Keizai Shinbunsha, 1976.

—— et al., eds. *Nihon no kigyōka* (Japanese entrepreneurs), Vol. III, Shōwa. Tokyo, Yūhikaku, 1978.

Morita Kazuo. "Noguchi Jun no Fusenkō kaihatsu" (Noguchi's development of the Pujon). In Takanashi Kōji, ed., *Noguchi Jun o tsuikairoku.* Ōsaka, Noguchi Jun Tsuikairoku Hensankai, 1952.

Munekata Eiji. "Agochi (Sekitan ekika) kōjō no omoide" (Memories of the Haoji [coal liquifaction] plant). In Kamata Shōji. ed., *Nippon Chisso shi e no shōgen,* Vol. IV. Tokyo, 1978.

——. "Noguchi-san hassō no sekitan ekika ni eikō are" (The glory of coal liquifaction pioneered by Noguchi). In Kamata Shōji, ed., *Nippon Chisso shi e no shōgen,* Vol. XXVI. Tokyo, 1985.

Myers, Ramon H. "Post World War II Japanese Historiography on Japan's Formal Colonial Empire." In Ramon H. Myers and Mark R. Peattie, eds., *The Japanese Colonial Empire, 1895–1945.* Princeton, Princeton University Press, 1984.

Nagasawa Yasuakai. "Nitchitsu Kasare-hō kōgyōka ni tsuite no ichikōsatsu: Denki Kagaku Kōgyō (KK) to no hikaku ni oite" (Examination of Nitchitsu's industrial application of the Casale method: A comparison with Denka), *Kansai Gakuin shōgaku ronkyū* 23.1–2:67–81 (December 1975).

Nagatsuka Riichi. *Kubota Yutaka den* (Biography of Kubota Yutaka). Tokyo, 1978.

Nahm, Andrew C., ed. *Korea under Japanese Colonial Rule.* Western Michigan University Press, 1973.

Nakagawa Keiichirō, Morikawa Hidemasa, and Yui Tsunehiko, eds. *Kindai Nihon keieishi no kisō chishiki* (Basic information about modern Japanese business history). Tokyo, Yūhikaku, 1974.

Nakagawa Keiichirō. *Nihonteki keiei* (Japanese-style management). Nihon keieishi kōza, Vol. V. Tokyo, Nihon Keizai Shinbunsha, 1977.

Nakamura Chūichi. *Nihon kagaku kōgyōshi* (History of the Japanese chemical industry). Tokyo, Tōyō Keizai Shinbunsha, 1959.

——. *Nihon sangyō no kigyōshiteki kenkyū* (Business historical research on Japanese industry). Tokyo, Yūkonsha, 1965.

Nakamura, James I. "Incentives, Productivity Gaps, and Agricultural Growth Rates in Prewar Japan, Taiwan and Korea." In Bernard S. Silberman and H. D. Harootunian, eds., *Japan in Crisis: Essays in Taishō Democracy.* Princeton, Princeton University Press, 1974.

Nakamura Seishi. *Shinkō zaibatsu no hatten* (Development of the new zaibatsu). Tokyo, Yūhikaku, 1976.

——. "Kyōdai denryoku kagaku konbināto no kensetsu" (Construction of the giant electrical and chemical conglomerates). In Morikawa Hidemasa, et al., *Nihon no kigyōka,* Vol. III, Shōwa. Tokyo, Yūhikaku, 1978.

Nakamura Takafusa. *Nihon keizai: sono seichō to kōzō* (The Japanese economy: Its growth and structure). Tokyo, Tōkyō Daigaku Shuppankai, 1978.

——, Itō Takashi, and Hara Akira. *Gendaishi o tsukuru hitobito* (The people who made modern history). 4 vols. Tokyo, Mainichi Shinbunsha, 1971.

Nakamura Takeshi. "The Contributions of Foreigners," *Journal of World History* 9 (1965).

Nakano Hitoshi. "Kōnan maguneshiumu kōjō ki" (Hungnam magnesium plant records). In Kamata Shōji, ed., *Nippon Chisso shi e no shōgen,* Vol. XI. Tokyo, 1980.

Nakayama Shigeru. "The Role Played by Universities in Scientific and Technological Development in Japan," *Journal of World History* 9 (1965).

Napier, Ron Wells. "Prometheus Absorbed: The Industrialization of the Japanese Economy 1905–1937." PhD dissertation, Harvard University, 1979.

Nenryō Konwakai, eds. *Nihon kaigun nenryōshi, Jō* (History of fuel in the Japanese Navy, Vol. I). Tokyo, Genshobō, 1972.

Nihon Kagakukai, eds. *Nihon no kagaku hyakunenshi: Kagaku to kagaku kōgyō no ayumi* (Hundred-year history of chemistry in Japan: Development of chemistry and the chemical industry). Tokyo, Nihon Kagaku Dōjin, 1978.

Nihon Kagakushi Gakkai, eds. *Nihon kagaku gijutsu shi taikei: Kagaku kōgyō* (Historical outline of Japanese science and technology: Chemical industry). Vol. XXI. Tokyo, Daiichi Hōki Shuppan KK, 1964.

Nihon Kasei Hiryō Kyōkai, eds. *Rinsan hiryō kōgyō no ayumi* (Development of the phosphate fertilizer industry). Tokyo, 1972.

Nihon Ryūan Kōgyō Kyōkai, eds. *Nihon ryūan kōgyōshi* (History of Japan's ammonium sulfate industry). Tokyo, 1963.

"Nippon Chisso sanjūnen kinen zadankai" (Nitchitsu's thirtieth anniversary symposium). In Kamata Shōji, ed., *Nippon Chisso shi e no shōgen,* Vol. XIV. Tokyo, 1981.

Nissan Kagaku Shashi Hensan Iinkai, eds. *Hachijūnenshi* (Eighty-year history). Tokyo, Nissan Kagaku Kōgyō, 1969.

Noda Keizai Kenkyūjo, eds. *Senjika no kokusaku kaisha* (Wartime National-Policy companies). Tokyo, Noda Keizai Kenkyūjo Shuppanbu, 1940.

Noguchi Jun. "Nihon Kai ni kiriotoshita Fusenkō no suiden jigyō" (Hydroelectric industry on the Pujon River which flows into the Sea of Japan). In Shinogi Itsuo, ed., *Chōsen no denki jigyō o kataru.* Seoul, Chōsen Denki Kyōkai, 1937.

Noguchi Tasuku, ed. *Mitsubishi kontsuerun: Keiei to zaimu no sōgō bunseki* (The Mitsubishi enterprise: General analysis of its management and finances). Tokyo, Shinhyōron, 1968.

Ogawa Masao. "Noguchi san to yushi kōgyō" (Noguchi and the fats and oils industry). In Kamata Shōji. ed., *Nippon Chisso shi e no shōgen,* Vol. IX. Tokyo, 1980.

Ogura, Takekazu. *Agricultural Development in Modern Japan.* Tokyo, Fuji Publishing Company, 1963.

Ōishi Kaichiro and Miyamoto Ken'ichi. *Nihon shihonshugi no hattatsushi no kisō chishiki* (Basic information about the development of Japanese capitalism). Tokyo, Yuhikaku, 1975.

Okamoto, Yasuo. "The Grand Strategy of Japanese Business." In Kazuo Sato and Yasuo Hoshino, *The Anatomy of Japanese Business.* Armonk, M. E. Sharpe, 1984.

Okimoto, Daniel I. "Regime Characteristics of Japanese Industrial Policy." In Hugh Patrick, *Japan's High Technology Industries: Lessons and Limitations of Industrial Policy.* Seattle, University of Washington Press, 1986.

——, Takuo Sugano, and Franklin B. Weinstein, eds. *Competitive Edge: The Semiconductor Industry in the U.S. and Japan.* Stanford, Stanford University Press, 1984.

Okochi, Akio and Hoshimi Uchida. *Development and Diffusion of Technology: Electrical and Chemical Industries.* International Conference on Business History, Vol. VI. Tokyo, University of Tokyo Press, 1980.

Ōshima Mikiyoshi. "Nitchitsu Nenryō–Ryūkō kōjō" (Nitchitsu Fuels–the Longxing plant). In Kamata Shōji, ed., *Nippon Chisso shi e no shōgen,* Vol. IV. Tokyo, 1978.

Ōshio Takeshi. "Nippon Chisso Hiryō Kabushiki Kaisha ni yoru hensei ryūan seizō kigyōka no katei" (The process of Nitchitsu's commercial production of cyanamide-method ammonium sulfate), *Keizai ronshū* 26:253–271 (March 1977).

——. "Nitchitsu kontsuerun no seiritsu to kigyō kin'yū" (The founding and company finances of the Nitchitsu Konzern), *Keizai kenkyū* 27:61–127 (March 1977).

——. "Fujiwara-Bosch kyōteian to Nihon no ryūan kōgyō" (The proposed Fujiwara-Bosch agreement and Japan's ammonium sulfate industry), *Keizai kenkyū* 49–50:45–62 (1978).

——. "Nitchitsu Kontsuerun to Chōsen Chisso Hiryō" (The Nitchitsu Konzern and Chōsen Chisso), *Keizai kenkyū* 49–50:63–77 (1978).

——. "Nippon Chisso Hiryō to Chōsen Chisso Hiryō no kigyō kin'yū" (Company finances of Nitchitsu and Chōsen Chisso), *Keizai kenkyū* 60:1–22 (March 1981).

——. "Shinkō kontsuerun" (New Konzerns), *Shakai keizai shigaku* 47.6:71–90 (1981).

——. "Nitchitsu Kontsuerun no kigyō haichi" (Company structure of the Nitchitsu Konzern), *Keizai kenkyū* 70:63–83 (November 1984).

——. "Chōsen Chisso Hiryō Kabushiki Kaisha no shūeki ni kansuru ichikōsatsu" (Examination of Chōsen Chisso's profits), *Keizai kenkyū* 72:97–124 (1985).

——. "Chōshinkō kaihatsu o meguru Nitchitsu to Mitsubishi no tairitsu ni tsuite" (Antagonism between Nitchitsu and Mitsubishi over development of the Changjin), *Shakai kagaku tōkyū* 89:167–185 (1985).

——. "Nitchitsu kankei kaisha no seiritsu to idō" (Establishment and change in Nitchitsu-related companies), *Keizai kenkyū* 73 (September 1985).

——. "Nitchitsu Kontsuerun no kin'yū kōzō" (Financial structure of the Nitchitsu Konzern), *Keizai kenkyū* 75:113–246 (March 1986).

Ouchi, William G. *Theory Z: How American Business Can Meet the Japanese Challenge.* Reading, Addison-Wesley, 1981.

Pascale, Richard Tanner and Anthony G. Athos. *The Art of Japanese Management.* New York, Simon and Schuster, 1981.

Patrick, Hugh. "Japanese High Technology Policy in Comparative Perspective." In Hugh Patrick, ed., *Japan's High Technology Industries: Lessons and Limitations of Industrial Policy.* Seattle, University of Washington Press, 1986.

Peattie, Mark R. "Introduction." In Ramon H. Myers and Mark R. Peattie, eds., *The Japanese Colonial Empire, 1895–1945.* Princeton, Princeton University Press, 1984.

Peck, Merton J., Richard C. Levine, and Akira Goto. "Picking Losers: Public Policy Toward Declining Industries in Japan," *Journal of Japanese Studies* 13:1 (Winter 1987).

Reader, William J. *Imperial Chemical Industries: A History–The Forerunners 1870–1926.* London, Oxford University Press, 1970.

Reischauer, Edwin O., and Albert M. Craig. *Japan: Tradition and Transformation.* Boston, Houghton Mifflin, 1978.

Sagami Teruo. "Chisso Sekken no omoide" (Reminiscences of Chisso Soap). In Kamata Shōji, ed., *Nippon Chisso e no shōgen,* Vol. XXII. Tokyo, 1984.

Sakudō Yōtarō. *Sumitomo Zaibatsu* (The Sumitomo Zaibatsu). Tokyo, Nihon Keizai Shinbunsha, 1982.

Sangyō Seisakushi Kenkyūjo, eds. *Waga kuni dai kigyō no keisei hatten katei* (Process of structural development of large Japanese companies). Tokyo, 1976.

Satō Kanji. *Hiryō mondai kenkyū* (Fertilizer problems research). Tokyo, Nihon Hyōronsha, 1930.

Sato, Ryuzo. "Nothing New? An Historical Perspective of Japanese Technology Policy." In Toshio Shishido and Ryuzo Sato, eds., *Economic Policy and Development: New Perspectives.* Dover, Auburn House Publishing, 1985.

Saxonhouse, Gary R. "Industrial Policy and Factor Markets: Biotechnology in Japan and the United States." In Hugh Patrick, *Japan's High Technology Industries: Lessons and Limitations of Industrial Policy.* Seattle, University of Washington Press, 1986.

Shibamura Yōgo. *Nihon kagaku kōgyōshi* (History of the Japanese chemical industry). Tokyo, Kurita Shoten, 1943.

——. *Kagaku hiryō* (Chemical fertilizer). Tokyo, Yūhikaku, 1959.

——. *Kigyō no hito Noguchi Jun den: Denryoku, kagaku kōgyō no paionia* (Biography of Noguchi Jun: Pioneer in the electrical and chemical industries). Tokyo, Yūhikaku, 1981.

Shibata Kenzō. "Suginishi 50 nenkan no hansei" (Reflections on more than fifty years). In Kamata Shōji, ed., *Nippon Chisso shi e no shōgen, Dokukan,* Vol. I (Tokyo, 1987).

Shibuya Reiji. *Chōsen Ginkō nijūgonenshi* (Twenty-five-year history of the Bank of Chōsen). Seoul, Chōsen Ginkō, 1934.

——, ed. *Chōsen no kigyō to sono shigen* (Korean companies and their resources). Seoul, Chōsen Kōgyō Kyōkai, 1937.

Shimotani Masahiro. "Dai Nihon Jinzō Hiryō torasuto to karinsan sekkai kōgyō" (Dai Nihon Artificial Fertilizer Trust and the superphosphate industry), *Nihon shi kenkyū* 146:32–60 (October 1974).

——. "Hensei ryūan, sekkai chisso kōgyō to Denki Kagaku Kōgyō KK no seiritsu: Waga kuni kagaku kōgyō ni okeru dokusen keiseishi (2)" (The ammonium sulfate and calcium cyanamide industry and the establishment of Denka: Monopolistic structure in Japan's chemical industry [2]), *Ōsaka Keidai ronshū* 106:25–53 (July 1975).

——. "Nippon Chisso Hiryō KK to takakuka no tenkai: Waga kuni kagaku kōgyō ni okeru dokusan keiseishi (3)" (Nitchitsu and the development of diversification: Monopolistic structure in Japan's chemical industry [3]), *Ōsaka Keidai ronshū* 112:94–124 (July 1976).

——. "Nitchitsu kontsuerun to gōsei ryūan kōgyō: Waga kuni kagaku kōgyō ni okeru dokusen keiseishi (4)" (Nitchitsu Konzern and the synthetic ammonia industry: Monopolistic structure in Japan's chemical industry [4]), *Ōsaka Keidai ronshū* 114:58–89 (November 1976).

——. *Nihon kagaku kōgyō shiron* (Historical analysis of Japan's chemical industry). Tokyo, Ochanomizu Shobō, 1982.

Shinoda Keiji. "Idai naru rōhi: Kichirin Jinzō Sekiyu" (A great waste: Jilin Artificial Oil). In Kamata Shōji, ed., *Nippon Chisso shi e no shōgen,* Vol. XI. Tokyo, 1980.

Shinogi Itsuo. *Chōsen no denki jigyō o kataru* (Speaking about Korea's electrical industry). Seoul, Chōsen Denki Kyōkai, 1937.

Shiobara Matasaku. *Takamine Hakushi* (Dr. Takahashi). Yokohama, Ōkawa Insatsujo, 1926.

Shiraishi Muneshiro. "Kōnan kōjō no gaisetsu" (Brief discussion of the Hungnam plant), *Kagaku kōgyō* January 1951:32–39.

Shiraishi Muneshiro Kankōkai, ed. *Shiraishi Muneshiro.* Tokyo, Pashifiku Ripro Sabisu, 1978.

Shishido, Toshio. "Japanese Technological Development." In Toshio Shishido and Ryuzo Sato, *Economic Policy and Development: New Perspectives.* Dover, Auburn House Publishing, 1985.

Shoji Tsutomu. *Jinzō hiryō kōgyō* (Artificial fertilizer industry). Tokyo, Kyoritsu-sha, 1933.

——. *Nihon sōda kōgyōshi* (History of the Japanese soda industry). Tokyo, Sōda Sarashiko Dōgyōkai, 1938.

Shōwa Denkō Kabushiki Kaisha Shashi Hensanshitsu, eds. *Shōwa Denkō 50 nen-shi* (Fifty-year history of Shōwa Denkō). Tokyo, Shōwa Denkō KK, 1978.

Sugimoto Toshio. "Ōtsu, Nobeoka kōjō no kaiko: Noguchi san no jinken kōgyō kaihatsu e no kaiko" (Memories of Ōtsu and Nobeoka: Memories of Noguchi's founding of the rayon industry). In Kamata Shōji, ed., *Nippon Chisso shi e no shōgen,* Vol. V. Tokyo, 1978.

Sugita Motomu. "Nippon Chisso no omoide" (Memories of Nitchitsu). In Kamata Shōji, ed., *Nippon Chisso shi e no shōgen,* Vol. XXVI. Tokyo, 1985.

Suh, Sang-chul. *Growth and Structural Changes in the Korean Economy 1910–1940.* Cambridge, Council on East Asian Studies, Harvard University, 1980.

Sumitomo Kagaku Kōgyō Kabushiki Kaisha, eds. *Sumitomo Kagaku Kōgyō Kabushiki Kaisha shi* (History of Sumitomo Chemical). Tokyo, 1981.

Suzuki Otokichi. "Kyūnenkan no Kōnan seikatsu dampen" (Bits of my nine years of living in Hungnam). In Kamata Shōji, ed., *Nippon Chisso shi e no shōgen,* Vol. XXVIII. Tokyo, 1980.

Suzuki Tsuneo. "Nihon ryūan kōgyō no jiritsuka katei" (The independent development of Japan's ammonium sulfate industry), *Shakai keizai shigaku* 43.2:66–91 (1978).

——. "Daiichiji taisenki Nitchitsu, Denka no tōshi to shikin chōtatsu" (Nitchitsu's and Denka's investments and capital formation during World War I), *Kurume Daigaku Shōgakubu seiritsu 30 shūnen kinen ronbunshū* August 1980.

Takahashi Kamekichi. *Nihon zaibatsu no kaibō* (Analysis of Japan's zaibatsu). Tokyo, Chūō Kōronsha, 1930.

Takahashi Takeo. *Kagaku kōgyōshi* (History of the chemical industry). Tokyo, Sangyō Zusho, 1973.

Takanashi Kōji, ed. *Noguchi Jun o tsuikairoku* (Memorial record of Noguchi Jun). Ōsaka, Noguchi Jun Tsuikairoku Hensankai, 1952.

Takashima Suekichi. *Kindai kagaku sangyō shiryō* (Data on the modern chemical industry). Tokyo, Sangyō Keizai Kenkyūjo, 1939.

Takeoka Kōji. "Kagami kōjō to Shin'etsu Chisso no kaisō" (Reflections on the Kagami plant and Shin'etsu Nitrogenous). In Kamata Shōji, ed., *Nippon Chisso shi e no shōgen,* Vol. IX. Tokyo, 1980.

Tamaki Shōji, former President, Japan Consulting Engineers Association, and long-time high ranking manager at Nitchitsu. Interview, 10 August 1983.

——. "Hungnam kōjō to Chōsen no daikibō hatsuden" (The Hungnam plant and Korea's large-scale electric generation), *Kagaku kōgyō* January 1951.

Tashiro Saburō. "Kōnan kenkyūjo no koto" (About the Hungnam research labs). In Kamata Shōji, ed., *Nippon Chisso shi e no shōgen*, Vol. III. Tokyo, 1978.

——. "Noguchi Kenkyūjo sōritsu gakuya hanashi" (Behind-the-scenes talk about the founding of the Noguchi Research Laboratory). In Kamata Shōji, ed., *Nippon Chisso shi e no shōgen*, Vol. XII. Tokyo, 1981.

Tatsuno, Sheridan. *The Technopolis Strategy: Japan, High Technology, and the Control of the Twenty-first Century*. New York, Prentice Hall, 1986.

Taylor, Graham D., and Patricia Sudnik, *Du Pont and the International Chemical Industry*. Boston, Twayne Publishers, 1984.

Togai Yoshio. "Zaibatsu to iu kotoba: Seishō to tomo ni Nihonsei" (The word *zaibatsu*: Politics and business and Made-in-Japan). In Nakagawa Keiichirō, et al., *Kindai Nihon keieishi no kisō chishiki*. Tokyo, Yūhikaku, 1974.

Tōkyō Gasu Kabushiki Kaisha, eds. *Tōkyō Gasu shichijūnenshi* (Seventy-year history of Tokyo Gas). Tokyo, Tōkyō Gasu KK, 1956.

Tōkyō Kōgyō Shikenjo, eds. *Tōkyō Kōgyō Shikenjo rokujūnenshi* (Sixty-year history of the Tokyo Industrial Experimental Laboratories). Tokyo, 1960.

Tuge, Hideomi. *Historical Development of Science and Technology in Japan*. Tokyo, Kokusai Bunka Shinkokai, 1961.

Uchiyama, S. "On the Manurial Effect of Calcium Cyanamide under Different Conditions," *Bulletin of the Central Agricultural Station* 1.2:93–103 (October 1907).

Udagawa Masaru. *Shōwashi to shinkō zaibatsu* (Shōwa history and the new zaibatsu). Tokyo, Kyōikusha, 1982.

——. *Shinkō zaibatsu* (New zaibatsu). Tokyo, Nihon Keizai Shinbunsha, 1984.

Ugaki Kazushige. "Noguchi Jun-kun o omou" (Thinking of Noguchi Jun). In Takanashi Kōji, ed., *Noguchi Jun o tsuikairoku*. Ōsaka, Noguchi Jun Tsuikairoku Hensankai, 1952.

——. "Ugaki Kazushige nikki ni kakareta Noguchi-san" (Mr. Noguchi as written about in Ugaki Kazushige's diary). In Kamata Shōji, ed., *Nippon Chisso shi e no shōgen*, Vol. XXIII. Tokyo, 1984.

Watanabe Minoru. "Japanese Students Abroad and the Acquisition of Scientific and Technical Knowledge," *Journal of World History* 9 (1965).

Watanabe Tokuji. *Gendai Nihon sangyō kōza: Kagaku kōgyō* (Lectures on modern Japanese industry: Chemical industry), Vol. IV. Tokyo, Iwanami Shoten, 1959.

——. *Gendai Nihon sangyō hattatsushi: Kagaku kōgyō* (History of the industrial development of modern Japan: The chemical industry), Vol. XIII. Tokyo, Gendai Nihon Sangyō Hattatsushi Kenkyūkai, 1968.

—— and Hayashi Yujirō. *Nihon no kagaku kōgyō* (Japan's chemical industry). Tokyo, Iwanami Shoten, 1974.

von Weiher, Sigfrid. "The Rise and Development of Electrical Engineering and Industry in Germany in the Nineteenth Century: A Case Study–Siemens and Halske." In Akio Okochi and Hoshimi Uchida, eds., *Development and Diffusion of Technology: Electrical and Chemical Industries.* Tokyo, Tokyo University Press, 1980.

Wray, William D. *Mitsubishi and the N.Y.K, 1870–1914: Business Strategy in the Japanese Shipping Industry.* Cambridge, Council on East Asian Studies, Harvard University, 1984.

Yamada Yutaka. "Chisso seikatsu 23 nen no omoide" (Reminiscences of 23 years of life at Chisso). In Kamata Shōji, ed., *Nippon Chisso shi e no shōgen,* Vol. XII. Tokyo, 1981.

Yamamoto Shigeru. *Dai Tō-A kagaku kōgyō ron* (Analysis of the chemical industry in Greater East Asia). Tokyo, Kokusai Nihon Kyōkai, 1942.

Yamamoto Tomio. *Nippon Chisso Hiryō jigyō taikan* (Overview of the Nitchitsu company; Official company history). Ōsaka, Nippon Chisso Hiryō, 1937.

Yamamura, Kozo. "The Japanese Economy, 1911–1930: Concentration, Conflicts, and Crises." In Bernard S. Silberman and H. D. Harootunian, eds., *Japan in Crisis: Essays in Taisho Democracy.* Princeton, Princeton University Press, 1974.

——. "Success Illgotten? The Role of Meiji Militarism in Japan's Technological Progress," *Journal of Economic History* 37.1:113–135 (March 1977).

Yamasaki, Kakujiro and Gataro Ogawa. *The Effects of the World War upon the Commerce and Industry of Japan.* New Haven, Yale University Press, 1929.

Yamashita Yukio. "Kagaku kōgyō no kusawaketachi" (Path-breakers in the chemical industry). In Nakagawa Keiichirō, et al., eds., *Kindai Nihon keieishi no kisō chishiki.* Tokyo, Yūhikaku, 1974.

Yasuda Shigeaki. *Mitsui Zaibatsu* (The Mitsui Zaibatsu). Tokyo, Nihon Keizai Shinbunsha, 1982.

Yasuoka Shigeaki. *Zaibatsu no keieishi* (Business history of the zaibatsu). Tokyo, Nihon Keizai Shinbunsha, 1978.

——, ed., *Zaibatsushi kenkyū* (Research on zaibatsu history). Tokyo, Nihon Keizai Shinbunsha, 1979.

Yokota Shigeru. "Taiwan Chisso no omoide" (Memories of Taiwan Nitrogenous). In Kamata Shōji, ed., *Nippon Chisso shi e no shōgen,* Vol. II. Tokyo, 1977.

Yoshida Mitsukuni. *Zaikaijin no gijutsuken* (Technical ability of men in the business world). Tokyo, 1969.

Yoshioka Kiichi. *Noguchi Jun.* Tokyo, Fuji Intanashiyonaru Konsarutanto, 1962.

Yuasa Mitsutomo. *Kagakushi* (History of chemistry). Tokyo, Keizai Shinposha, 1961.

Yukizawa Kenzō and Maeda Shōzō. *Nihon bōeki no chōki tōkei: Bōeki kōzōshi kenkyū no kisō sagyō* (Long-term statistics for Japanese trade: Basic history of the structure of trade). Tokyo, Dōhōsha, 1978.

Index

Harvard East Asian Monographs

1. Liang Fang-chung, *The Single-Whip Method of Taxation in China*
2. Harold C. Hinton, *The Grain Tribute System of China, 1845–1911*
3. Ellsworth C. Carlson, *The Kaiping Mines, 1877–1912*
4. Chao Kuo-chün, *Agrarian Policies of Mainland China: A Documentary Study, 1949–1956*
5. Edgar Snow, *Random Notes on Red China, 1936–1945*
6. Edwin George Beal, Jr., *The Origin of Likin, 1835–1864*
7. Chao Kuo-chün, *Economic Planning and Organization in Mainland China: A Documentary Study, 1949–1957*
8. John K. Fairbank, *Ch'ing Documents: An Introductory Syllabus*
9. Helen Yin and Yi-chang Yin, *Economic Statistics of Mainland China, 1949–1957*
10. Wolfgang Franke, *The Reform and Abolition of the Traditional Chinese Examination System*
11. Albert Feuerwerker and S. Cheng, *Chinese Communist Studies of Modern Chinese History*
12. C. John Stanley, *Late Ch'ing Finance: Hu Kuang-yung as an Innovator*
13. S. M. Meng, *The Tsungli Yamen: Its Organization and Functions*
14. Ssu-yü Teng, *Historiography of the Taiping Rebellion*
15. Chun-Jo Liu, *Controversies in Modern Chinese Intellectual History: An Analytic Bibliography of Periodical Articles, Mainly of the May Fourth and Post-May Fourth Era*
16. Edward J. M. Rhoads, *The Chinese Red Army, 1927–1963: An Annotated Bibliography*
17. Andrew J. Nathan, *A History of the China International Famine Relief Commission*
18. Frank H. H. King (ed.) and Prescott Clarke, *A Research Guide to China-Coast Newspapers, 1822–1911*
19. Ellis Joffe, *Party and Army: Professionalism and Political Control in the Chinese Officer Corps, 1949–1964*
20. Toshio G. Tsukahira, *Feudal Control in Tokugawa Japan: The Sankin Kōtai System*

21. Kwang-Ching Liu, ed., *American Missionaries in China: Papers from Harvard Seminars*
22. George Moseley, *A Sino-Soviet Cultural Frontier: The Ili Kazakh Autonomous Chou*
23. Carl F. Nathan, *Plague Prevention and Politics in Manchuria, 1910–1931*
24. Adrian Arthur Bennett, *John Fryer: The Introduction of Western Science and Technology into Nineteenth-Century China*
25. Donald J. Friedman, *The Road from Isolation: The Campaign of the American Committee for Non-Participation in Japanese Aggression, 1938–1941*
26. Edward Le Fevour, *Western Enterprise in Late Ch'ing China: A Selective Survey of Jardine, Matheson and Company's Operations, 1842–1895*
27. Charles Neuhauser, *Third World Politics: China and the Afro-Asian People's Solidarity Organization, 1957–1967*
28. Kungtu C. Sun, assisted by Ralph W. Huenemann, *The Economic Development of Manchuria in the First Half of the Twentieth Century*
29. Shahid Javed Burki, *A Study of Chinese Communes, 1965*
30. John Carter Vincent, *The Extraterritorial System in China: Final Phase*
31. Madeleine Chi, *China Diplomacy, 1914–1918*
32. Clifton Jackson Phillips, *Protestant America and the Pagan World: The First Half Century of the American Board of Commissioners for Foreign Missions, 1810–1860*
33. James Pusey, *Wu Han: Attacking the Present through the Past*
34. Ying-wan Cheng, *Postal Communication in China and Its Modernization, 1860–1896*
35. Tuvia Blumenthal, *Saving in Postwar Japan*
36. Peter Frost, *The Bakumatsu Currency Crisis*
37. Stephen C. Lockwood, *Augustine Heard and Company, 1858–1862*
38. Robert R. Campbell, *James Duncan Campbell: A Memoir by His Son*
39. Jerome Alan Cohen, ed., *The Dynamics of China's Foreign Relations*
40. V. V. Vishnyakova-Akimova, *Two Years in Revolutionary China, 1925–1927,* tr. Steven I. Levine
41. Meron Medzini, *French Policy in Japan during the Closing Years of the Tokugawa Regime*
42. *The Cultural Revolution in the Provinces*
43. Sidney A. Forsythe, *An American Missionary Community in China, 1895–1905*
44. Benjamin I. Schwartz, ed., *Reflections on the May Fourth Movement: A Symposium*
45. Ching Young Choe, *The Rule of the Taewŏn'gun, 1864–1873: Restoration in Yi Korea*
46. W. P. J. Hall, *A Bibliographical Guide to Japanese Research on the Chinese Economy, 1958–1970*
47. Jack J. Gerson, *Horatio Nelson Lay and Sino-British Relations, 1854–1864*
48. Paul Richard Bohr, *Famine and the Missionary: Timothy Richard as Relief Administrator and Advocate of National Reform*

49. Endymion Wilkinson, *The History of Imperial China: A Research Guide*
50. Britten Dean, *China and Great Britain: The Diplomacy of Commercial Relations, 1860–1864*
51. Ellsworth C. Carlson, *The Foochow Missionaries, 1847–1880*
52. Yeh-chien Wang, *An Estimate of the Land-Tax Collection in China, 1753 and 1908*
53. Richard M. Pfeffer, *Understanding Business Contracts in China, 1949–1963*
54. Han-sheng Chuan and Richard Kraus, *Mid-Ch'ing Rice Markets and Trade, An Essay in Price History*
55. Ranbir Vohra, *Lao She and the Chinese Revolution*
56. Liang-lin Hsiao, *China's Foreign Trade Statistics, 1864–1949*
57. Lee-hsia Hsu Ting, *Government Control of the Press in Modern China, 1900–1949*
58. Edward W. Wagner, *The Literati Purges: Political Conflict in Early Yi Korea*
59. Joungwon A. Kim, *Divided Korea: The Politics of Development, 1945–1972*
60. Noriko Kamachi, John K. Fairbank, and Chūzō Ichiko, *Japanese Studies of Modern China Since 1953: A Bibliographical Guide to Historical and Social-Science Research on the Nineteenth and Twentieth Centuries, Supplementary Volume for 1953–1969*
61. Donald A. Gibbs and Yun-chen Li, *A Bibliography of Studies and Translations of Modern Chinese Literature, 1918–1942*
62. Robert H. Silin, *Leadership and Values: The Organization of Large-Scale Taiwanese Enterprises*
63. David Pong, *A Critical Guide to the Kwangtung Provincial Archives Deposited at the Public Record Office of London*
64. Fred W. Drake, *China Charts the World: Hsu Chi-yü and His Geography of 1848*
65. William A. Brown and Urgunge Onon, translators and annotators, *History of the Mongolian People's Republic*
66. Edward L. Farmer, *Early Ming Government: The Evolution of Dual Capitals*
67. Ralph C. Croizier, *Koxinga and Chinese Nationalism: History, Myth, and the Hero*
68. William J. Tyler, tr., *The Psychological World of Natsumi Sōseki*, by Doi Takeo
69. Eric Widmer, *The Russian Ecclesiastical Mission in Peking during the Eighteenth Century*
70. Charlton M. Lewis, *Prologue to the Chinese Revolution: The Transformation of Ideas and Institutions in Hunan Province, 1891–1907*
71. Preston Torbert, *The Ch'ing Imperial Household Department: A Study of its Organization and Principal Functions, 1662–1796*
72. Paul A. Cohen and John E. Schrecker, eds., *Reform in Nineteenth-Century China*
73. Jon Sigurdson, *Rural Industrialism in China*
74. Kang Chao, *The Development of Cotton Textile Production in China*
75. Valentin Rabe, *The Home Base of American China Missions, 1880–1920*
76. Sarasin Viraphol, *Tribute and Profit: Sino-Siamese Trade, 1652–1853*

77. Ch'i-ch'ing Hsiao, *The Military Establishment of the Yuan Dynasty*
78. Meishi Tsai, *Contemporary Chinese Novels and Short Stories, 1949–1974: An Annotated Bibliography*
79. Wellington K. K. Chan, *Merchants, Mandarins, and Modern Enterprise in Late Ch'ing China*
80. Endymion Wilkinson, *Landlord and Labor in Late Imperial China: Case Studies from Shandong by Jing Su and Luo Lun*
81. Barry Keenan, *The Dewey Experiment in China: Educational Reform and Political Power in the Early Republic*
82. George A. Hayden, *Crime and Punishment in Medieval Chinese Drama: Three Judge Pao Plays*
83. Sang-Chul Suh, *Growth and Structural Changes in the Korean Economy, 1910–1940*
84. J. W. Dower, *Empire and Aftermath: Yoshida Shigeru and the Japanese Experience, 1878–1954*
85. Martin Collcutt, *Five Mountains: The Rinzai Zen Monastic Institution in Medieval Japan*

STUDIES IN THE MODERNIZATION OF THE REPUBLIC OF KOREA: 1945–1975

86. Kwang Suk Kim and Michael Roemer, *Growth and Structural Transformation*
87. Anne O. Krueger, *The Developmental Role of the Foreign Sector and Aid*
88. Edwin S. Mills and Byung-Nak Song, *Urbanization and Urban Problems*
89. Sung Hwan Ban, Pal Yong Moon, and Dwight H. Perkins, *Rural Development*
90. Noel F. McGinn, Donald R. Snodgrass, Yung Bong Kim, Shin-Bok Kim, and Quee-Young Kim, *Education and Development in Korea*
91. Leroy P. Jones and Il SaKong, *Government, Business and Entrepreneurship in Economic Development: The Korean Case*
92. Edward S. Mason, Dwight H. Perkins, Kwang Suk Kim, David C. Cole, Mahn Je Kim, et al., *The Economic and Social Modernization of the Republic of Korea*
93. Robert Repetto, Tai Hwan Kwon, Son-Ung Kim, Dae Young Kim, John E. Sloboda, and Peter J. Donaldson, *Economic Development, Population Policy, and Demographic Transition in the Republic of Korea*
106. David C. Cole and Yung Chul Park, *Financial Development in Korea, 1945–1978*
107. Roy Bahl, Chuk Kyo Kim, and Chong Kee Park, *Public Finances during the Korean Modernization Process*

94. Parks M. Coble, *The Shanghai Capitalists and the Nationalist Government, 1927–1937*
95. Noriko Kamachi, *Reform in China: Huang Tsun-hsien and the Japanese Model*

96. Richard Wich, *Sino-Soviet Crisis Politics: A Study of Political Change and Communication*
97. Lillian M. Li, *China's Silk Trade: Traditional Industry in the Modern World, 1842–1937*
98. R. David Arkush, *Fei Xiaotong and Sociology in Revolutionary China*
99. Kenneth Alan Grossberg, *Japan's Renaissance: The Politics of the Muromachi Bakufu*
100. James Reeve Pusey, *China and Charles Darwin*
101. Hoyt Cleveland Tillman, *Utilitarian Confucianism: Ch'en Liang's Challenge to Chu Hsi*
102. Thomas A. Stanley, *Ōsugi Sakae, Anarchist in Taishō Japan: The Creativity of the Ego*
103. Jonathan K. Ocko, *Bureaucratic Reform in Provincial China: Ting Jih-ch'ang in Restoration Kiangsu, 1867–1870*
104. James Reed, *The Missionary Mind and American East Asia Policy, 1911–1915*
105. Neil L. Waters, *Japan's Local Pragmatists: The Transition from Bakumatsu to Meiji in the Kawasaki Region*
108. William D. Wray, *Mitsubishi and the N.Y.K., 1870–1914: Business Strategy in the Japanese Shipping Industry*
109. Ralph William Huenemann, *The Dragon and the Iron Horse: The Economics of Railroads in China, 1876–1937*
110. Benjamin A. Elman, *From Philosophy to Philology: Intellectual and Social Aspects of Change in Late Imperial China*
111. Jane Kate Leonard, *Wei Yuan and China's Rediscovery of the Maritime World*
112. Luke S. K. Kwong, *A Mosaic of the Hundred Days: Personalities, Politics, and Ideas of 1898*
113. John E. Wills, Jr., *Embassies and Ilusions: Dutch and Portuguese Envoys to K'ang-hsi, 1666–1687*
114. Joshua A. Fogel, *Politics and Sinology: The Case of Naitō Konan (1866–1934)*
115. Jeffrey C. Kinkley, ed., *After Mao: Chinese Literature and Society, 1978–1981*
116. C. Andrew Gerstle, *Circles of Fantasy: Convention in the Plays of Chikamatsu*
117. Andrew Gordon, *The Evolution of Labor Relations in Japan: Heavy Industry, 1853–1955*
118. Daniel K. Gardner, *Chu Hsi and the* Ta Hsueh: *Neo-Confucian Reflection on the Confucian Canon*
119. Christine Guth Kanda, *Shinzō: Hachiman Imagery and its Development*
120. Robert Borgen, *Sugawara no Michizane and the Early Heian Court*
121. Chang-tai Hung, *Going to the People: Chinese Intellectual and Folk Literature, 1918–1937*
122. Michael A. Cusumano, *The Japanese Automobile Industry: Technology and Management at Nissan and Toyota*
124. Steven D. Carter, *The Road to Komatsubara: A Classical Reading of the Renga Hyakuin*

125. Katherine F. Bruner, John K. Fairbank, and Richard T. Smith, *Entering China's Service: Robert Hart's Journals, 1854–1863*
126. Bob Tadashi Wakabayashi, *Anti-Foreignism and Western Learning in Early Modern Japan: The* New Theses *of 1825*
127. Atsuko Hirai, *Individualism and Socialism: The Life and Thought of Kawai Eijirō (1891–1944)*
128. Ellen Widmer, *The Margins of Utopia:* Shui-hu hou-chuan *and the Literature of Ming Loyalism*
129. R. Kent Guy, *The Emperor's Four Treasuries: Scholars and the State in the Late Ch'ien-lung Era*
130. Peter C. Perdue, *Exhausting the Earth: State and Peasant in Hunan, 1500–1850*
131. Susan Chan Egan, *A Latterday Confucian: Reminiscences of William Hung (1893–1980)*
132. James T. C. Liu, *China Turning Inward: Intellectual-Political Changes in the Early Twelfth Century*
133. Paul A. Cohen, *Between Tradition and Modernity: Wang T'ao and Reform in Late Ch'ing China*
134. Kate Wildman Nakai, *Shogunal Politics: Arai Hakuseki and the Premises of Tokugawa Rule*
135. Hans Ulrich Vogel, *Chinese Central Monetary Policy and Yunnan Copper Mining in the Early Qing (1644–1800)*
136. Jon L. Saari, *Legacies of Childhood: Growing Up Chinese in a Time of Crisis, 1890–1920*
137. Susan Downing Videen, *Tales of Heichū*
138. Heinz Morioka and Miyoko Sasaki, *Rakugo: The Popular Narrative Art of Japan*
139. Joshua A. Fogel, *Nakai Ushikichi in China: The Mourning of Spirit*
140. Alexander Barton Woodside, *Vietnam and the Chinese Model: A Comparative Study of Vietnamese and Chinese Government in the First Half of the Nineteenth Century*
141. George Elison, *Deus Destroyed: The Image of Christianity in Early Modern Japan*
142. William D. Wray, ed., *Managing Industrial Enterprise: Cases from Japan's Prewar Experience*
143. T'ung-tsu Ch'ü, *Local Government in China under the Ch'ing*
144. Marie Anchordoguy, *Computers Inc.: Japan's Challenge to IBM*
145. Barbara Molony, *Technology and Investment: The Prewar Japanese Chemical Industry*
146. Mary Elizabeth Berry, *Hideyoshi*
147. Laura E. Hein, *Fueling Growth: The Energy Revolution and Economic Policy in Postwar Japan*
148. Wen-hsin Yeh, *The Alienated Academy: Culture and Politics in Republican China, 1919–1937*